Intelligent Systems Reference Library

Volume 210

The aim of this series is to publish a Reference Library, including novel advances and developments in all aspects of Intelligent Systems in an easily accessible and well structured form. The series includes reference works, handbooks, compendia, textbooks, well-structured monographs, dictionaries, and encyclopedias. It contains well integrated knowledge and current information in the field of Intelligent Systems. The series covers the theory, applications, and design methods of Intelligent Systems. Virtually all disciplines such as engineering, computer science, avionics, business, e-commerce, environment, healthcare, physics and life science are included. The list of topics spans all the areas of modern intelligent systems such as: Ambient intelligence, Computational intelligence, Social intelligence, Computational neuroscience, Artificial life, Virtual society, Cognitive systems, DNA and immunity-based systems, e-Learning and teaching, Human-centred computing and Machine ethics, Intelligent control, Intelligent data analysis, Knowledge-based paradigms, Knowledge management, Intelligent agents, Intelligent decision making, Intelligent network security, Interactive entertainment, Learning paradigms, Recommender systems, Robotics and Mechatronics including human-machine teaming, Self-organizing and adaptive systems, Soft computing including Neural systems, Fuzzy systems, Evolutionary computing and the Fusion of these paradigms, Perception and Vision, Web intelligence and Multimedia.

Indexed by SCOPUS, DBLP, zbMATH, SCImago.

All books published in the series are submitted for consideration in Web of Science.

More information about this series at http://www.springer.com/series/8578

Pardeep Kumar · Ahmed Jabbar Obaid · Korhan Cengiz · Ashish Khanna · Valentina Emilia Balas
Editors

A Fusion of Artificial Intelligence and Internet of Things for Emerging Cyber Systems

Springer

Editors
Pardeep Kumar
Department of Software Engineering
Quaid-e-Awam University of Engineering
Science and Technology (QUEST)
Nawabshah, Pakistan

Ahmed Jabbar Obaid
Department of Computer Science
University of Kufa
Najaf, Iraq

Korhan Cengiz
Department of Electrical and Electronics
Engineering
Trakya University
Edirne, Turkey

Ashish Khanna
Department of Computer Science
and Engineering
Maharaja Agrasen Institute of Technology
New Delhi, India

Valentina Emilia Balas
Department of Automatics and Applied
Software
University of Arad
Arad, Romania

ISSN 1868-4394 ISSN 1868-4408 (electronic)
Intelligent Systems Reference Library
ISBN 978-3-030-76652-8 ISBN 978-3-030-76653-5 (eBook)
https://doi.org/10.1007/978-3-030-76653-5

This Springer imprint is published by the registered company Springer Nature Switzerland AG
The registered company address is: Gewerbestrasse 11, 6330 Cham, Switzerland

Contents

Chapter 1
IoT for Better Mobile Health Applications

Rehab A. Rayan, Christos Tsagkaris, and Imran Zafar

Abstract Using big data in health is booming. Recently, several novel innovations and techniques applying big data to Bioinformatics and health research are released. Nowadays, devices such as smartphones, sensors; individuals like patients, healthcare professionals, or investigators and collective entities such as healthcare facilities and institutions are regularly creating and collecting a vast quantity of health data that can diagnose and treat new disorders. The genuine dilemma in applying mobile health is how to identify, collect, analyze, and distribute information to make better and simpler lives for individuals via early predicting risk factors. For instance, researchers have built many techniques to enable managing chronic diseases. Medical devices for ongoing monitoring health indicators or detecting timely online healthcare data like patient self-administering physical therapy are greatly needed. Today, for accessing high-speed internet connections on mobile phones, several smart patients are using smartphone applications (apps) to monitor routinely various daily health demands. Such smartphone apps and devices are increasingly utilized and embedded with e-health and telemedicine through the Internet of Things (IoT). However, confidentiality and safety are major concerns that require cooperation with policymakers and prompt communication of potential hazards. This work examines the applications of the IoT and m-health in healthcare. It also describes innovative methods for improving health via computational technologies and techniques for integrating IoT and m-health, including a proposed model for diabetes self-management.

Keywords IoT · Big data · E-health · Mobile health · M-health · Telemedicine

---

R. A. Rayan (✉)
Department of Epidemiology, High Institute of Public Health, Alexandria University, Alexandria, Egypt
e-mail: rayanr@alexu.edu.eg

C. Tsagkaris
Faculty of Medicine, University of Crete, Heraklion, Greece

I. Zafar
Department of Bioinformatics and Computational Biology, Virtual University of Pakistan, Lahore, Pakistan

P. Kumar et al. (eds.), *A Fusion of Artificial Intelligence and Internet of Things for Emerging Cyber Systems*, Intelligent Systems Reference Library 210,
https://doi.org/10.1007/978-3-030-76653-5_1

1.1 Introduction

The Internet of things (IoT), a modern approach, is a promising solution worldwide for being applied largely in health. Remarkably, IoT is linking almost 22 billion devices together and expected to reach 50 billion devices by 2030 [1]. IoT connects digital devices and enables exchanging information among networked devices to address certain issues. Such IoT-derived inter-networking has turned people's lives simpler. Nowadays, the world is confronting health-related serious issues [2]. The healthcare sector, using smartphones and IoT-based devices, could assist patients in controlling their daily health-related conditions and in urgent situations as well [3]. Hence, mobile health (m-health) could enable quality health services as needed by each patient [4].

Based on the IoT, the m-health technology could tackle care entirely for a patient, and such innovative technologies are adapting a patient's circumstances and their metrics could be altered per condition. Therefore, under the IoT, the m-health technology would control the current and future health state of severely ill patients. Applying IoT to m-health programs has a high potential to grow the capabilities of the primary healthcare systems, and thus it would be readily available to have regular communications between patients and health professionals involving physicians. For instance, a wearable could instantly relate to a tool for big data analytics to identify directly and deal with any data fluctuations. Another wearable sensor could check blood pressure through the skin, meanwhile, a contact lens could examine and show the levels of sugar in tears [2]. Lately, Dell has undertaken an experimental initiative to monitor and investigate chronic conditions. Hence, a patient who has diabetes could be dynamically observed while offering daily meal alerts and recommendations. The USA El Camino Hospital has declared that they have adopted a telemedicine-integrated assessment program successfully to determine at-high-risk patients and their m-health program could directly suggest the best solution [5].

Recently, telemedicine-integrated assessing health programs have arisen, however, the more growth in innovative digital health techniques offer novel solutions to collect and exchange timely a vast amount of data on health. Therefore, it provides great potential to grow the capacities of health systems to enhance communication between the patient and health caregivers and decrease risks. Such interventions could maintain patient comprehensive care that would decrease the probabilities of avoidable risks. In healthcare facilities, the IoT is built upon healthcare datasets, Internet Protocol (IP) connectivity, sensors, and clinical devices that could remotely operate electronic health records. Combining IP with enterprise service hubs and other clinical devices allows transferring data among patients and healthcare workers, which are known as the Internet of Healthy Things (IoHT). The IoHT could provide real-time precise clinical data and healthcare resources for the healthcare providers in vital situations demanding prompt care (Fig. 1.1). The clinical devices' configurations are getting further complicated; therefore, they are challenging the experts in information technology to incorporate m-health and the IoT into the health systems. The demand for data in establishing an effective and well-structured m-health system has

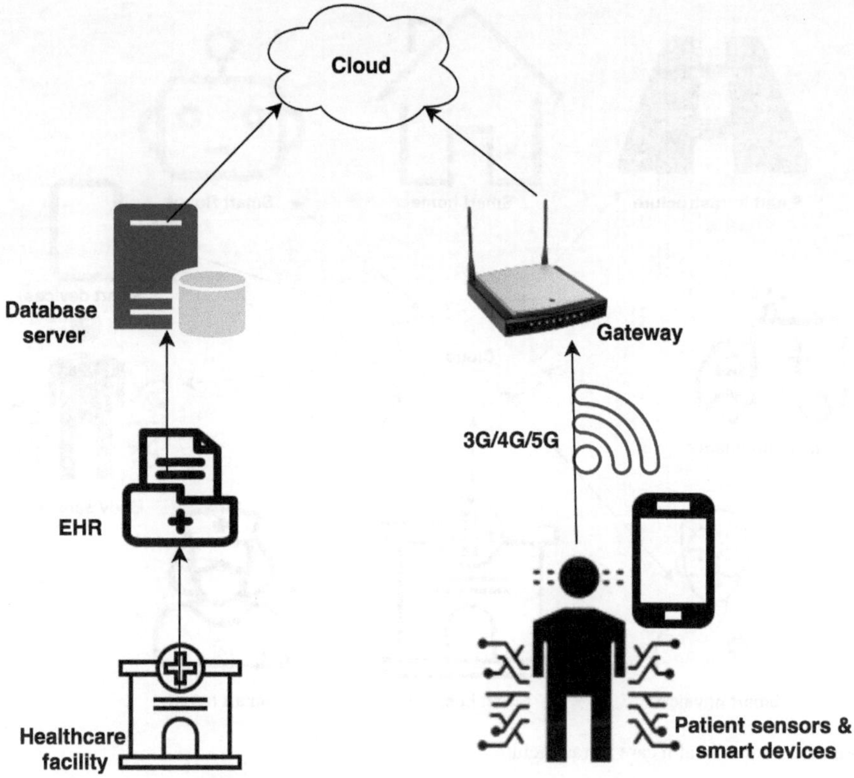

Fig. 1.1 The IoHT framework. EHR: Electronic health record

added usually more concerns. Hence, it is recommended regarding smart business for enhancing IoT-integrated m-health big data systems' functionality [6].

Value-derived m-health targets incentivizing health caregivers according to delivering timely and affordable patient care [7]. Figure 1.2 outlines the needed infrastructure for the healthcare system delivering IoT-integrated m-health services. The findings obviously reveal the required group support for both the healthcare workers and the patients to operate efficiently [8]. Incorporating wearables with m-health and the IoT technology is promising in more advanced availability and access to affordable health services [9]. The IoT underlines a universal network of internet-connected services and devices anytime and anywhere where the IoHT or the Internet of Patients (IoP) reflects affordable and simple communications between healthcare professionals and patients via protected connectivity [10].

This chapter examines the applications of the IoT and m-health in healthcare. It also describes innovative methods for improving health via computational technologies and techniques for integrating IoT and m-health, including a proposed model for diabetes self-management.

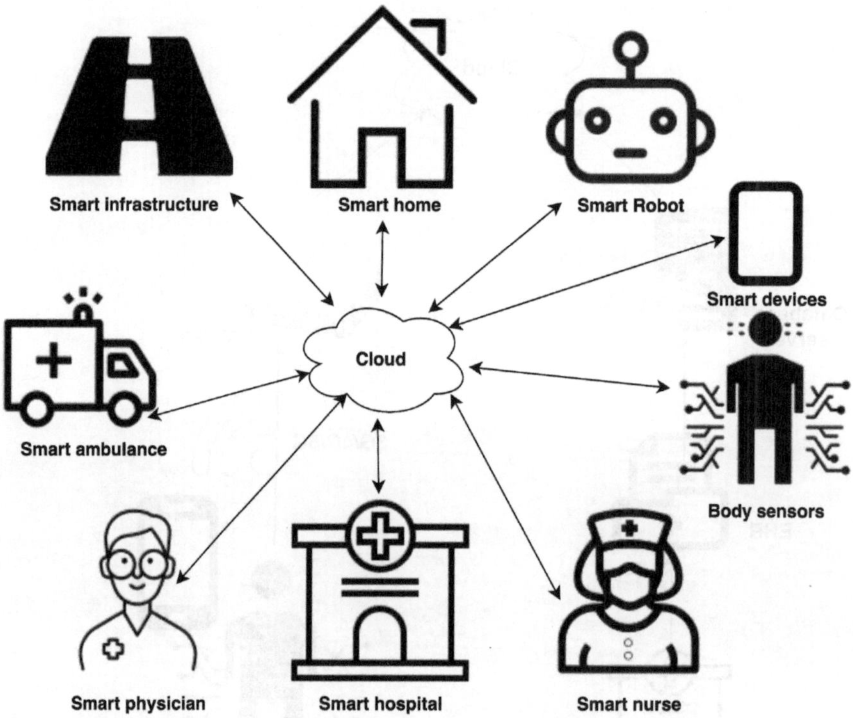

Fig. 1.2 Smart health care infrastructure

1.2 Literature Survey

An m-health-based monitoring system was proposed to house services and data where an innovative technique was proposed for arranging and handling online data and managing IoT-integrated big data applications where the deployed technique got patient data from the IoT devices and sent it successfully to the cloud [11]. Another technique emphasized carrying the received m-health data from the shared IoT devices to the cloud where the dataset controlling system allowed the IoT devices to gather timely consumer's vital data [12]. The cloud maintained rapid handling of the housed data therefore, the consumer could receive alerts quickly in critical conditions. Another model was designed to process and display data by adequately deploying cloud-based architecture [13].

One model applied encryption for securing exchangeable clinical data [14]. Another example could identify safe and trusted terminals, involving m-health hospitals and healthcare professionals, to exchange clinical data by connecting with patients. A patient-specific cyber-physical framework approved efficient and m-health services by collecting data systematically in parallel processing and distributed housing centers [15]. A digital m-health system was developed to monitor patients distantly focusing on accelerating connectivity and processing [9]. A Cloud-based and e-health integrated calorie-monitoring system that classified accurately various foods in a meal and aggregate meal-energy content [16]. An effective technique, based on artificial intelligence and deep learning principles, was proposed to forecast, and precisely classify the heart rate via a smartphone-integrated sensor called an accelerometer [17].

An IoT-Cloud-derived m-health intervention was proposed to precisely detect the patient's voice where it uses a smart machine learning-based voice pathology monitoring tool [18]. Another IoT-Cloud intervention was described for monitoring physical activity applying embedded clinical sensors, cloud architecture, and Extensible Markup Language (XML) web services for simple, fast, and secure retrieving and exchanging of data where it notifies the patient about problems or abnormalities during physical activities [19]. The Health Industrial IoT (HealthIIoT) principle was proposed for the timely monitoring of the disabled and geriatrics [20].

A health control and monitoring model was proposed that could monitor affordably and effectively heart-disease patients. The model includes intelligent gateways and a power-efficient IoT-based sensor where the sensor could gather data on the temperature of the body, the respiration levels, and the Electrocardiography (ECG) output; and send this data to the gateways for digital analysis and interpretations in a wireless setting with a negligible loss [21]. An m-Health intervention was presented that could operate as a mediating level between the IoT terminals and the cloud where it applies a Cloud Access Security Broker (CASB) for enhancing the confidentiality and protection of the data [22].

Another model was proposed, which could deal with lag-sensitive m-health data [23]. A service-centered framework was described to approve and examine raw clinical data collected by IoT terminals via resource-limited integrated digital instances [24]. An intelligent e-health gateway was discussed that could support IoT-based m-health services where the gateways are at various sites in the distributed network and each could operate and manage many IoT terminals freely linked with both health caregivers and the patients [25]. Another intelligent ehealth gateway was described that could offer timely patient data housing, refining, and analysis along with an IoT-based early warning score (EWS) to test functionality [8].

Big data is a key element in m-health since clinical sensors produce big health data, therefore, a scalable cloud model-embedded sensor was introduced which could manage all the m-health-based sensors and work timely on big datasets [26]. In a smart city context, personal behavior was examined precisely by applying big data analytics into social IoT via approaches of integrated screening [27, 28].

1.3 Integrating IoT with Mobile Health

Figure 1.3 shows integrating m-health with the IoT for monitoring diseases or daily activities and educating patients along with rehabilitation and surveillance purposes. The IoT enables networking between patients and health professionals such as general practitioners, nurses, specialists, and others. Therefore, the patient could get

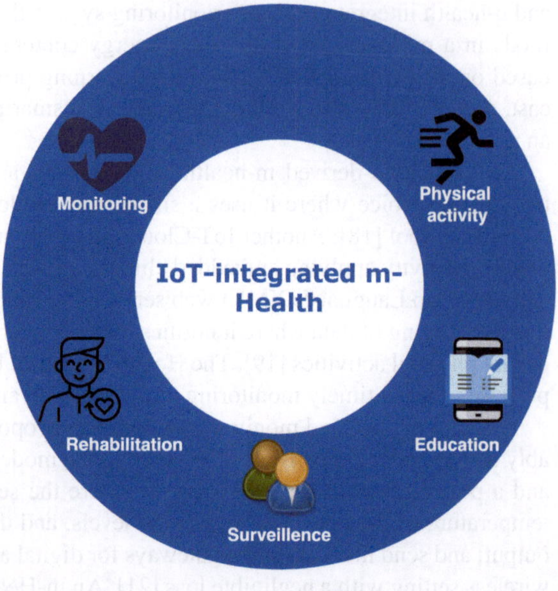

Fig. 1.3 IoT-integrated m-health framework

appointed to healthcare providers or receive critical care services either at home or in healthcare facilities via IoT-integrated m-health applications as mentioned below:

1.3.1 Appointment Reminders

The patient would receive either a voice or text Short Message Service (SMS) from the healthcare facility to notify about confirmed, rescheduled, or deleted appointments with the healthcare provider, covering immunization dates, tests' findings, and other services. Now, the mobile phone's SMS is the major appointment reminder in many countries worldwide.

1.3.2 Telemedicine

The lack of healthcare providers is the major driver towards telemedicine where the patient could instantly communicate or consult with healthcare providers via mobile devices-bases text, voice, images, or video calls. Telemedicine facilities could provide care for a chronically diseased patient at home especially networking between health professionals in cities and healthcare workers and the patient in the countryside promoting affordable and quality healthcare services [29].

1.3.3 Monitoring and Empowerment

Patients' care could be monitored remotely with the IoT-integrated m-health applications where distant sensors and imaging devices connected to mobile phones limiting face-to-face encounters while allowing interacting with healthcare providers and sending data.

1.4 A Proposed Model for Diabetes Self-management

A novel IoT-integrated m-health model could enable self-management and offer multidimensional healthcare for diabetes, as an example of chronic disease requiring periodic monitoring, through remotely gathering and tracking patient data and returning tailored and individualized feedback on a smartphone. This model would provide self-management of diabetes and facilitate remote and timely clinical interaction and customized feedback including advice, decision support, or lifestyle modification based on the individual needs of a patient, using newly collected and previous patient data.

The model could detect whether the level of a patient's activities follow the desired regimen for personal care, using clinical parameters (blood pressures, blood sugar levels, weight, etc.), and produce suitable helping or alerting feedback advice. The physical layer of the model integrates wireless hubs including a group of biosensors-based devices (blood pressure meter, glucometer, weight scale, etc.) connected wirelessly to a smartphone which is also connected to a web-based application layer via the available telecommunication infrastructure (Fig. 1.4). The next step is to pilot this model for clinical evaluation to examine functionality, the precision of the shared data, security, patient-provider acceptability, and further improvements, which are among the ongoing work of the researcher.

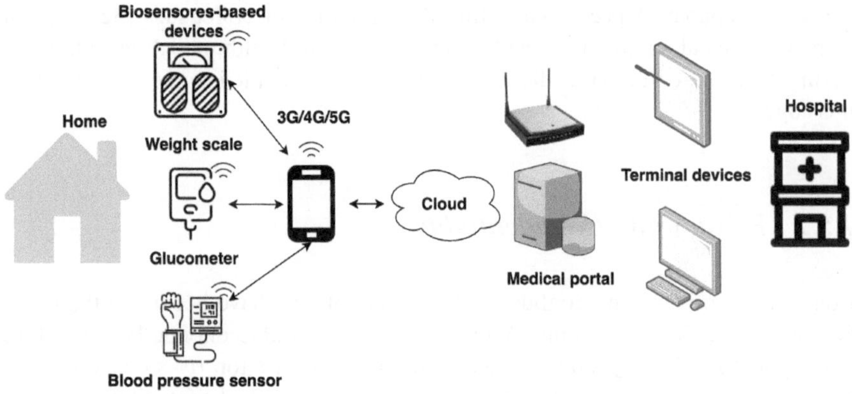

Fig. 1.4 A proposed IoT-integrated m-health model for diabetes self-management

1.5 Discussions

1.5.1 *Feasibility of Implementation*

IoT is a new and evolving concept that provides connectivity to the Internet via sensing devices. M-health, a variant of healthcare defined as the practice of medicine and public health with the support of mobile devices, is probably one of the most IoT compatible paradigms of healthcare. From the level of concept to the level of services, there are technical, financial, and human factors affecting the feasibility of IoT-integrated m-health in the third decade of the twenty-first century. As far as technical factors are concerned, a wide variety of (bio)sensors and a sound infrastructure of communications. Explaining the technicalities of the sensors and the network technologies is beyond the scope of this chapter; however, it is important to realize that such equipment should be widely available, easily accessible, and easy to use. The experience with sensors so far indicates that they need to be as comfortable as possible given that patients – let alone people with minor health nuisances–will not tolerate heavy or annoying devices in their environment let alone on their bodies [30, 31].

Education concerning such technological modalities has long debated under the term of "digital literacy". Medical universities and professional boards are expected to train physicians and allied health professionals. Health professionals will need to act as multipliers, disseminating knowledge to their patients. Sometimes, authorized representatives of tech providers could intervene. There is similar experience in fields such as Diabetes mellitus, where tech professionals collaborate with physicians to educate patients in terms of wearables or glucose monitoring devices. Civil society can also contribute to digital literacy, providing simplified information concerning hot topics such as security or fundamentals of use. There are many examples of civil society entities (Non-governmental organizations (NGOs), scientific associations, etc.) delivering workshops or vocational training sessions. Depending on the rate of IoT-integrated m-health implementation, it will be important to train people as educators capable of peer – educating their families or colleagues [32, 33]. The response of local communities and professionals to m-health modalities introduced during the novel coronavirus disease (COVID-19) pandemic is quite encouraging at this edge.

1.5.2 *Privacy and Confidentiality*

Concerns on privacy and confidentiality on the internet have become a thing long before IoT itself came to being. According to the "World Economic Forum" global risk report for 2019, cyberattacks and data violations are top risks concerning the world today [30, 34]. 2020, the year of the pandemic, marked a transition to remote healthcare services with physicians and patients sharing information about new or preexisting health conditions as well as prescriptions and test results online. For instance, popular communication platforms such as Zoom have faced security issues and although there are no specific figures on how many medical centered meetings were intercepted, this has been alarming. Conventional internet networks have been the "crime scene" of such cyberattacks, but it is possible that IoT suffers from the same problem. Fortunately, IoT has higher security standards. Developers are addressing its security weaknesses, while features such as blockchain revolutionize the concept of online confidentiality [35, 36].

Despite the technical superiority of IoT, there are two factors increasing its vulnerability to health-oriented cyberattacks. First, a prominent part of IoT has been built for the sake of healthcare. There is an ample number of publications, including this very chapter, presenting various aspects of IoT based healthcare. Frankly, its assets have been praised and its weaknesses have been pointed out, providing cyber-attackers with the technical knowledge they would otherwise struggle to find for themselves. IoT is technically facilitating health data transmission and storage. With Magnetic Resonance Images (MRI) scans and even trial datasets processed in a minimal amount of time, it is almost expected that IoT channels of health data communication can become prey for scammers and abusers [37, 38]. Furthermore, the value of health-related data is high. According to the World Economic Forum report, on the illegal market, "Protected Health Information" (PHI) is more significant than a credit card and online banking credentials or even personally identifiable information. Hence, cybercriminals are highly motivated to gain access to medical databases, to trade the PHI, or adapt it for their benefit [35].

Finally, when IoT will become widely available, its safety will become operator-dependent up to an important extent. In recent years, both physicians and patients have been tricked to install credential-stealing malware. This way patients and physicians themselves serve as insiders who either systematically or accidentally unveil PHI [39, 40]. Keeping IoT-integrated m-health safe will not only take to design sophisticated security systems but also educate their users. Choosing IoT-integrated m-health modalities with a low learning curve and educating the users repeatedly about security issues will contribute significantly to safeguarding health data. Apart from technology and healthcare providers, civil society could play an important role in spreading the word and increasing awareness about safety [30].

1.6 Future Insights

Ultimately, using mobile devices in promoting healthcare would play a key role in managing m-health facilitating accessing and using electronic health records via mobile devices whenever needed. The IoT integrated applications for smooth and ongoing monitoring of critical patients could minimize health-related crises, reduce healthcare expenses, and competently use health resources. IoT integrated technologies, the blessing of the new era, could highly minimize chronically diseased patients' healthcare expenses with enhanced health monitoring and patient empowerment towards a healthy lifestyle hence, enabling financially rewarding patients showing better lifestyle behaviors [41]. Professionals expect that the IoT- integrated m-health monitoring platforms could reduce the daily load of hypertension, diabetes, and anemia conditions, and hence improve the community economy with better incomes for healthcare providers. Therefore, advancing human–machine interaction-enabling techniques leveraging the heterogeneous, complex health big data with m-health modalities would deliver better healthcare with the help of healthcare experts, technologists, and researchers [42].

1.7 Conclusions

Integrating m-health and big data with the IoT is promising in providing affordable and accessible healthcare, however, it is also highly challenging and requires more investigations. Using competent biosensors and computers or smartphones-installed health applications could manage IoT-integrated m-health data where a distributed model could incorporate bio-sensing, monitoring, analyzing, and delivering quality m-health services for a distant patient. Leveraging big data with the IoT involving biosensors and other wearables could robustly and smoothly diagnose a critically diseased patient or monitor a chronically diseased one according to signs and symptoms and medical records, thus exchanging technical resources to manage a distant patient. Using big data opens the window for discovering novel knowledge and developing innovative technologies for better standards and quality of healthcare delivery at lower costs. Nowadays, healthcare devices are highly required for timely and ongoing monitoring of clinical indicators or online health-related data. Today, for mobile phones' smooth access to high-speed Internet connectivity, several smart patients use apps for routinely monitoring various health needs from day-to-day. Such devices and apps are increasingly utilized and incorporated with both e-health and telemedicine through the IoT. This work examined the applications and challenges of the IoT and m-health in healthcare. It also described innovative methods for improving health via computational technologies and techniques for integrating IoT and m-health. It contributed with a proposed IoT-integrated m-health model for diabetes self-management combining multidimensional aspects of care, turning the focus from the classical provider-oriented to a patient-oriented one that needs to be piloted clinically.

References

1. Statista Research Department. (2019, May). Number of connected devices worldwide 2030. *Statista*, https://www.statista.com/statistics/802690/worldwide-connected-devices-by-access-technology/. Accessed June 20, 2020.

2. Mishra, K. N., & Chakraborty, C. (2020). A novel approach towards using big data and IoT for improving the efficiency of m-health systems. In D. Gupta, A. E. Hassanien, & A. Khanna (Eds.), *Advanced Computational intelligence techniques for virtual reality in healthcare* (pp. 123–139). Cham: Springer International Publishing.
3. Andriopoulou, F., Dagiuklas, T., & Orphanoudakis, T. (2017). Integrating IoT and fog computing for healthcare service delivery. In G. Keramidas, N. Voros, & M. Hübner (Eds.), *Components and services for IoT platforms: Paving the way for IoT standards* (pp. 213–232). Cham: Springer International Publishing.
4. Dong, B., Yang, J., Ma, Y., & Zhang, X. (2016). Medical monitoring model of internet of things based on the adaptive threshold difference algorithm. *International Journal Multimedia Ubiquitous Engineering, 11*(5), 75–82.
5. AbuKhousa, E. (2017). Analytics and Telehealth emerging technologies: The path forward for smart primary care environment. *Journal Healthcare Communication, 2*(4). https://doi.org/10.4172/2472-1654.100108.
6. Chakraborty, C., Gupta, B., & Ghosh, S. K. (2014). Mobile metadata assisted community database of chronic wound images. *Wound Medicine, 6*, 34–42. https://doi.org/10.1016/j.wndm.2014.09.002
7. Mahmud, R., Koch, F. L., & Buyya, R. (2018). Cloud-fog interoperability in IoT-enabled healthcare solutions. In *Proceedings of the 19th International Conference on Distributed Computing and Networking* (pp. 1–10), Varanasi, India. https://doi.org/10.1145/3154273.3154347.
8. Rahmani, A. M., et al. (2018). Exploiting smart e-Health gateways at the edge of healthcare Internet-of-Things: A fog computing approach. *Future Generation Computer System, 78*, 641–658. https://doi.org/10.1016/j.future.2017.02.014
9. Fazio, M., Celesti, A., Márquez, F. G., Glikson, A., & Villari, M. (2015). Exploiting the FIWARE cloud platform to develop a remote patient monitoring system. In *2015 IEEE Symposium on Computers and Communication (ISCC)* (pp. 264–270). https://doi.org/10.1109/ISCC.2015.7405526.
10. Hassanalieragh, M., et al. (2015). Health monitoring and management using Internet-of-Things (IoT) sensing with cloud-based processing: Opportunities and challenges. In *2015 IEEE International Conference on Services Computing* (pp. 285–292). https://doi.org/10.1109/SCC.2015.47.
11. Doukas, C., & Maglogiannis, I. (2012). Bringing IoT and cloud computing towards pervasive healthcare. In *2012 Sixth International Conference on Innovative Mobile and Internet Services in Ubiquitous Computing* (pp. 922–926). https://doi.org/10.1109/IMIS.2012.26.
12. Tsiachri Renta, P., Sotiriadis, S., & Petrakis, E. G. M. (2017). Healthcare sensor data management on the cloud. In *Proceedings of the 2017 Workshop on Adaptive Resource Management and Scheduling for Cloud Computing* (pp. 25–30). Washington, DC, USA. https://doi.org/10.1145/3110355.3110359.
13. Mahmud, S., Iqbal, R., & Doctor, F. (2016). Cloud enabled data analytics and visualization framework for health-shocks prediction. *Future Generation Computer System, 65*, 169–181. https://doi.org/10.1016/j.future.2015.10.014
14. Chen, M., Qian, Y., Chen, J., Hwang, K., Mao, S., & Hu, L. (2016). Privacy protection and intrusion avoidance for cloudlet-based medical data sharing. *IEEE Transactions Cloud Computing*, 1–1. https://doi.org/10.1109/TCC.2016.2617382.
15. Zhang, Y., Qiu, M., Tsai, C.-W., Hassan, M. M., & Alamri, A. (2017). Health-CPS: Healthcare cyber-physical system assisted by cloud and big data. *IEEE Systems Journal, 11*(1), 88–95. https://doi.org/10.1109/JSYST.2015.2460747
16. Peddi, S. V. B., Kuhad, P., Yassine, A., Pouladzadeh, P., Shirmohammadi, S., & Shirehjini, A. A. N. (2017). An intelligent cloud-based data processing broker for mobile e-health multimedia applications. *Future Generation Computer System, 66*, 71–86. https://doi.org/10.1016/j.future.2016.03.019
17. Jindal, V. (2016). Integrating mobile and cloud for PPG signal selection to monitor heart rate during intensive physical exercise. In *2016 IEEE/ACM International Conference on Mobile Software Engineering and Systems (MOBILESoft)* (pp. 36–37). https://doi.org/10.1109/MobileSoft.2016.027.

18. Muhammad, G., Rahman, S. M. M., Alelaiwi, A., & Alamri, A. (2017). Smart health solution integrating IoT and cloud: A case study of voice pathology monitoring. *IEEE Communications Magazine, 55*(1), 69–73. https://doi.org/10.1109/MCOM.2017.1600425CM
19. Gupta, P. K., Maharaj, B. T., & Malekian, R. (2017). A novel and secure IoT based cloud centric architecture to perform predictive analysis of users activities in sustainable health centres. *Multimedia Tools Applications, 76*(18), 18489–18512. https://doi.org/10.1007/s11042-016-4050-6
20. Hossain, M. S., & Muhammad, G. (2016). Cloud-assisted industrial Internet of Things (IIoT)—Enabled framework for health monitoring. *Computer Networks, 101*, 192–202. https://doi.org/10.1016/j.comnet.2016.01.009
21. Nguyen Gia, T., et al. (2017). Low-cost fog-assisted health-care IoT system with energy-efficient sensor nodes. In *2017 13th International Wireless Communications and Mobile Computing Conference (IWCMC)* (pp. 1765–1770). https://doi.org/10.1109/IWCMC.2017.7986551.
22. Ahmad, M., Amin, M. B., Hussain, S., Kang, B. H., Cheong, T., & Lee, S. (2016). Health Fog: A novel framework for health and wellness applications. *The Journal of Supercomputing, 72*(10), 3677–3695. https://doi.org/10.1007/s11227-016-1634-x
23. Chakraborty, S., Bhowmick, S., Talaga, P., & Agrawal, D. P. (2016). Fog networks in healthcare application. In *2016 IEEE 13th International Conference on Mobile Ad Hoc and Sensor Systems (MASS)* (pp. 386–387). https://doi.org/10.1109/MASS.2016.065.
24. Dubey, H., Yang, J., Constant, N., Amiri, A. M., Yang, Q., & Makodiya, K. (2016). Fog data: Enhancing telehealth big data through fog computing. *ArXiv160509437 Cs*. https://doi.org/10.1145/2818869.2818889.
25. Negash, B., et al. (2018). Leveraging fog computing for healthcare IoT. In A. M. Rahmani, P. Liljeberg, J.-S. Preden, & A. Jantsch (Eds.), *Fog computing in the internet of things: Intelligence at the edge* (pp. 145–169). Cham: Springer International Publishing.
26. Vuppalapati, C., Ilapakurti, A., & Kedari, S. (2016). The role of big data in creating sense EHR, an integrated approach to create next generation mobile sensor and wearable data driven electronic health record (EHR). In *2016 IEEE Second International Conference on Big Data Computing Service and Applications (BigDataService)* (pp. 293–296). https://doi.org/10.1109/BigDataService.2016.18.
27. Ahmad, A., Rathore, M. M., Paul, A., & Rho, S. (2016). Defining human behaviors using big data analytics in social internet of things. In *2016 IEEE 30th International Conference on Advanced Information Networking and Applications (AINA)* (pp. 1101–1107). https://doi.org/10.1109/AINA.2016.104.
28. Ahmed, E., & Rehmani, M. H. (2017). Introduction to the special section on social collaborative internet of things. *Computers & Electrical Engineering, 58*, 382–384. https://doi.org/10.1016/j.compeleceng.2017.04.023
29. Mora, H., Gil, D., Terol, R. M., Azorín, J., & Szymanski, J. (2017). An IoT-based computational framework for healthcare monitoring in mobile environments. *Sensors, 17*(10). https://doi.org/10.3390/s17102302.
30. Jagadeeswari, V., Subramaniyaswamy, V., Logesh, R., & Vijayakumar, V. (2018). A study on medical Internet of Things and big data in personalized healthcare system. *Health Information Science System, 6*(1), 14. https://doi.org/10.1007/s13755 018-0049-x
31. Zhang, P., Schmidt, D., White, J., & Mulvaney, S. (2018). Towards precision behavioral medicine with IoT: Iterative design and optimization of a self-management tool for type 1 Diabetes. In *2018 IEEE International Conference on Healthcare Informatics (ICHI)* (pp. 64–74). https://doi.org/10.1109/ICHI.2018.00015.
32. Dimitrov, D. V. (2016). Medical Internet of Things and big data in healthcare. *Healthcare Information Research, 22*(3), 156–163. https://doi.org/10.4258/hir.2016.22.3.156
33. Georga, E. I., Protopappas, V. C., Bellos, C. V., & Fotiadis, D. I. (2014). Wearable systems and mobile applications for diabetes disease management. *Health Technology, 4*(2), 101–112. https://doi.org/10.1007/s12553-014-0082-y

34. Baldini, G., Botterman, M., Neisse, R., & Tallacchini, M. (2018). Ethical design in the Internet of Things. *Science and Engineering Ethics, 24*(3), 905–925. https://doi.org/10.1007/s11948-016-9754-5
35. Islam, S. M. R., Kwak, D., Kabir, M. D. H., Hossain, M., & Kwak, K.-S. (2015). The Internet of Things for health care: A comprehensive survey. *IEEE Access, 3*, 678–708. https://doi.org/10.1109/ACCESS.2015.2437951.
36. Baker, S. E., Xiang, W., & Atkinson, I. M. (2017). Internet of Things for smart healthcare: Technologies, challenges, and opportunities. *IEEE Access*. https://doi.org/10.1109/ACCESS.2017.2775180
37. Gopal, G., Suter-Crazzolara, C., Toldo, L., & Eberhardt, W. (2019). Digital transformation in healthcare—Architectures of present and future information technologies. *Clinical Chemistry and Laboratory Medicine, 57*(3), 328–335. https://doi.org/10.1515/cclm-2018-0658.
38. Han, S. H., et al. (2010). Implementation of medical information exchange system based on EHR standard. *Healthcare Information Research, 16*(4), 281–289. https://doi.org/10.4258/hir.2010.16.4.281
39. Alkhatib, S., Waycott, J., Buchanan, G., & Bosua, R. (2018). Privacy and the Internet of Things (IoT) monitoring solutions for older adults: A Review. *Studies Health Technology Information, 252*, 8–14.
40. Wright, J. P., & Boisvert, D. (2009). What biosocial criminology offers criminology. *Criminal Justice and Behavior*. https://doi.org/10.1177/0093854809343140
41. Joyia, G. J., Liaqat, R. M., Farooq, A., & Rehman, S. (2017). Internet of Medical Things (IOMT): Applications, benefits and future challenges in healthcare domain. *Journal of Communication*. https://doi.org/10.12720/jcm.12.4.240-247
42. Kubo. (2014). The research of IoT based on RFID technology. In *2014 7th International Conference on Intelligent Computation Technology and Automation* (pp. 832–835). https://doi.org/10.1109/ICICTA.2014.199.

Chapter 2
Energy Efficient Hybrid IoT System for Ambient Living

Atif Saeed, Abdul Basit Aftab, and Faraz Junejo

Abstract This chapter aims to highlight the usage of IoT in the current world scenario and how it is extended and connected to the devices that we use in our daily life, making them smart and creating a global infrastructure for ambient living. The IoT offered making our life's easier; hence, increasing the quality of life through ambient living environments. The chapter also presents an in-detail knowledge regarding different IoT protocols available and current state of the art for ambient living using IoT. The chapter will also give its readers a future direction to the topic by including a case study regarding the 'Smart Geyser'. The case study concluded that implementing IoT on simple gas geyser is not only easy but it could give an energy savings up to 15%.

2.1 Introduction

2.1.1 What Is IoT?

IoT is the short form for the widely used term of Internet of Things. It explains the synchronization of computing and smart devices with physical hardware [16]. This connection between the physical and the cyber systems allows a seamless transfer of data (through network connectivity) without any interference from users. There are numerous examples of the system of IoT in the present world and the scope of IoT is showing meteoric rise than ever before. An example of this could be to

A. Saeed (✉) · A. B. Aftab · F. Junejo
Shaheed Zulfiqar Ali Bhutto Institute of Science and Technology (SZABIST), Clifton Block 5, Karachi, Pakistan
e-mail: m.atif@szabist.edu.pk

A. B. Aftab
e-mail: abdul.basit@szabist.edu.pk

F. Junejo
e-mail: faraz.junejo@szabist.edu.pk

P. Kumar et al. (eds.), *A Fusion of Artificial Intelligence and Internet of Things for Emerging Cyber Systems*, Intelligent Systems Reference Library 210,
https://doi.org/10.1007/978-3-030-76653-5_2

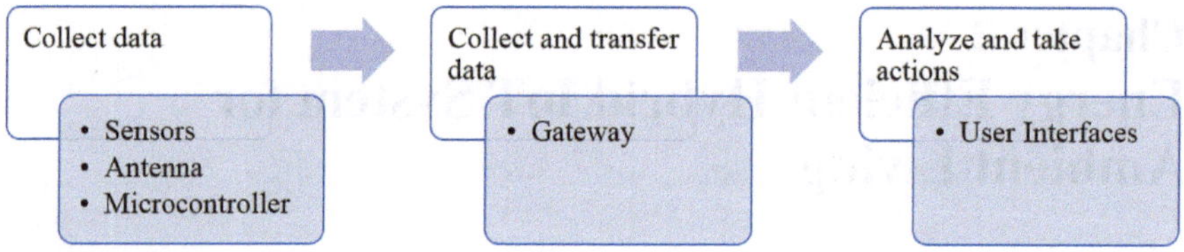

Fig. 2.1 IoT framework

acquire heart rate data using a simple heart rate monitor and connecting it through a network to seamlessly transfer data to the cloud or any web-based server [17]. The two prerequisites in this would be to a parameter that could be assigned an IP (Internet Protocol) and a connection network that could transmit that parameter for remote access. The system and application of IoT are sought out by companies more and more to increase the efficiency as well as increase the customer base [24]. All of this would eventually result in better decision-making and overall increase the effectiveness of the system. The IoT framework as shown in Fig. 2.1 is mostly based on web-based devices that are able to process or record some sort of data. Sensors, processors and communicating devices form the very skeleton of the IoT framework to gather data from their respective locations or data points and then transfer it to the desired internet protocols or server. All of this communication is done when the sensor which is recording the data connects to an internet portal for its logging to be done on cloud servers. Cloud servers are often programmed with algorithms to provide a better context to the incoming data. These devices are not only limited to transfer the data to the cloud server but often are able to communicate within them to provide a better input to the server. Human interaction is mostly limited but not prohibited [2]. Users can provide specific inputs to the framework which could in the form of data access or instructions. The type of sensors, connectivity capabilities are often dictated by the purpose or application of the IoT system. With the growing prowess in data analytics such as Machine Learning (ML) and Artificial intelligence (AI) which not only makes the collection of data easier but also the analysis of the corresponding data [25].

The workforce and general public are greatly benefitting from the progress made in the field of IoT. IoT has undoubtedly improved the standard of living as well as the convenience in people's lives. Business corporations are also not far behind in taking advantage of the IoT framework [32]. Business owners and managers are able to monitor and visualize the progress of their companies and even physical process at all times with zero lag or delay. Human dependance has also been brought to an all-time low as the processes can be easily automated, saving the companies extra labor costs. This creates a chain effect and eventually results in lesser down- times and delivery periods overall increasing the customer satisfaction and outreach of the company. It could easily be said that this rise of IoT based applications is only the start and this would only pick up more pace with the recent technological advancements [26].

From cottage industries to large-scale manufacturing, IoT has widened the scope for each sector of business. Below are listed some of the more shared advantage to the business:

- Continuous monitoring of business operations.
- Improvement in customer satisfaction (CX).
- Time-efficient and cost-effective solution.
- Improvement in employee potency.
- More practical business models .
- More sales and eventually profit.
- Better business deals.

The main motive of the IoT framework is to provide companies and businesses with a new, alternate strategy to empower them with unique ideas to increase their revenue and outreach. IoT has tremendous potential in the manufacturing and transportation industry as these departments are inherently remote and involve a variety of processes. As IoT is booming with extreme pace across all industries, it has found great appreciation and development in the agricultural sector as well as the home automation processes. Advanced sensors are available which can record any type of physical quantity or parameter such as the amount of rainfall or soil contents and even the humidity in the environment. These are only few of the examples, IoT could be very well be adapted by the construction industry for gauging different stresses and factor in the overall safety of the buildings [23]. As with all cases, IoT can be extremely beneficial to home automation setup with providing remote access of all the appliances of the house in the palm of our hands. These applications of IoT would encourage the reduction of waste especially paper waste. IoT can encompass every industry in the modern world including that of manufacturing, agriculture, health and finance [30].

The novelty or the newness presented in this research work is the implementation of a geyser system (operating on natural gas) using the IoT protocols. Very few working or consumer-type models have been presented in the industry with energy consumption data to compare of a conventional consumer geyser (operating on gas). This IoT model is previously implemented only on electric geysers which are practical in first-world or developed countries where power shortage is not a problem. Countries that are still developing enough infrastructure and generation like Pakistan have to rely upon their natural resources. Almost 90% of the geysers operated in Pakistan are operated on Natural gas [18]. Implementing IoT protocol on a conventional geyser is tricky with precise control of the valve controlling the gas influx to the heating flame or element.

2.1.2 History of IoT

The expression Internet of Things or IoT was coined back almost two decades ago. The actual concept of such a framework to incorporate cyber and physical systems has

Fig. 2.2 Kevin Ashton, inventor of the Internet of Things [29]

been around from at least the1970s. It was known by the term of 'embedded internet' back then. Kevin Ashton (Fig. 2.2) while working at Procter Gamble, supply chain division came up with this terminology. His main aim behind this term was to attract the higher management in the company and internet being the buzzword in the late nineties grasped their attention. Kevin was successful in gaining attention of a few but Internet of Things rose to real prominence after a decade.

In the middle of 2010, the idea of IoT gained much more momentum when it was leaked that Google Maps and Street View service was in addition to the photographs was also storing data regarding people's WIFI networks. This was a combination of storing both the physical as well as the cyber world. It was then picked up by the Chinese government and included IoT in their next five-year plan.

"Internet of Things" was picked up by one other company named Gartner at the start of the previous decade (2011) and included it in their emerging phenomena's list. Gartner is also attributed with the invention of 'Hype-cycle for emerging technologies. The very next year Europe saw the conference be a massive hit gathered around the concept of Internet of Things. Simultaneously other media magazines and outlets also added the term IoT to their vocabularies.

In late 2013, IDC only estimated the potential market and worth that IoT was able to collect. It was estimated that it would be at a whopping 8.9 trillion dollars by the end of this decade. By 2014, the term had become widely popular and was the mass public was very much accustomed to it. Tech giants like Google announced their pilot projects such as Nest for 3.2 billion dollars. CES (Consumer Electronic Show) was also prepared around the theme of IoT.

2.1.3 Energy Efficiency Using IoT

The use of IoT would help in lowering the costs as well as bring sustainability to the way businesses are being carried out. 2018 saw the rates of energy reach a five-year high to 10.67 cents per Kilowatt, even worryingly these rates were the second highest in the decade of 2010–2020 [36].

To better understand how IoT can play a part in reducing the cost is to look at how the whole framework works. The IoT is based on sensors, which collect data on whatever they are placed in and whatever parameter is required off them. This could be used to gauge electricity usage as to monitor and cancel any unnecessary usage. Once the trend or analytics of how electricity is being used in a particular facility, different strategies or plans could be implemented that could offer us valuable benefits both in terms of lowering the costs as well as sustainability's.

For example, with this information, high power consuming devices could be replaced with smart, green electronics. A traditional light bulb could be replaced by a LED. Overhead costs could be removed by effective and efficient monitoring through IoT systems. IoT based systems could be incorporated into newer renewable plants to achieve maximum throughput of both the facilities. Many other methods are being explained in the following paragraphs [22].

2.1.3.1 Ventilate Only When Necessary

This is pretty novel and logical. Sensors are placed in the environment to carefully monitor the quality of the air. One major indicator of the quality of the air is the amount of Carbon Dioxide in the air. The ASHRAE (American Society of Heating, Refrigerating and Air Conditioning Engineers) are also of the opinion to set standards and prevent over-ventilation in their facility. As more air is being replenished, it further needs cooling or heating which results in extra usage of electrical energy.

Application of a control ventilation system would help in the monitoring of Carbon Dioxide levels as well as prevent over ventilation. Only when the air quality is worse enough, will the air be replenished resulting in saving energy costs.

2.1.3.2 Reduce Your Energy Usage During Peak Times

Energy being one of the most vital resources available to man is often rated or calculated according to its demand and usage. This is why its demand and eventually its rates vary with different times of a day. In hot climate, the rates must be higher than that for a cooler season.

IoT devices can also curb the additional costs deliberated by these peak hours by using the phenomenon of peak shaving. In this the IoT devices carefully look at the trend of each and every appliance or device in a facility and determine which of those are more likely to operate in peak hours. Then adjustments are made to minimize these costs.

Smart meters are applying this concept very effectively as they collect the power usage of a particular facility and then carry out the whole concept of peak shaving.

2.1.3.3 Take Advantage of renewable Energy

Sustainability cannot be achieved without making the switch from fossil fuels to natural resources. Incorporating these new streams of inputs would not only help in reduction in costs but also improve the brand image in the eyes of mass public. IoT can be used to tap into numerous renewable resources including those of thermal, wind or solar energy.

Solar energy is one of the most abundant in the world. IoT can be used to setup solar power monitoring through the period of sunshine and other system parameters. IoT sensors can gauge the levels of these parameters and this would optimize the performance of the whole setup.

These measures are impactful but their real impact can be seen when they are put in combination with each other such as renewable resources combined with the monitoring power of IoT to reach the optimum level of performance.

2.2 IoT Protocols and Standards

2.2.1 IoT Protocols

One of the most important and critical components of the IoT framework are the IoT protocols. In the absence of these protocols, physical systems will be of no use to the whole system as they would not be able to exchange any kind of data between them and also to the cloud. The basic foundation of data interchange is based on this aspect and protocol of IoT. The protocols allow for the transfer of data packets and specifically the useful information to the end user resulting in a financially viable implementation of the IoT device management. As mentioned before, IoT involves a lot of two way information between the sensors, cloud, devices and users and that is why robust protocols need to be followed to make sure that the vital information is passed on without any corruption and without any leakage [1].

2.2.1.1 Constrained Application Protocol (CoAP)

This protocol provides a separate, limited and restricted protocol for the working of IoT devices. The current state of the art network facilities is free and can be used anyone. This makes it vulnerable to hacking as well as power consuming for the use of IoT. The CoAP converts HTTP models to work in restrictive environments. This protocol is highly dependent on the User Datagram Protocol (UDP) to deploy a secure and smooth communication path between the ends. This protocol also includes the ability to broadcast simultaneously to multiple endpoints resulting in the ability to maintain a connection between various hosts and networks. The communication speed is adequate due to the low bandwidth resulting in this protocol being suitable for

Fig. 2.3 CoAP protocol

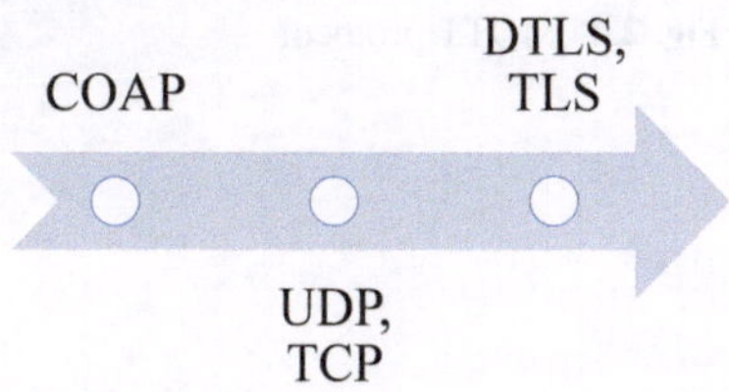

the wireless networks and especially in the M2M environments where the resources are limited. CoAP also uses a RESTful architecture which is based on request and response interaction between different devices. The basic flow of CoAP is shown in Fig. 2.3.

The protocol being discussed also provides a greater degree of control on the messages being sent and tag them as confirmable or not. This makes it easier for the devices as they could acknowledge. To add to these features, the CoAP includes the attribute of resource discovery and content negotiation. CoAP is an extremely easy and convenient protocol that meets the demands of a very basic IoT system and is often preferred to lightweight devices. CoAP is a very viable system that helps in IoT systems especially those already existing web-based clients [41].

2.2.1.2 Message Queuing Telemetry Transport (MQTT)

This protoçol (MQTT) is one of the most used and viable standards in the deployment of Internet of Things (IoT), especially in the industrial environment. This application is a very thin protocol used for messages. As it is made for portable devices, its structure has been kept rather fine and convenient to make sure that the power drainage is not immense. It is based on TCP/IP protocol which means that it takes into account any disturbances in communication networks [20].

The model that the MQTT adopts is a three-prong model with a subscriber, publisher and broker. The publisher is able to gather information and transmit it to the subscriber and the broker in between acts as a bridge. The main purpose of the broker is to make sure that the security standards are being followed to ensure no data leakage. The various modes of assuring quality and standard with in this protocol are listed below:

- QoS0 = is the fastest method as the confirmation is not required from the subscriber.
- QoS1 = is ensures only one confirmation message but the messages sent from the publisher might be multiple.
- QoS2 = is the most reliable method as confirmation is received and the number of duplicate messages is controlled.

MQTT has found wide applications in various devices such as electric meters, detectors and industrial machinery. MQTT has the apt number of features to fulfill the following needs:

Fig. 2.4 MQTT protocol

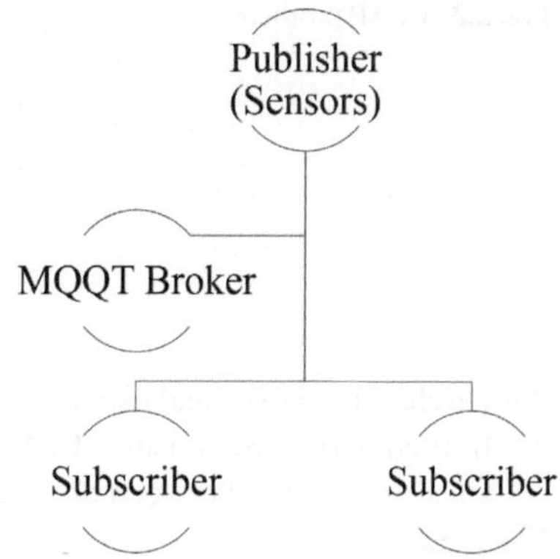

- Limited usage of bandwidth.
- Wireless connectivity.
- Low power drain.
- Reliable.
- Less load on processing power and memory.

MQTT does have its limitations and one of them is that the environment that this protocol offers is the a very restrictive environment. MQTT-SN solves this problem by using a UDP protocol and supports name indexing. MQTT also is not well suited to data visualization and device management. This is why the main onus is on the device to manage and represent data as shown in Fig. 2.4.

2.2.1.3 WiFi

WIFI is one of the most common and important technology in the contemporary world. Wi-Fi networks are based on sending signals over the air to various devices which may be smartphones, computers or even cloud servers. Routers are one of the most common devices which are responsible to disperse the network signals all throughout the house and eventually provide a network connection to the devices in the home. With modern smartphones, they can be converted into WIFI hotspots which results in mobile phone sharing their networks even if the routers are not operating or present.

The main technology that is used to transmit wireless signals is radio waves. Radio waves have the frequency either 2.4 or 5 GHz. These frequencies are in the range in which different devices can work. These frequencies are far apart to make sure that the load is distributed and not burdened on only one frequency.

A WIFI connection has a typical reach of about 100 m. However, the most common WIFI routers are only up to 35 m. There are several factors that play an important part in the transmission of WIFI signals which include the antenna strength as well as the frequency of transmission. Other external factors also play a very important role which concludes the environment. One simple rule applies that the closer the device to the main transmission device, the stronger the wireless connection between the two is established [34].

2.2.1.4 ZigBee

ZigBee is especially designed for low power applications and devices. Its throughput is comparatively low as compared to others at only 250 kbps. The connectivity reach for ZigBee is 100 m between access points. The most common implementation of the ZigBee protocol is for the sensors and personal WPAN. This makes it suitable for low power home automation and monitoring systems. In 2003, its specifications were classified as IEEE standard but its widescale adoption by devices did not start by the start of 2006.

The main purpose behind the discovery of ZigBee was to find a protocol for short range devices that had the ability to adapt and change automatically without any user input. It was supposed to be used mainly in communication devices such as different sensors and monitoring devices. This protocol is very convenient to setup and not prone to corruption of information. Its prerequisites are relatively cheap and simple like microcontroller and sensors to gather information.

Features like self-assembly and self-healing grid topology help the ZigBee to be convenient and scalable to multiple and several nodes making it one of the most preferred choices for suppliers and developers [14].

2.2.1.5 Bluetooth

Another technology that enables wireless connection between devices is the Bluetooth technology and it is widely available in mobile phones and smart devices. It is an open-standard specification with IEEE. The reach of the Bluetooth technology varies according to the source antenna and could be ranging from 1 to 100 m in open space. Typically, the most common one is with the range of 10 m and is widely used in portable devices. Connection between devices on different floors as well as different rooms can be made using the 2.4 GHz frequency.

Bluetooth technology makes use of data packets to transfer data and does this in about 79 channels. This standard has been carried out from the very first generation of Bluetooth (Bluetooth 1.0) and a bandwidth of 1 MHz. All of these specifications make the maximum data transfer speed at 721 kbit/s. However, the latest Bluetooth standard of 4.0 has pumped the transfer speed to a whopping 3.0 Mb/s. Each iteration of Bluetooth standard is also backward compatible which makes legacy applications and devices still workable.

Bluetooth is a patented technology and most of its technology is free and products could incorporate Bluetooth to make it Bluetooth compatible [9].

2.2.1.6 Extensible Messaging and Presence Protocol (XMPP)

The credit of introduction of this protocol goes to the Jabber open source community. It was first introduced in 1999 and its main purpose was for real-time communication through messages. It is based on the XML language and enables exchange of messages and other elaborate data at real-time between multiple devices.

Ever since XMPP has been introduced it has been adopted more by the communication frameworks. As it is a lightweight client, XMPP-IoT is used more and more in the context of Internet of Things. In addition to being lightweight, it also has the advantage of being able to scale up for different applications which is why developers are preferring it over other protocols, providing a much better experience at the consumer end.

It does have a few limitations which includes no quality assurance neither any encryption which is imperative to maintain privacy. This is the reason that some consumers are wary of using it as they believe that their sensitive data might be leaked. The protocol needs to work on its end-to-end encryption to make sure that it is widely accepted in the industrial landscape and is often replaced with the protocols such as MQTT or LwM2M [19].

2.2.1.7 Data-Distribution Service (DDS)

DDS framework of technology makes use of the publish-subscribe methodology. Object Management Group (OMG) is responsible for the introduction of this protocol. This protocol enables multiple new benefits that were not possible before such as the ability to communicate in real time between connected devices with robust results and a structure that could be scaled to different sizes. Other positives include that it reinforces broker less architecture and multicasting to ensure greater standards and operations.

The basic skeleton of the DDS framework is based on the DCPS (Data Centric Publish-Subscribe) layer and the DLRL (Data-Local Reconstruction Layer). Subscribers are given a resourceful, efficient system that could be scaled to different sizes by the DCPS Later whereas the DLRL offers an outlet to the DCPS layer for data exchange between different Internet of Things devices.

The DDS protocol could not be counted as run-of-the-mill Internet of Things solution but it is still able to hold its place among the various other options available. Its application has been beneficial in the management of smart grid as well as the new talk of the town, autonomous vehicles and robotics. In addition to the before mentioned applications, the DDS framework can also be applied to the data interchange management between smaller applications and devices and scaled up to almost higher applications. Data exchange from the cloud is also possible using this framework [47].

2.2.1.8 Advanced Message Queuing Protocol (AMQP)

This is an open standard protocol with the ability to publish and subscribed. It was introduced in 2003 and builds upon the works of financial services sector. This technology is still primitive or in its early stages when in talked in context of the Internet of Things (IoT) Industry. This protocol has picked up considerable momentum in the information communication department. The single most important benefit of the AMQP is the durability and quality it provides in the communication model. It assures proper transactions. IoT applications are not always looking for these qualities but are nevertheless important in data exchange. It is not preferred in the application of sensors with confined resources that may include bandwidth, processing power or network connectivity.

2.2.1.9 Lightweight M2M (LwM2M)

The basic difference between the various other protocols from this one is that this protocol is specifically designed for IoT and its resource handling and for those devices that have certain limitations in terms of their resource. It originated in 2014 by the Open Mobile Alliance (OMA). It is a comprehensive platform for IoT data exchange and device management.

2.2.2 Importance of Different IoT Protocols

These different IoT protocols are the reason and difference between a smart device and an ordinary device. These protocols enable the device to actually report the cause of any failure to other or 'mother' devices. This is the basic trait along with being sync to other devices and communicate that offers any device any ability to be smart. This communication can only happen when there is a common medium in between the two devices, these protocols provide the basis for these mediums and result in effective communications between the two devices. These protocols could be the existing ones already in use or newly developed for the purpose of communication between these devices.

Furthermore, the presence of specific protocols for communication between devices results in the elimination or at least the minimization of clutter and fragmentation that would arise if different frameworks were being used for each new single device. Different frameworks also pose a threat to the devices and overall system as these makes them make vulnerable to security failures and threats.

As the issue of unification of protocols is pretty logical and acknowledged by all, very few efforts have been put in this direction to come to a uniform platform or protocol for the IoT industry. In recent times, a boom of new IoT protocols is witnessed and each of the protocols has taken the challenge to offer flexibility as well as security benefits to the IoT industry. OMA lightweight M2M is one of those protocols that has greatly reduced the needs of other protocols as it results in greater security benefits as well as the ability to provide a wide range of solutions to the IoT applications.

Fragmentation in IoT application is a reality because of the very decentralized nature of any IoT application. This is because multiple devices are used to connect to each other and with each new device comes its different caveats and frameworks. This creates a very diverse ecosystem of IoT devices with each devices only needing certain features in their protocols. IoT protocols are often divided with their functionality as the connectivity infrastructure (6LowPAN) , data transmission (MQTT, CoAP) and communications (Wi-Fi, Bluetooth).

2.3 IoT in Smart Living

The main purpose of introducing the concept of connected devices and an overall system where each and every process happens in sync to the other devices is to make life easier for the common person. One other purpose that it serves is that it results in a smart ecosystem that is aware of the resources that it has to use and eventually uses its resources mindfully for a sustainable future [11]. Below are mentioned some of the smart home applications that can help in the convenience as well as the sustainability of our future lives [42].

2.3.1 Features of IoT Based Smart Living

2.3.1.1 Communicate and Interact with the User

The concept home is a very basic and interesting phenomena. The fact that each and every appliance of a house is connected to each other and the user and at real time communicate within would really help in making our lives easier. All the different sensors, actuators and other controllers in the house are linked to one other through a central connection and the user can program different scenarios as well as different configurations according to different time and situations. This would save immensely on the energy costs as for example all the lights be dimmed when its night time or all lights turned off for unoccupied rooms. Several smart devices are available in the market that provide a better contextual awareness to the sensor that would further make energy management system look like a walk in the park.

2.3.1.2 Using a Wi-Fi Based Wireless Sensor Network

The type of automation proposed in this system is that the user could control the appliances of the home remotely. The control panel appears on his/her phone or PC and he/she could perform the required action. This automation could also be done using the application of motion or proximity sensors to detect the presence of people and then turn the desired appliance on or off.

2.3.1.3 Personal Assistance

One other form of home automation could be the use of virtual assistants that are made in charge of the house affairs. These assistants then using artificial intelligence (AI) and Machine learning (ML) run the tasks of the house and notify the user with the help of speakers and screens.

2.3.1.4 Saving People Time and Energy

All of these home automation techniques are used for two basic purposes that is to conserve energy and time. By using all of these devices, we are putting all the redundant tasks of our daily life in the hands of the computer to handle using smart hubs and smart homes.

2.3.2 IoT Based Smart Products for Ambient Living

Internet of things is based upon internet. The more devices that are compatible with the internet, the more we can widen the scope of IoT applications. Already smartphones, computers, laptops are fully connected to the internet stream and make intercommunication between them possible and efficient. These devices also provide remote control of any such system based on IoT. It is a staggering fact that the number of connected devices has outnumbered the human population on earth.

The production of IoT devices is seeing a meteoric rise. These products are greatly in demand including smart gadgets, watches, laptops and mobile phones. Even vehicles are being atomized and digitized and are wanted by the people even more.

2.3.2.1 IoT Products Life Cycle

The list continuous to grow and many more IoT based devices are introduced that made human life much easier. However, the purpose of this chapter is to put a light on device that is not much discussed and considered to be a new in IoT based product line [44]. The basic cycle can be seen in Fig. 2.5.

IoT products follow a very simple development cycle. Deployment is the first step and the processes of monitoring, servicing and managing follow after that. Few of the very popular internet of things devices are mentioned below:

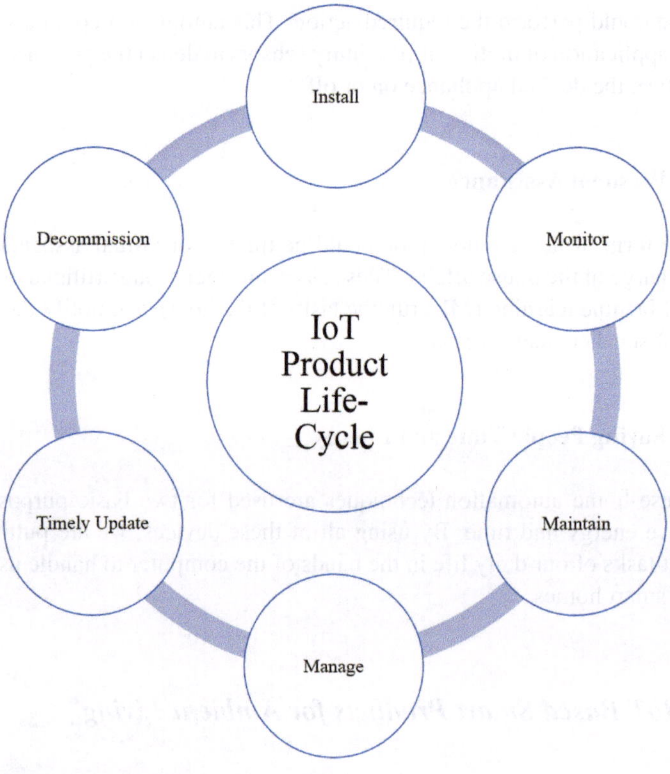

Fig. 2.5 IoT product lifecycled

- Google Home Voice Controller
- Amazon Echo Plus Voice Controller
- Amazon Dash Button
- August Doorbell Cam
- August Smart Lock
- Kuri Mobile Robot
- Belkin WeMo Smart Light Switch
- Flow by Plume Labs Air Pollution Monitor
- Nest Smoke Alarm
- Nest T3021US Learning Thermostat Easy Temperature Control
- Philips Hue Bulbs and Lighting System
- Bitdefender BOX IoT Security Solution
- IoT based air conditioners.

2.4 State of the Art in IoT Technologies for Ambient Living

The research areas of smart devices for ambient assisted living (AAL) holds a major chunk of IoT and isn't restricted to smart home, controlled environment, Adaptive and self-decision-making algorithms and sensing and actuating technologies. Previously, many published researches showcased enough work related to above mentioned filed highlighting its current state of the art and advance proposed features to improve the quality of life and work. The research [17] presented the combination of intelligent personal assistants, cloud networking and IoT based smart sensing network to provide a smart environment for older or disabled people. The research articles [4, 6, 43] summarizes many subtopics and features related to ambient assisted living like human health monitoring through wear- able gadget, smart activity recognition or alert system, mood monitoring, adaptive devices and user interfaces, application of various algorithms on collected data, etc. Also, [5] gives brief review on IoT based ecosystem that could be the thing of future. As discussed, usually the AAL is related to smart homes or environment as presented in [38, 45] but the concept of smart buildings was bought up in [13]. For controlling, sensing and triggering purposes [15] presented an approach to sense human presence using wireless sensor networks (WSN), the protocol adjustment of such sensing network is usually based on scattered mobile terminals [3]. Home automation using emotion and mood recognition were also presented in research [28] however, still work is required on it for its large-scale implementation. Another prominent area to think of is IoT based smart security-based system, this emerging field could bring a revolution since many IoT devices does have basic security protocols but doesn't have a one window solution for all, the researchers tries to overcome this gap [8, 21]; however, still various firms are still underway of its development. Yet another work on the concept of smart cities and services is carried out in [10] and IoT is implemented for educational and entertainment purposes and also smart museums [35] system based on localization technique. Some of the researches also presented systems like autonomous parking stations [31], a WSN based navigation system [37], Safety Systems [33] and AI based activity prediction system [27], all of these researched are meant to make our life's easier and create an environment of ambient assisted living AAL.

2.5 Future Direction for Ambient Assisted Living

2.5.1 Case Study: Smart Geysers

2.5.1.1 Need of Automation

Because of the rising competition in society, the value of time has increased. Moreover, as everyone has a busy schedule, we prefer appliances that require minimum effort and saves time. People have their homes automated to easily control their appliances with their smartphones, or using voice recognition. Appliances such as tank gas geyser need to be automated as well because of their complex operation and the time they take. The normal tank gas geysers take up time, are not easy to turn on and off, waste quite an amount of gas, and are not ecofriendly. Since they are not very convenient to turn on and off, people tend to keep them on for a longer period of time and during the winters, they usually keep it on throughout the season, this results in gas consumption and unnecessary pollution. With an alarming brink of global warming [12], the least we can do is reduce this consumption and automation does not only help us with this but save time as well. Usually, with tank gas geysers, we do not have constant supervision on gas leakage and flame/temperature control, this can be fixed as well and gas leakage can be detected using automation [40].

2.5.1.2 What Is Smart Geyser?

Smart Geyser machine kit is a control box kit that can be easily integrated to any tank water geyser making the device communicate with a smartphone through iot, this makes it easier for the user to control their geyser and turn it on or turn is off with just a push of button, they can monitor their geyser parameters like temperature, water presence, flame status, gas leakage. There is time scheduled turn on and turn off in the device too so that it will perform its action daily on the fixed scheduled time.

2.5.1.3 Working Framework of Smart Geyser

Once the geyser is turned on the pilot flame is always on producing some pollution and wasting gas and a lot of population use gas water geyser and throughout the winters it always stays on so had to come up with some technique that lights the geyser up only when want and turn it off when don't without wasting a significant amount of gas, not only needed a convenient way to control the geyser too, therefore had chosen iot to control the whole hardware and monitor it's parameters on a single mobile screen. The pilot performs three major functions:

- Heat the thermocouple to create small electric current that triggers the operation of gas control.
- Signals gas control valve that pilot is lit,
- Ignites the gas from the main burner when it is released [7].

If the pilot light is ON and thermocouple is under the influence of the pilot flame, and the thermocouple is working and making electrical connection to the gas control, then the electromagnet is energized. The electromagnet holds open the safety valve for as long as the pilot flame is heating up the thermocouple. The block diagram of the system is given below in Fig. 2.6.

On the app, have a control button and timer input function as well now when the user input the specific time interval from a real time clock from the phone the

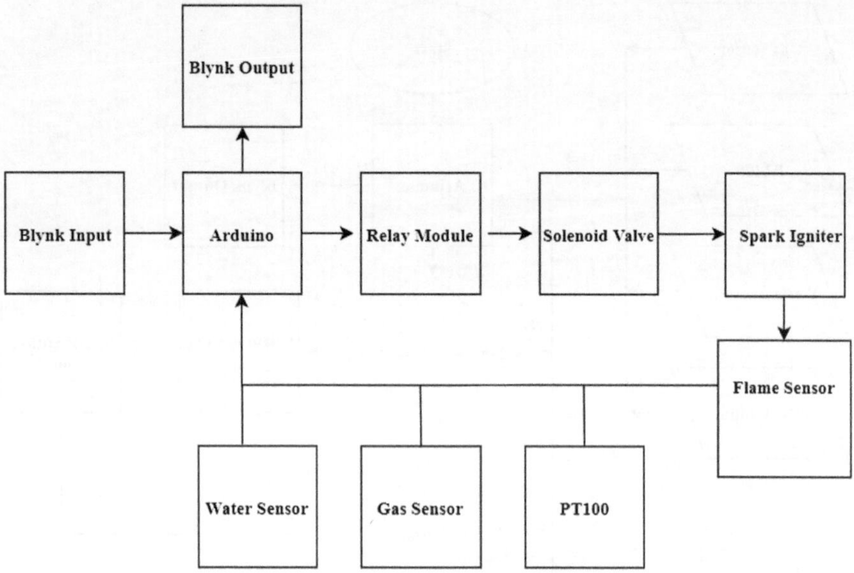

Fig. 2.6 Smart Geyser block diagram

MCU checks if the real time matches with the input put time range if yes then MCU sends signal to relay to switch the solenoid valve on and the igniter as well and since already discussed that the flame sensor and the spark igniter are in close loop so the MCU will continuously check if the flame is on to turn off the igniter, the geyser parameters like water status, gas leakage and flame statuses are displayed on the app and the temperature of the water is displayed in the form of numeric value and graph as well. The process flow of system is given below in Fig. 2.7 explains how it works.

2.5.1.4 Mobile Application for User Interface

Blynk is a communication protocol for mobile to device interface using internet or Bluetooth, With an ease of access and template formation blynk can connect several boards like nodemcu, Arduino, rasberry pi, etc. can control hardware remotely with ease, also can monitor sensor data, and can do various tasks from controlling to monitoring [39]. Blynk server is a bridge that connects the smartphones and IoT devices. The GUI can be seen in Fig. 2.8 User has the authority to use it under local network. Being an open-source application, and have ability to handle thousands of devices and can even work with raspberry Pi. The final working prototype of the system can be seen in Fig. 2.9.

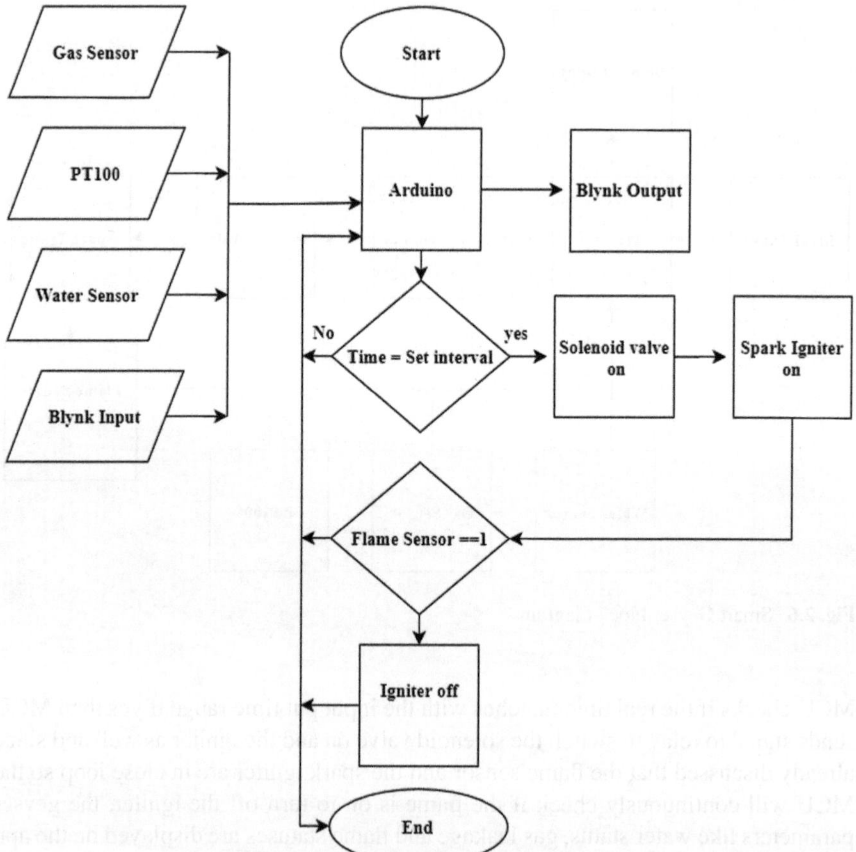

Fig. 2.7 Smart Geyser process flow diagram

2.5.1.5 Energy Savings Using Smart Geyser

The IoT based device help user conserve energy by continuously monitoring the temperature of water and automatically turn on or off the supply of gas when geyser didn't want it. In most of the household this is a very common practice to turn on geyser before using it and then after reaching the specific temperature the pilot mode got turn on and geyser continuously consumes energy even if it's in small quantity. The smart geyser will not only turn off the gas when a certain temperature reached but also automatically turns it on when temperature falls below the set threshold by user. The smart geyser was observed for a period of 1year in order to observe the gas savings and it was observed that over the period of one year the gas savings was around 15 %, the readings can be seen in Table 2.1 and its graphical representation is given in Fig. 2.10.

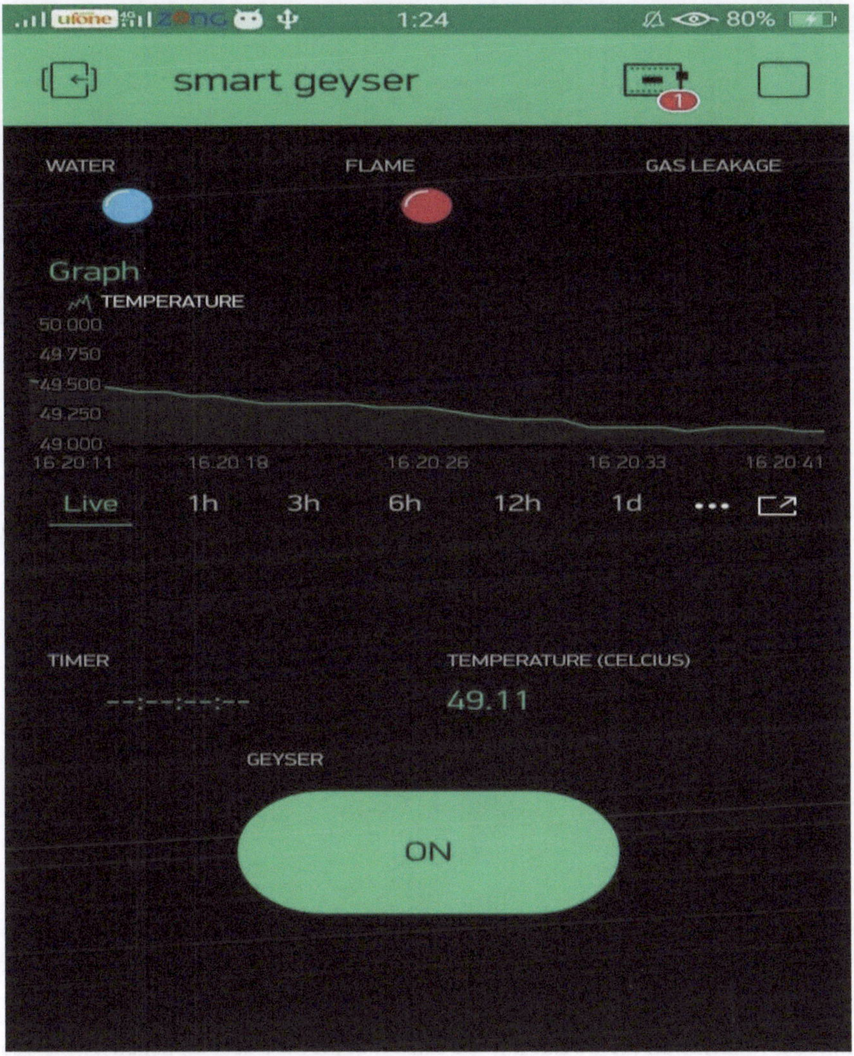

Fig. 2.8 Smart Geyser user interface

2.5.1.6 Is Smart Geyser Sustainable?

The sustainable development goals are the plan layout to achieve a preferable, steady, and a better future for all. It confronts the global challenges are being faced or would face if action is not taken at this very stage [46]. Issues such as climate changes, unemployment, poverty, inequality, pollution, environmental degradation, peace, and justice are challenged through this layout. There are around 17 highlighted issues that need urgent attention and action; these highlighted issues are all interconnected.

Fig. 2.9 Final working prototype

In order to tackle all these issues, it is a must that they are accomplished by the year 2030 or else it can become quite impossible for us to face these issues, The goals are:

- GOAL 1: No Poverty
- GOAL 2: Zero Hunger
- GOAL 3: Good Health and Well-being
- GOAL 4: Quality Education
- GOAL 5: Gender Equality
- GOAL 6: Clean Water and Sanitation
- GOAL 7: Affordable and Clean Energy
- GOAL 8: Decent Work and Economic Growth
- GOAL 9: Industry, Innovation and Infrastructure

Table 2.1 Monthly energy savings using smart Geyser

Month	Units consumed without smart Geyser (Mcf)	Units consumed with smart Geyser (Mcf)
1	300	252
2	325	266.5
3	312	277.68
4	338	270.4
5	330	287.1
6	223	189.55
7	240	216
8	220	187
9	180	144
10	210	178.5
11	212	180.2
12	256	212.48

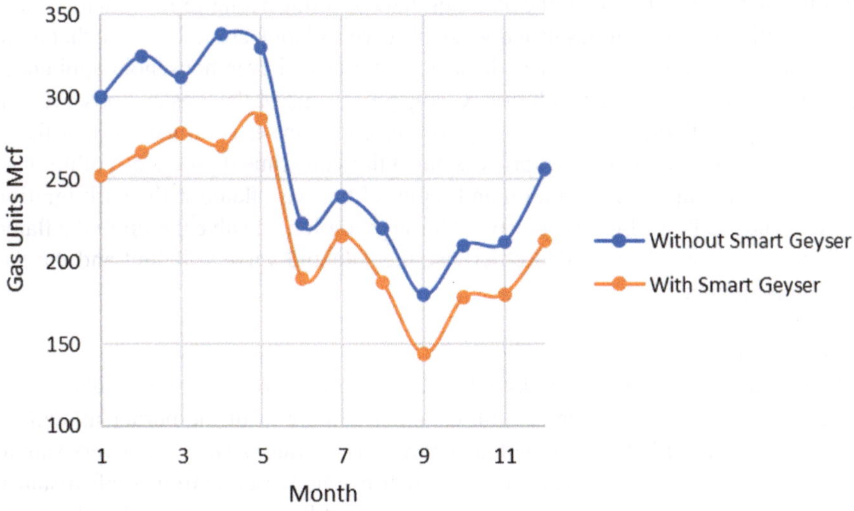

Fig. 2.10 Energy savings using smart Geyser

- GOAL 10: Reduced Inequality
- GOAL 11: Sustainable Cities and Communities
- GOAL 12: Responsible Consumption and Production
- GOAL 13: Climate Action
- GOAL 14: Life Below Water
- GOAL 15: Life on Land
- GOAL 16: Peace and Justice Strong Institutions
- GOAL 17: Partnerships to achieve the Goal

Smart Geyser covers three main goals from the above

- GOAL 9: Industry, Innovation and Infrastructure
- GOAL 13: Climate Action
- GOAL 15: Life on Land

Industry, Innovation and Infrastructure
Mechanical thermostat tank water gas geyser was taken and upgraded it with electronic kit that will make it work more efficiently and will conserve gas. The kit is very easy to install the sensors are very easy to integrate in a normal house hold gas geyser.

Climate Action
Air pollution affects virtually every country in the world and every part of society. Only one in ten people lives in a city that meets WHO air quality guidelines. Air pollution in homes is an important risk factor for an estimated 2.9 billion people worldwide, especially those in low- and middle-income countries where biomass and carbon fuels are often burned for cooking and heating. In total, approximately 2 billion children live in areas that exceed WHO's annual fine particle guideline of 10 μ g/m^3 (the concentration of fine particles that constitute a long-term inhalation risk) so basically this a serious issue and when observe and look around and see that there are countless number of vehicles, devices, factories and even household appliances that are responsible for air pollution. Gas geyser is one of them as can observe that there is a pilot flame that always stays on once the geyser is turned on, that flame controls the geyser combustion process but at the same time it produces pollution so therefore came up with a solution and replaced the pilot flame with spark ignition system and the flow of gas is controlled through a solenoid valve therefore the flame will only turn on when needed otherwise the solenoid valve will shut and the gas will stop flowing.

Life on Land
The operation of a geyser is risky when want to turn it on, sometimes when there is gas leaking inside the combustion chamber and try to lit the burner up can do some serious hazard, thus can say that a normal tank water geyser is not very safe to operate. Smart geyser gives you an option to turn your geyser on from a safe distance it gas feedback sensors and conditional sensors that keeps you updated with geyser status and parameters.

2.6 Conclusion

The use of AAL technologies is increasing day by day with the increasing population of older people. IoT and WSN made the implementation of AAL easy and has played a vital role in its increasing popularity. Support for active and independent living requires addressing the main challenges of IoT such as data veracity, integration, interoperability, privacy, reliability, security, and usability.

Various recent research showcased the challenges faced by IoT based AAL technologies such as the advanced coding to develop AI based algorithms to make system intelligent enough for decision making, control, domain related issues, user friendly interfaces and SOPs to develop wearables. However, the long-term benefits for the implementation of AAL are listed below.

- Improved life quality in terms of office, home, spare time activities, social gatherings through innovative wearable technologies;
- Smart cities and digital transportation systems;
- Easy to access public sectors through the implementation of smart e-services;
- Better security;
- Advanced data gatherings and statistical analysis for public benefits;
- Stable economy through smart and digital currencies.

References

1. Al-Fuqaha, A., Guizani, M., Mohammadi, M., Aledhari, M., & Ayyash, M. (2015). Internet of things: A survey on enabling technologies, protocols, and applications. *IEEE Communications Surveys & Tutorials*, *17*(4), 2347–2376.
2. Amin, I., & Saeed, A. (2018). 5.10 wireless technologies in energy management
3. Beibei, T., & Yi, L. (2015). Upgraded application of intelligent environment monitoring system in iot smart home. In *2015 Sixth International Conference on Intelligent Systems Design and Engineering Applications (ISDEA)* (pp. 916–919). IEEE.
4. Bravo, J., Hervás, R., & Rodríguez, M. (2012). *Ambient Assisted Living and Home Care: 4th International Workshop, IWAAL 2012, Vitoria-Gasteiz, Spain, December 3–5, 2012, Proceedings*, (Vol. 7657). Springer.
5. Calvaresi, D., Cesarini, D., Sernani, P., Marinoni, M., Dragoni, A. F., & Sturm, A. (2017). Exploring the ambient assisted living domain: a systematic review. *Journal of Ambient Intelligence and Humanized Computing*, *8*(2), 239–257.
6. Catarinucci, L., De Donno, D., Mainetti, L., Palano, L., Patrono, L., Stefanizzi, M., et al. (2015). An iot-aware architecture for smart healthcare systems. *Ieee Internet Things*, *2*(6), 515–526.
7. Catherine, Q., Wheeler, J., Wilkinson, R., & de Jager, G. (2012). Hot water usage profiling to improve geyser efficiency. *Journal of energy in Southern Africa*, *23*(1), 39–45.
8. Chakrabarty, S., & Engels, D. W. (2016) A secure iot architecture for smart cities. In *2016 13th IEEE Annual Consumer Communications & Networking Conference (CCNC)* (pp. 812–813). IEEE.
9. Chang, K. H. (2014). Bluetooth: a viable solution for iot?[industry perspectives]. *IEEE Wireless Communications*, *21*(6), 6–7.
10. Chianese, A., Piccialli, F., & Valente, I. (2015). Smart environments and cultural heritage: A novel approach to create intelligent cultural spaces. *Journal of Location Based Services*, *9*(3), 209–234.
11. Dohr, A., Modre-Opsrian, R., Drobics, M., Hayn, D., & Schreier, G. (2010). The internet of things for ambient assisted living. In *2010 Seventh International Conference on Information Technology: New generations* (pp. 804–809). Ieee.
12. Drake, F. (2014). *Global warming*. Routledge.
13. Ghayvat, H., Mukhopadhyay, S., Gui, X., & Suryadevara, N. (2015). Wsn-and iot-based smart homes and their extension to smart buildings. *Sensors*, *15*(5), 10350–10379.
14. Gill, K., Yang, S. H., Yao, F., & Lu, X. (2009). A zigbee-based home automation system. *IEEE Transactions on consumer Electronics*, *55*(2), 422–430.

15. Gračanin, D., Matković, K., & Wheeler, J. (2015). An approach to modeling internet of things based smart built environments. In *2015 Winter Simulation Conference (WSC)* (pp. 3208–3209). IEEE.
16. Gregorio, F., González, G., Schmidt, C., Cousseau, J. (2020). Internet of things. In *Signal Processing Techniques for Power Efficient Wireless Communication Systems* (pp. 217–245). Springer.
17. Gubbi, J., Buyya, R., Marusic, S., & Palaniswami, M. (2013). Internet of things (iot): A vision, architectural elements, and future directions. *Future generation computer systems*, *29*(7), 1645–1660.
18. Hassan, M. S., Tahir, M. N., Wajid, A., Mahmood, H., & Farooq, A. (2018). Natural gas consumption and economic growth in pakistan: production function approach. *Global Business Review*, *19*(2), 297–310.
19. Hornsby, A., Belimpasakis, P., & Defee, I. (2009). Xmpp-based wireless sensor network and its integration into the extended home environment. In *2009 IEEE 13th International Symposium on Consumer Electronics* (pp. 794–797). IEEE.
20. Hunkeler, U., Truong, H. L., Stanford-Clark, A. (2008). Mqtt-s—publish/subscribe protocol for wireless sensor networks. In *2008 3rd International Conference on Communication Systems Software and Middleware and Workshops (COMSWARE'08)* (pp. 791–798). IEEE.
21. Inukollu, V. N., Arsi, S., & Ravuri, S. R. (2014). Security issues associated with big data in cloud computing. *International Journal of Network Security & Its Applications*, *6*(3), 45.
22. Jo, H., & Yoon, Y. I. (2018). Intelligent smart home energy efficiency model using artificial tensorflow engine. *Human-centric Computing and Information Sciences*, *8*(1), 1–18.
23. Kazim, M. F., Hameed, S., Hamza Khan, M., Bilal Taak, M., & Azfar Khan, A. (2018). *Indian Journal of Science and Technology*, *11*, 46.
24. Khan, M. A., & Salah, K. (2018). Iot security: Review, blockchain solutions, and open challenges. *Future Generation Computer Systems*, *82*, 395–411.
25. Khan, R., Khan, S. U., Zaheer, R., & Khan, S. (2012). Future internet: the internet of things architecture, possible applications and key challenges. In *2012 10th International Conference on Frontiers of Information Technology* (pp. 257–260). IEEE.
26. Khatua, P. K., Ramachandaramurthy, V. K., Kasinathan, P., Yong, J. Y., Pasupuleti, J., & Rajagopalan, A. (2020). Application and assessment of internet of things toward the sustainability of energy systems: Challenges and issues. *Sustainable Cities and Society*, *53*, 101957.
27. Kim, B., Kang, S., Ha, J. Y., & Song, J. (2015). Agatha: Predicting daily activities from place visit history for activity-aware mobile services in smart cities. *International Journal of Distributed Sensor Networks*, *11*(12), 867602.
28. Konstantinidis, E. I., Billis, A., Savvidis, T., Xefteris, S., & Bamidis, P. D. (2017). Emotion recognition in the wild: Results and limitations from active and healthy ageing cases in a living lab. In *eHealth360°* (pp. 425–428). Springer.
29. Kramp, T., Van Kranenburg, R., & Lange, S. (2013). Introduction to the internet of things. *Enabling Things to Talk* (pp. 1–10). Berlin, Heidelberg: Springer.
30. Krotov, V. (2017). The internet of things and new business opportunities. *Business Horizons*, *60*(6), 831–841.
31. Lee, C., Han, Y., Jeon, S., Seo, D., & Jung, I. (2016). Smart parking system using ultrasonic sensor and bluetooth communication in internet of things. *KIISE Transactions on Computing Practices*, *22*(6), 268–277.
32. Lee, I., & Lee, K. (2015). The internet of things (iot): Applications, investments, and challenges for enterprises. *Business Horizons*, *58*(4), 431–440.
33. Lopes, N. V., Santos, H., & Azevedo, A. I. (2015). Detection of dangerous situations using a smart internet of things system. In *New Contributions in Information Systems and Technologies* (pp. 387–396). Springer.
34. Maier, M., & Ghazisaidi, N. (2011). *FiWi access networks*. Cambridge University Press.
35. Mighali, V., Del Fiore, G., Patrono, L., Mainetti, L., Alletto, S., Serra, G., & Cucchiara, R. (2015). Innovative iot-aware services for a smart museum. In *Proceedings of the 24th International Conference on World Wide Web* (pp. 547–550)

36. Pan, J., Jain, R., Paul, S., Vu, T., Saifullah, A., & Sha, M. (2015). An internet of things framework for smart energy in buildings: designs, prototype, and experiments. *IEEE Internet of Things Journal*, *2*(6), 527–537.
37. Paredes, H., Fernandes, H., Sousa, A., Fernandes, L., Koch, F., Fortes, R., Filipe, V., & Barroso, J. (2015). Exploring smart environments through human computation for enhancing blind navigation. In *International Workshop on Multiagent Foundations of Social Computing* (pp. 66–76). Springer.
38. Perumal, T., Datta, S. K., & Bonnet, C. (2015). Iot device management framework for smart home scenarios. In *2015 IEEE 4th Global Conference on Consumer Electronics (GCCE)* (pp. 54–55). IEEE.
39. Rif'an, M, et al. (2019). Internet of things (iot): Blynk framework for smart home. *KnE social sciences* (pp. 579–586).
40. Ryan, J. (1989). Home automation. *Electronics & Communication Engineering Journal*, *1*(4), 185–192.
41. Shelby, Z., Hartke, K., & Bormann, C. (2014). The constrained application protocol (coap).
42. Sun, Y., Song, H., Jara, A. J., & Bie, R. (2016). Internet of things and big data analytics for smart and connected communities. *IEEE Access*, *4*, 766–773.
43. Suryadevara, N. K., & Mukhopadhyay, S. C. (2014). Determining wellness through an ambient assisted living environment. *IEEE Intelligent Systems*, *29*(3), 30–37.
44. Tao, F., Wang, Y., Zuo, Y., Yang, H., & Zhang, M. (2016). Internet of things in product life-cycle energy management. *Journal of Industrial Information Integration*, *1*, 26–39.
45. Wang, D., Lo, D., Bhimani, J., & Sugiura, K. (2015). Anycontrol–iot based home appliances monitoring and controlling. In *2015 IEEE 39th Annual Computer Software and Applications Conference* (Vol. 3, pp. 487–492). IEEE.
46. Water, U. (2018). *Sustainable development goal 6 synthesis report on water and sanitation* (p. 10017). New York: Published by the United Nations New York.
47. Yang, J., Sandström, K., Nolte, T., & Behnam, M. (2012). Data distribution service for industrial automation. In *Proceedings of 2012 IEEE 17th International Conference on Emerging Technologies & Factory Automation (ETFA 2012)* (pp. 1–8). IEEE.

Chapter 3
Analysis of Agriculture Production and Impacts of Climate Change in South Asian Region: A Concern Related with Healthcare 4.0 Using ML and Sensors

Rohit Rastogi, Sarvam Maheshwari, Priyanshi Garg, Mukund Rastogi, and Pardeep Kumar

Abstract The Effect of Global Warming and rapid changing climate in an indefinite manner is a major concern and all domain of science are trying to address it in their ways. It is not only creating challenges to food production, yet to the human health. The presented research work is all about the prediction of the yield of agriculture of the land without involving any activity of humans and this makes our procedure superfast and quite easy and reliable for humans and hence the name of the project "Predicting Agricultural Productivity". Main purpose of the research work includes the implementation and training of machine learning algorithms for the prediction of the yield of agriculture so that the error can get minimized and accuracy gets maximized. For training of the model, a collection of features from actual yield and pictures of satellite is extracted by us. After This phase, a suitable algorithm like Naive Bayes, NN and its variant are chosen and used as the mathematical way to learn the parameters that are based on the features of yield. Then during study, harvest of agriculture is prognoses for a separate set of data. Data that is prognosticated is compared in contrast to the actual land yield. The manuscript also focuses the different data sets which are obtained by satellite imaging and using remote sensing, the clear mapping of current condition is obtained which helps to predict the yield in better way.

Keywords Agriculture · Squared error · Machine learning · Re-projection · Masking · Modis · Test cases

---

R. Rastogi (✉)
Department of (Physics and CS), DEI Agra, Agra, India

R. Rastogi · S. Maheshwari · P. Garg · M. Rastogi
Department of CSE, ABES Engineering College Ghaziabad, U.P, India

P. Kumar
Department of Software Engineering, Quaid-e-Awam University of Engineering, Science and Technology (QUEST), Nawabshah, Pakistan
e-mail: pardeep.kumar@quest.edu.pk

P. Kumar et al. (eds.), *A Fusion of Artificial Intelligence and Internet of Things for Emerging Cyber Systems*, Intelligent Systems Reference Library 210,
https://doi.org/10.1007/978-3-030-76653-5_3

3.1 Introduction

The study content of this research manuscript gives a short explanation of what is explained in the project, for example usage of NASA's data of remote sensing that is publicly available and the application of certain algorithms such as Naive Bayes theorem and Neural Network theorem for implementation of the idea. Design of system and methodology of this work includes the design of the system that is depicting the flow of data in data flow diagram and use case diagram and also the diagram of activity. Result and implementation described in this manuscript gives the requirements of the system for implementation of project.

3.1.1 Problem Introduction

In today's world, there are many positive impacts by production of agricultural commodity on water resources, food security and economic stability. Without the use of any highly improved sensing technology, we are many times helpless like on Earth, humans are able to see a very tiny part of electromagnetic spectrum. In order to see complete range of complete electromagnetic spectrum, usage of gamma, microwave and infrared waves. For solving complex problems, all the data and required pictures acquired by these satellites are taken [1].

The key problems need to be addressed in remote sensing data area used in agriculture statistics with more accuracy are mainly for

- Monitoring of cover of Land.
- Construction of frame of area.
- Data collection with the help of data collected from Support field.
- Estimation of area of crop [2].
- Miniaturization of crops and Forecasting of yield.

3.1.1.1 Motivation

For prevention of famine and supporting of humanitarian efforts, monitoring of agriculture is used in developing countries. To subsidize production of farms and regulation of commodity prices, improved forecasts of yield of agriculture is used. These forecasts are also used in water use optimization and for the growers demanding better strategic management. Financial institutions use forecasts of production to trade the features of commodity [3].

3.1.1.2 Experimental Research Work Objective

There are many challenges in prediction of crop yield such as yield estimation which means the prediction of the crop yields before harvesting. We are able to present an accurate, inexpensive and the scalable methodology to predict the yields of crop using the publicly available data from remote sensing. We are also going to exhibit the traditional old way for the extraction of features in project and also wish to propose an approach that is based on the ideas fetched by learning of modern representation. In other case, a technique is also introduced for reducing dimension. This allows training a network such as LSTM or CNN. This also helps us to learn beneficial features when there is less work of labelling training data [4].

Since in India and most developing countries, the rural persons especially farmers are less computer literate, so the application has been designed for mobile platform as a prototype and after its success, more advanced version will be developed.

3.1.1.3 Scope of the Research Work

Forecasting of crop means to the prediction of the harvest of crop in tons/ha and the ability for defining the production before the harvesting of crop actually take place and the time period includes a couple or month or more. Production of crop yields such as corn, wheat, rice has always a very attractive area of research for earlier meteorologists. In international and national economic programming, prediction of crop yield may be proved useful sometimes. It can also be used for the estimation of crop production in long or brief term when a network which learns relations of effective climate factors is defined [5].

Satellite imagery are also used for the forecasting of agricultural yields in many previous studies, whereas traditional studies use only moving average techniques, physically based models and regressions. In present scenario, machine learning and IoT based techniques are also used for the prediction of yield. One example of such a study is "Predicting agricultural productivity in California using satellite data and machine learning" [6].

3.1.2 Key Areas of Our Concerns

3.1.2.1 Healthcare and Human Health in Global Scenario

Healthcare is basically an industry or a field on which people focus on themselves to be healthy by opting good and natural habits. People can take advice from the professionals in order to become healthy. The healthcare professionals are basically nursing and doctors. In order to become healthy, one must have good hygiene, regular exercise and yoga included in daily activities [7].

There are many areas of study in healthcare which includes Clinical Healthcare, Administrative Healthcare, Therapy and Rehabilitation and Public Health [8, 9].

In our global scenario, good global health governance is getting important day by day. Scenarios help us for exploring the possible health development future. Eight scenarios are considered by considering three criteria i.e., "global-scope", "long-range outlook" and "integration". After looking into the dimension of health in some of the collected scenarios, only 14 gave reasonable description out of 31. Nine of them described only some specific pressures with regard to certain aspects of human health [10, 11].

Only 15% of the scenarios describe health in a perfect adequate manner and in a proper adequate way. Two groups of Global Scenario described health as the pure outcome of environmental and socio-economic changes and only one of them gave genuine reasonable description. This indicates that in the current sets of scenarios, health is given no importance [12].

3.1.2.2 Big Data Analytics and its Approach for Healthcare

It actually refers to the systematic processing of the large volume of data that we get through the industry in order to produce the relevant information. The data includes the transactions of various business, records of an organization etc. Big data collects all the chunk of data and then do the analysis to produce information [13].

It is basically used to derive relations between the customers and their preferences so that they are able to know about customer's choice and behaviour in order to increase the sales by offering them a no. of varieties [14].

Whenever it comes to healthcare, we know that healthcare is implemented at many levels. In healthcare we need to process many data at a time in order to get the information. Data of healthcare includes no. of doctors and nurses, clinics, medicine, no. of surgical instruments, record of patient etc. In order to manage all the data, we use the big data analytics and it has a main role in maintaining the sustainability of healthcare system [1].

Big data basically works on three parameters: volume, variety and velocity. Here volume symbolizes the huge volume of data that we get to process information. Variety symbolizes the no. of parameters to relate the data to get accuracy and velocity summarizes the speed at which we are able to get the result of a particular query [15].

3.1.2.3 Big Data Analytics in Agriculture

In agriculture, big data is viewed as a perfect combo of technology and analysis which can be used to convert the data that is collected from various resources into useful information that is actionable. The process by which the big data analytics is applied on agriculture is called Data mining. It basically refers to the identification of certain patterns in the given data in order to get a behaviour which really helps businessman and many professionals [16].

Big data analytics is very useful in agriculture for controlling the weeds and for protection of crops. Some of the challenges that we have to face during big data analysis are such as collection, modelling and analysis of data. The ownership of the data is analysed via ethical matrix method so as to find out the concerns of the landlord of agriculture business, consumers etc. [4].

3.1.2.4 Annals of Agriculture Production in Indian Scenario

The rank of India is second worldwide in producing the outputs of farm while it ranks first in the area of cropped area. In India, agriculture and its sectors have 15.4% of the GDP but now due to the broad economical growth of the country, the agriculture's GDP is continuously decreasing. Most of agriculture exports of India serve the nations that are developing or very less developed. Indian foods are exported to approximate 130 countries [17].

In order to expand production and give a greater number of varieties, Crop diversification is done. In India, it is basically reviewed as transition from less grown to more remunerative crops. It takes place due to the policies of government such as TMO. Crop diversification needs to practice in areas which receives enough rainfall so that the chance of crop failure is less [5].

3.1.2.5 Correlation of Human Health with Agriculture Production

There has been tragic change in the field of agriculture since 1988. There are two types of human health that is associated with agriculture. Firstly, the people who are working on agriculture, their health has been improved by the use of improved technology and by some efforts such as the establishment of NIOSH which collaborates the professionals to provide knowledge to farmers. Secondly, the people who eat the processed food of the crops have also their health dependent on agriculture as what amount of pesticides and fertilizers are used, quality of crops etc. These all things can be regulated by the behaviour of weather on crops which can be further regulated by the analysis of data collected over years hence big data analytics is used [3].

3.1.2.6 Application of Machine Learning in Healthcare, Machine Vision and Medical Imaging

Machine Learning is basically the study of algorithms and the mathematical models of statistics that the systems use to perform the tasks. Machine learning is basically the subset of Artificial Intelligence. In the last few years, it has been extensively used as it needs relatively cheap systems and inexpensive memory. It is basically depending on the user that which algorithm to apply to get the desired information and accuracy, which also helps to take certain decisions. Many enhanced algorithms have been developed to improve the accuracy. Learning algorithms are also very agnostic to the domain. The actually value of the application of the machine learning depends on the way we use these algorithms to solve real world problems [7].

It is basically the technology and procedure used to provide the automatic inspection of the image in order to control the process or robot guidance etc. According to AIA, it has full scope over all the application whether industrial or not. All the applications mostly use same algorithms but the constraints are different. Systems basically rely on the digital sensors which are cameras that are protected from inside in industry so that the hardware and software of the computer can measure various characteristics according to the constraints. They also perform the even-handed measurements such as determining the sparking plug gap [18].

Artificial Intelligence means the intelligence that is shown by machines. It is also called as machine intelligence. Basically, it refers to the ability of the system to learn the behaviour of the environment that helps in increasing the accuracy of prediction steadily over a period of time. The goals of artificial intelligence are to include reasoning, representation of knowledge, perception etc. [19, 20].

It is basically a technique for creating the visual representation of the interior parts of the body and then the clinical analysis and medical intervention of that representation is done. It is also able to record the functioning of some organs and tissue in the visual representation. It is basically a part of the biological imaging and uses the imaging technologies of ultrasound, thermography, nuclear medicine functional imaging etc. Various types of medical imaging include Radiography, magnetic resonance imaging, nuclear medicine etc. [21].

3.2 Literature Survey and Related Previous Work

In this presented work, the author team have tried to measure precipitation by using Tropical Rainfall Measuring Mission (TRMM) [22– 26] satellite constellation and measures vegetation fluorescence using Moderate Resolution Spectro Radiometer (MODIS) [27] satellite sensor in order to access data of remote sensing that is publicly provided by NASA. Short analysis and learning the knowledge of working of Geographic Information Systems (GIS) is done [21].

GIS is basically a machine-based tool to analyse, store, manipulate and visualise the geographic info in map. Understanding key processes (re-projection, masking, and many more) and researching different products of MODIS is done. It is an important need to analyse the measurements of remote sensing and assess whether the target has green vegetation or not, Normalized Difference Vegetation Index (NDVI) which acts as simple graphical indicator. For the vegetation which has a strong signature. The spectral wavelengths have been included by using the subset of the MODIS data and then the per-pixel averages per month is calculated in order to degrade volumes. In 2000—present, monthly per-pixel sums are calculated so as to coincide with the MODIS data [24]. Using LULC data, all the non-agricultural areas were cut out from all images obtained by MODIS so as to stop the signal dampening of vegetation. Two methods are used to train the models which are as follows:

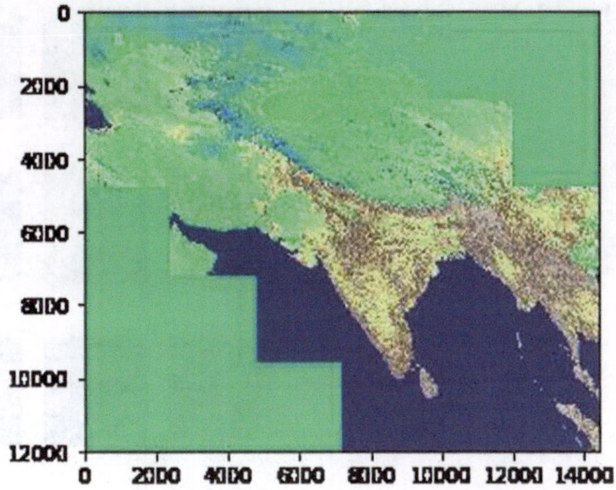

Fig. 3.1 Example image of vegetation fluorescence data obtained by MODIS [21]

3.2.1 The Naïve Bayes

It is basically a classification technique. It assumes that all the features present in the class are not dependent on any other. As all the features uniquely contribute to the probability of a certain result that's why this algorithm is called as 'Naïve'.

Laplace Smoothing and multinomial event model are used to implement Naïve Bayes implementation. TRMM precipitation and MODIS vegetation fluorescence are used as input features. The Fluorescence values of MODIS data for each year were binned in equal length (refer to Fig. 3.1). Binning of the data starts at 0 and moving over by 0.025 per bin, with 0.975 to 1 as the last fluorescence bin value of MODIS data. In each of the bin, the value is calculated as number of times a certain pixel along with the value of respective year's image obtained from MODIS (Huffman, G. J. Et al. 1997). Data of ten bins is equal to the TRMM data and then the data is scaled with the average of the data of all the bins and MODIS data of all the years. Discretization of yield data is done to map data from labs that are continuous to classes that are discrete (refer to Fig. 3.2).Every year's data of yield is paired with that year's data of MODIS and then used as an example of test and training set. Data of years from 2000 to 2019 is included for India from Ministry of Agriculture and Farmers Welfare, Govt. of India, New Delhi (https://aps.dac.gov.in/APY/Index.htm). Calculation of error was done by comparison of the mid value of predicted bin yield for the year which is left out with its true value.

3.2.2 The Neural Network

- It is basically a series of algorithms to recognize the underlying relationships in the data.

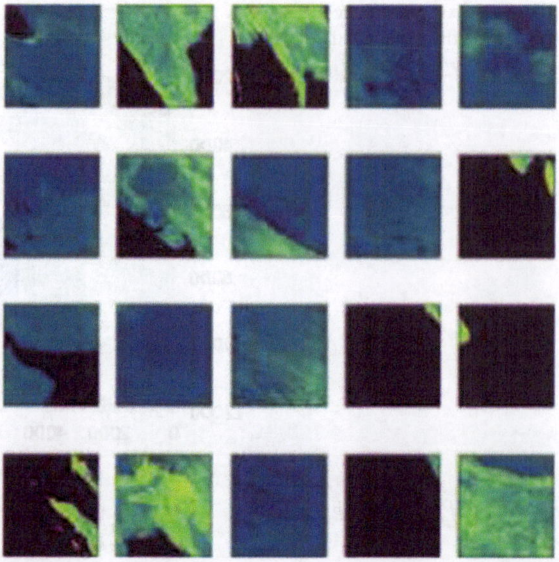

Fig. 3.2 Mosaic king the satellite image [21]

- These networks can adapt to changing input so that it is able to produce the best possible result.
- Varieties of feed forward neural nets are constructed and trained with the help of keras library for the purpose of non-linear regression. Testing and evaluation of different architectures and training functions are carried out by the cross validation of the data collected over years from India (refer to Fig. 3.3).

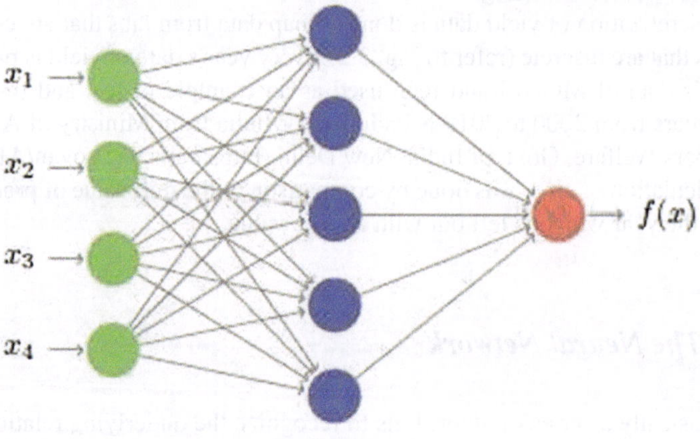

Fig. 3.3 Simple feed forward neural network [28]

- Levenberg–Marquardt (LM) optimization, Bayesian Regularization (BR) optimization and Gradient descent with momentum and adaptive learning rate back propagation (GDX) is used for the training of nets. BR is two orders slower than LM and GDX. But as the accuracies of LM and GDX are worse, so the best method chosen is BR.

3.2.3 Gist of Literature Survey

- In order to use less time and to measure efficiency, above described methods were fully provided with training and testing is also done on a sample of data for India. Training accuracies of 85 and 96% for feed forward neural network and 70 and 73% for Naïve Bayes are obtained with overall tests via the promising preliminary results of the cross validation.
- Neural network result of the training and test accuracies for final feed forward net is shown which is trained via BR optimization, [25] for the cross verification of the data for all the years for India as well as all the predicted production versus true production.
- To summarize, despite favourable initial performance by the models that are simple, improved performance would be attained by following some steps such as:
  - Optimization of model parameter needs to be done.
  - Pre-processing of input features need to be adjusted.
  - More input features such as additional satellites and satellite derived satellites should be added.

3.3 System Design and Methodology

3.3.1 System Design

System design refers to the process of the designing of the elements so that components, modules, interfaces and architecture of those components and its respective data goes through the system. The purpose of the process of System design is to enable the system elements in such a manner so that there is consistency between the entity of architecture and implementation.

3.3.2 System Architecture/Diagrammatical View

It is conceptual model also known as the systems model that defines the behaviour, more expanded views and structure of a system. Convolution neural network is used to extract the features of the images fetched from satellite [26]. As in the process of network adjusting with the weights respectively, they got tested with the data of test which is generally present in 80:20 ratio. Then a comparison is carried out between the Predicted yield and actual one in order to minimize mean squared error (refer to Fig. 3.4).

The detailed implementation and approach of this work can be understood by Use case Diagram, Data Flow Diagram and Activity Diagram followed in the experimentation and research execution which has been presented in below section.

3.3.2.1 Data Flow Diagram

Whenever one has to represent data, one has to follow some chronology. In order to follow chronology, Data Flow Diagram helps us. It is actually a way to represent the data in a certain flow. As shown in Fig. 3.2, a user may be able to predict the yield of agriculture either by doing selection of state or by selecting their own district [21].

These chosen district or state is given to the trained model resulting the prediction of yield in graphical form (refer to Fig. 3.5).

3.3.2.2 Use Case Diagram

In order to make our representation as simple as possible, one needs a tool. For this purpose, one can use the Use Case Diagrams. It can identify the different varieties of users and some use cases too. Shapes such as circles are used in making of this

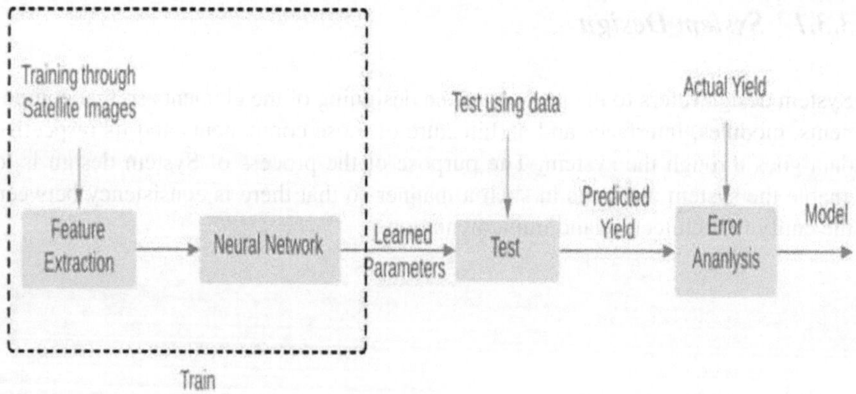

Fig. 3.4 Methodology of implementation of work

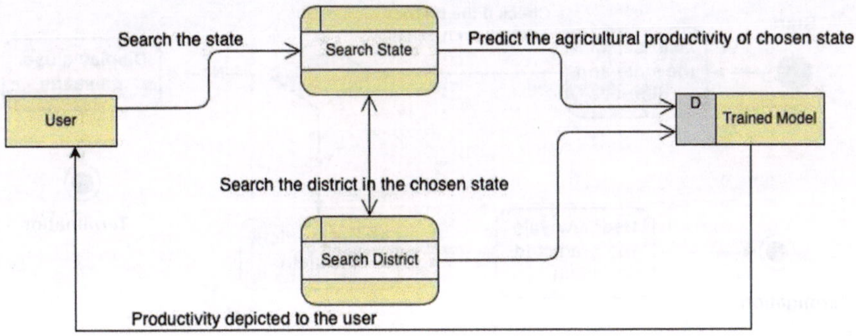

Fig. 3.5 Data flow diagram of research work

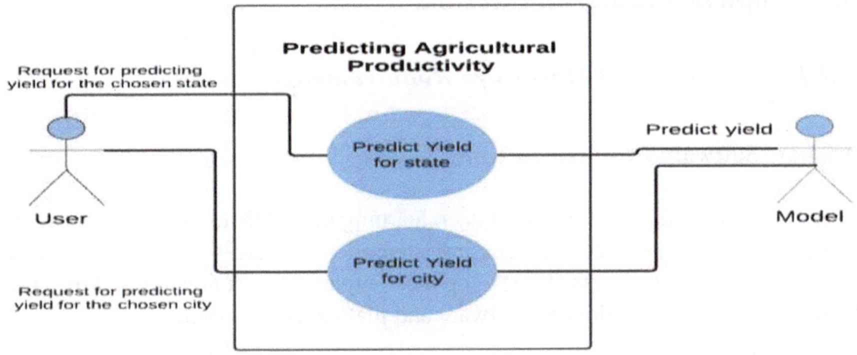

Fig. 3.6 Use case diagram

diagram. A user is also able to interact with the system in order to requesting for prediction of the agricultural productivity for any state or either for district (refer to Fig. 3.6).

3.3.2.3 Activity Diagram

Activity diagram basically depicts the behaviour of a certain activity [29]. The user selects whether he/she wants to know the data for state or district and then the system performs the check for its prediction and then it proceeds to predict the yield otherwise an error message is displayed.(refer to Fig. 3.7).

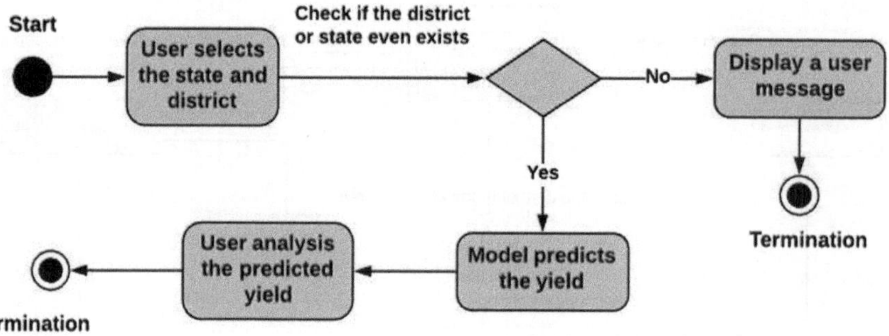

Fig. 3.7 Activity diagram

3.4 Implementation and Results

3.4.1 Software and Hardware Requirements

3.4.1.1 Software Requirements

Packages that contain algorithms of deep learning like CNN to execute consistently and for the purpose of image processing some more packages are required so as to fulfil software requirements. To develop the complete project and to run it on a user friendly android app, following software and platforms were used.

- TensorFlow: It is a software library which is free and is also open source. It is symbolic math library and also used in case of neural networks.
- Keras: It is also an open-source library. It is used to enable fast experimentation and it is quite extensible and user-friendly. It is written in Python.
- OpenCV: It is a cross platform library which consists of programming functions and their main objective is computer vision in real time.
- XCode: It is basically an IDE for macOS and it contains some software tools. It is written in C and C++.
- Google Earth Engine: It actually used to combine multi-petabyte catalogue of the images fetched from the satellites.

Other requirements for development are

- Anaconda: It is an IDE to code in Python & R. It is managed by conda and it is free and an open source distribution.
- Python: It is a language of programming developed by Guido Van Rossum and is high level and interpreted programming language.
- Swift: It is a Programming language developed by Apple and is designed to work with frameworks of Cocoa.

3.4.1.2 Hardware Requirements

- 20 GB RAM or above
- Tesla K80 GPU

The only edge case where 20 GB might do better is if you have a workload that needs over 16 GB but less than 20 GB of ram and Tesla K80 Accelerator features and benefits 4992 NVIDIA CUDA cores with a dual-GPU design. Up to 2.91 teraflops double-precision performance with NVIDIA GPU Boost is also available in market. Up to 8.73 teraflops single-precision performance with NVIDIA GPU Boost may give more efficient results. 24 GB of GDDR5 memory. 480 GB/s aggregate memory bandwidth will also work.

3.4.2 Assumptions and Dependencies

Here are certain assumptions that the authors' team has to make. The first is to suppose that the model is only able to work in case of land in Indian territories. The recent development that is going on in the states or districts is not considered in model training. The model of app is basically designed so as to work at district level. And it is also needed that the user who is going to use this app has the full knowledge of using app.

3.4.3 Implementation Details

3.4.3.1 Snapshots of Interfaces

The author team is also developing a mobile app with user friendly GUI where users can use search bar to search for his/her state and after successful selection of the state, the user may predict the productivity of agriculture (right) or may select the district. This app contains an easy and maintainable interface which makes it more user friendly. The scope of the app is state and district level (refer to Fig. 3.8).

3.4.3.2 Test Cases

Here are some of the samples that researchers have used in our project. Following are some other examples.

Some of them include:

- For prevention of vegetation dampening, all the non-agricultural areas are removed from the images obtained by MODIS.

Fig. 3.8 App home screen (left) and graphical representation (right)

- Data that is provided in LP DAAC is like granules i.e., in grid format (refer to Fig. 3.9).
- To form map, data is organized according to the values of grid, also known as Mosaic king.
- Sinusoidal projection of mosaic is converted into UTM (Mercator), also known as Re-Projection.
- Masking is performed by using respective shape files to crop the area that is re-projected (districts) (refer to Fig. 3.10).

Data is collected year wise for every district with its respective season and then this same data is used to train the convolution network and as a result of all this, the total feature yield (tons/ha) is target variable (refer to Table 3.1).

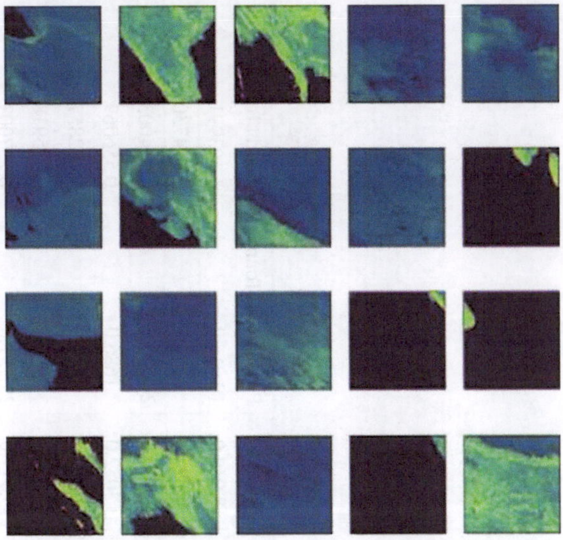

Fig. 3.9 LP DAAC satellite granules

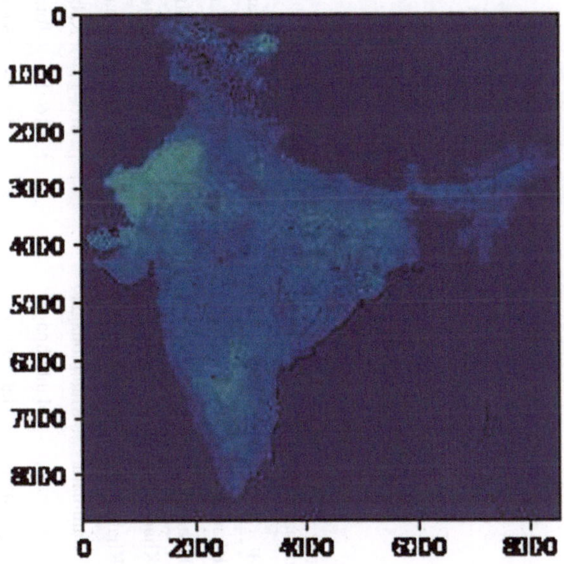

Fig. 3.10 Required processed image after masking

3.5 Conclusion

3.5.1 Performance Evaluation

The authors' team has used either Stochastic search or grid search to examine the aggregation scheme and for the sensitive analysis of each and every variable. One

Table 3.1 Corresponding seasonal crop data (Year wise and district wise)

	State	Crop	District	Year	Season	Area (Ha)	Production (Tonnes)	Yield (Tonnes/Ha)
180,014	Telangana	9. Khammam	Komaram bheem asifabad	2016–17	Kharif	18,040.0	19,916.0	1.103991
180,015	Telangana	9. Khammam	Komaram bheem asifabad	2016–17	Rabi	296.0	327.0	1.104730
180,016	Telangana	9. Khammam	Komaram bheem asifabad	2016–17	Total	18,336.0	20,243.0	1.104003
180,017	Telangana	9. Khammam	Mahbubnagar	2016–17	Kharif	3032.0	1686.0	0.556069
180,018	Telangana	9. Khammam	Mahbubnagar	2016–17	Rabi	35.0	19.0	0.542857
180,019	Telangana	9. Khammam	Mahbubnagar	2016–17	Total	3067.0	1705.0	0.555918
180,020	Telangana	9. Khammam	Mahbubnagar	2015–16	Kharif	115,917.0	29,791.0	0.257003
180,021	Telangana	9. Khammam	Mahbubnagar	2015–16	Rabi	60.0	15.0	0.250000

may choose other variants of it. Till now, when the team run this project on sample data, only accuracy of 62% was achievable. This 62% performance is purely based on the preliminary analysis of the data and Feed Forward Neural networks algorithm is used in it. The accuracy calculated for this training model is not much as expected but the author team is continuously working to gather more data and upgrade the dataset and consequently accuracy.

3.5.2 Future Directions and Limitations

- There are many aspects of this project which are remain untouched till now hence the study will be continued on various aspects. During the training, some additional variables, satellites and talukas will be considered.
- Testing of convolution neural networks will be done and feed forward neural will be compared with the further data.
- The requirements of the agricultural land may be minimum can be predicted by the more advanced training of the model.
- This model can be trained also to predict the requirements for the sustainable survival of agriculture in states and districts.
- In order to achieve the enhanced production of the model, some steps that need to be taken are as follows:

1. Optimization of model parameters is required.
2. Pre-processing of the input features need to be adjusted.
3. Some input features need to be added such as the additional satellites and some satellite derived variables.
4. More number of need to be added in order to improve the accuracy.
5. Also, since its beginning of the work, so application has been developed in standalone system. In future, with powerful GPUs, it can be well implemented in distributed system and with advanced networks.
6. High End GPU for handling big data and implementation on cloud with distributed platform may be thought of.

Due to limited computing power, expanded results can't be presented now. Process is completed and scripts were got checked but only on limited data. After whole data processing, expanded meaningful results are obtained.

Appendix

Data Format

Following is data used for testing and training has particular features listed below:

- Id

- State: It basically is used in order to find out the place of which the user wants to predict the agricultural yield.
- Crop: It is basically used to determine whether it is wheat or rice or maize etc.
- District: It basically is used in order to find out the place of which the user wants to predict the agricultural yield.
- Year: It is used to know that which year's data need to be accessed.
- Season: It is asked to user to know that which season's data need to be accessed.
- Area in hectare
- Production in tonnes
- Yield in tonnes/hectare
- MODIS Image Array.

TRMM Image Array.

Input Features: Satellite Data

a-MODIS: The **M**oderate **R**esolution Imaging Spectro radiometer is a payload imaging sensor built by Santa Barbara Remote Sensing that was launched into Earth orbit by NASA in **1999** on board the Terra Satellite, and in 2002 on board the Aqua satellite.

b-TRMM: The Tropical Rainfall Measuring Mission was a joint space mission between NASA and the Japan Aerospace Exploration Agency designed to monitor and study tropical rainfall. The term refers to both the mission itself and the satellite that the mission used to collect data.

A. MODIS satellite data are available twice daily at 500 m resolution for any area in the world from 2000—present.

B. TRMM satellite data are available every 3 h at 25 km resolution for any area in the world from 1997—present (Figs. 3.11, 3.12, 3.13, 3.14, 3.15, 3.16, 3.17, 3.18 and 3.19).

Global Vegetation Greenness: Vegetation greens can be determined by using remote sensing. Red and grey are regions with low NDVI below 0.1 and represent barren sand, rock or snow. Yellow and light green are regions with moderate NDVI between 0.2 and 0.3 and represent shrub and grassland. Dark green are regions with high NDVI and represent tropical rainforests.

Determine the amount of reflectivity.

1. The sun's energy is absorbed, reflected or re-emitted by vegetation depending on the amount of chlorophyll in the plant.
2. Satellites collect the reflected or emitted energy from the vegetation.

In green vegetation:

- Chlorophyll absorbs most of the visible ed and blue to make food
- Much of the near infrared is reflected.

Introducing a Normalized term:

- Normalized Difference Vegetative Index is a remote sensing method that allows one to display greens of vegetation.

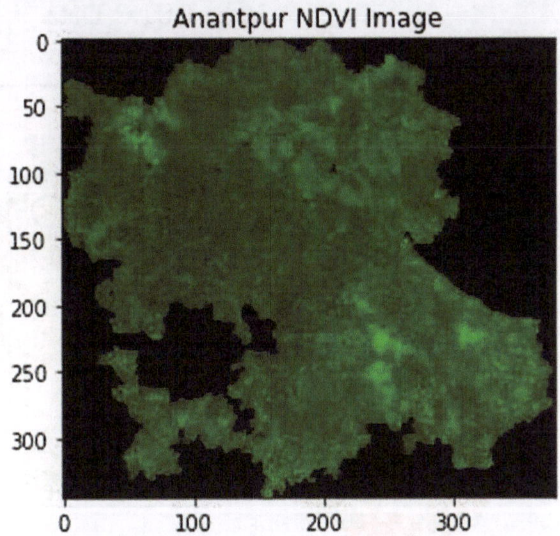

Fig. 3.11 Sample APY data

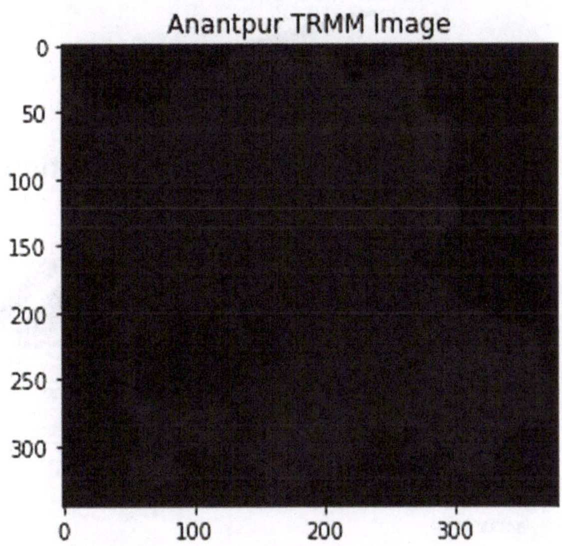

Fig. 3.12 Sample NDVI and TRMM images (Anantpur)

- NDVI compares reflectivity of NIR and Red wavelength bands.
- Image analysis using remote sensing allows one to determine NDVI using this formula (Fig. 3.20):

$$\text{NDVI} = (\text{NIR} - \text{Red})/(\text{NIR} + \text{Red})$$

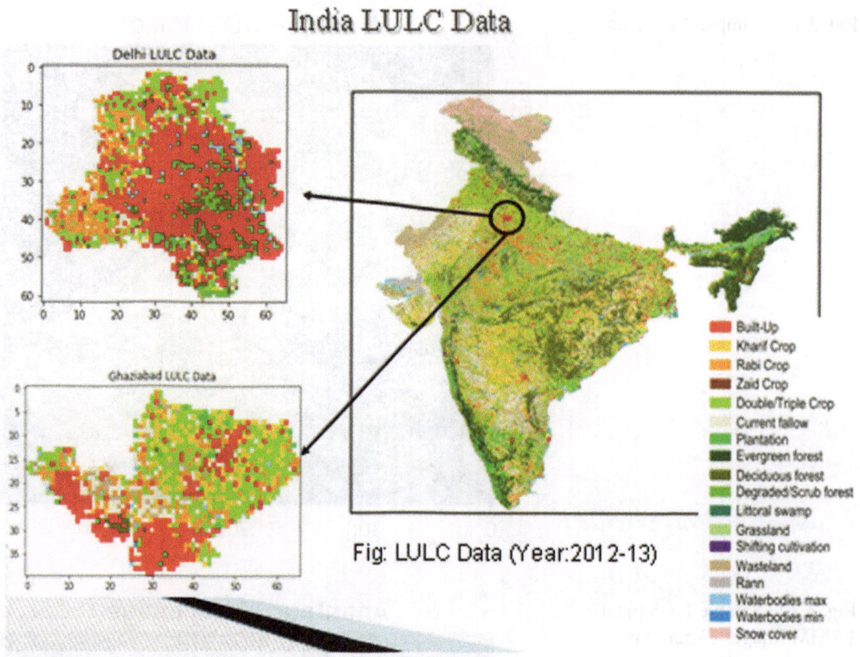

Fig. 3.13 India LULC data

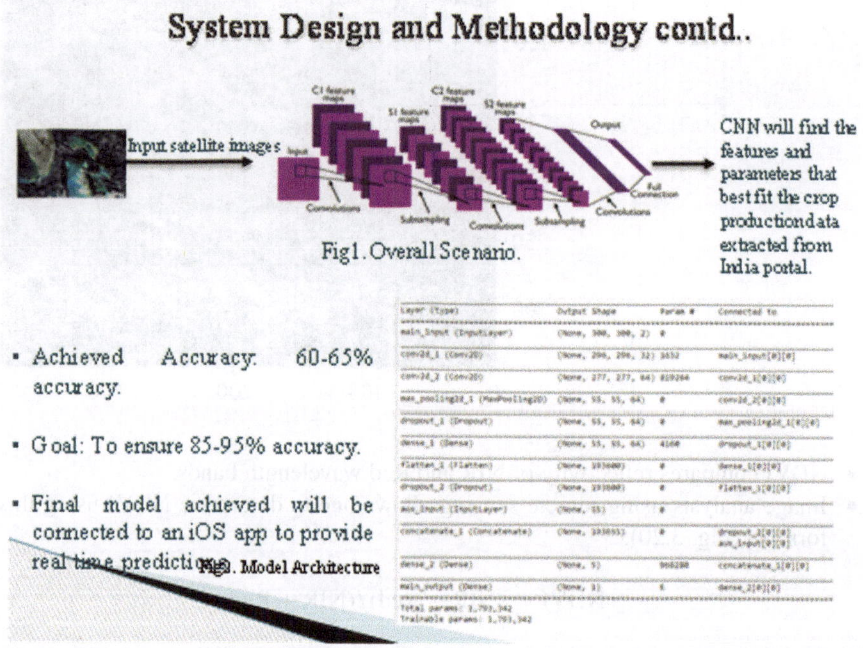

Fig. 3.14 Design and methodology at a glance

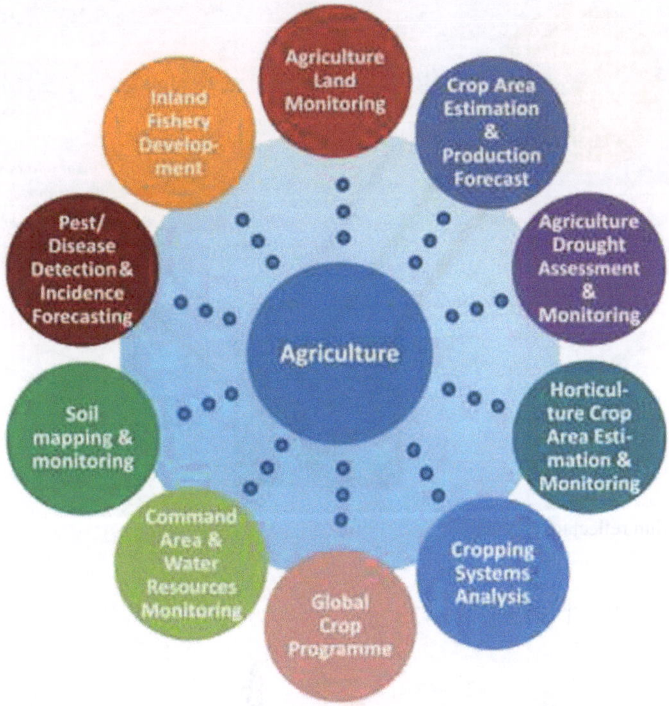

Fig. 3.15 Domains in agriculture. *Source* National Remote Sensing Centre

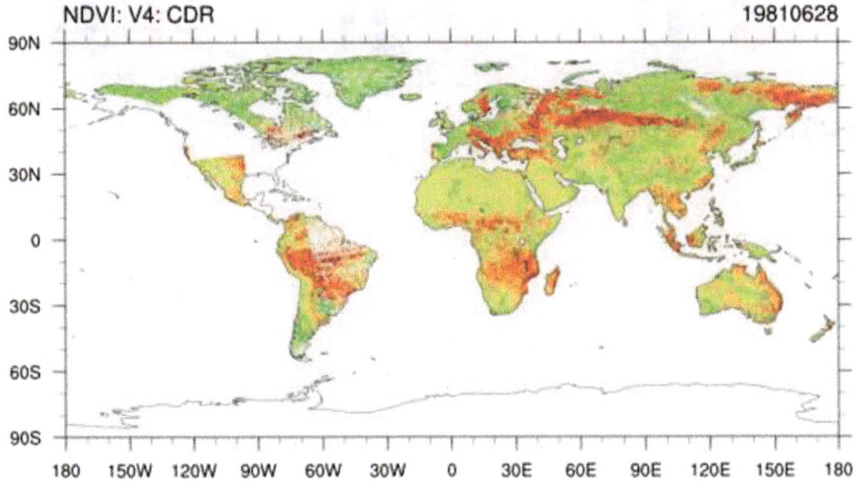

Fig. 3.16 Global vegetation greenness

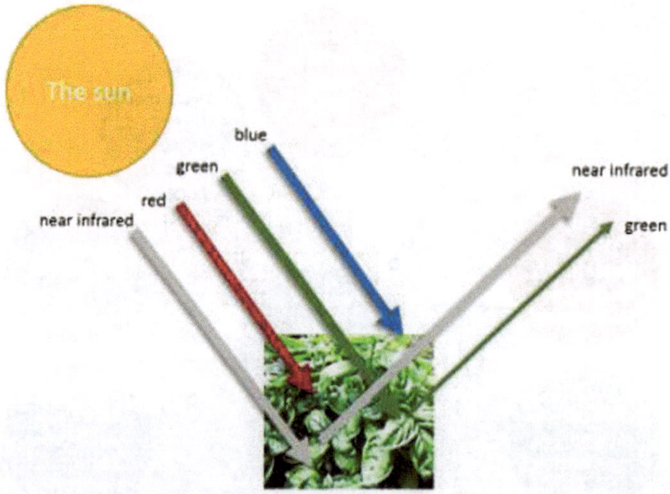

Fig. 3.17 Sun reflectance model

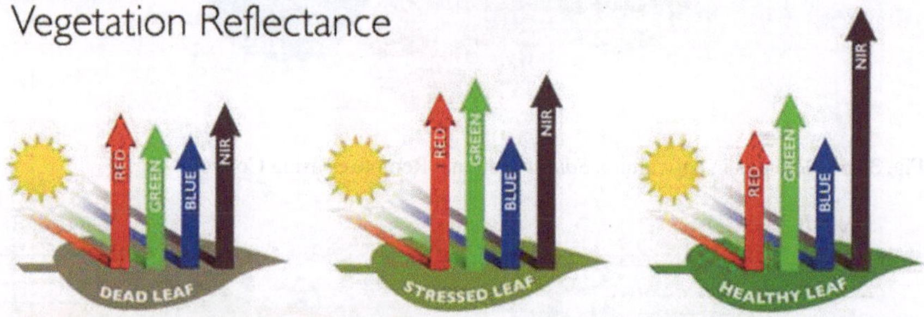

Fig. 3.18 Vegetation reflectance model

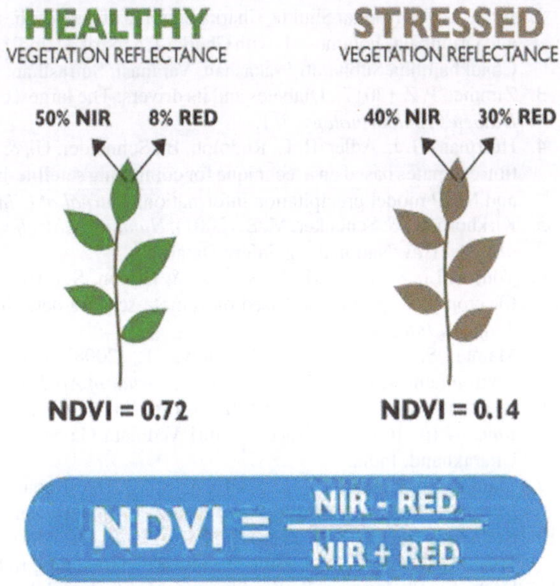

Fig. 3.19 Normalized difference vegetation index (NDVI)

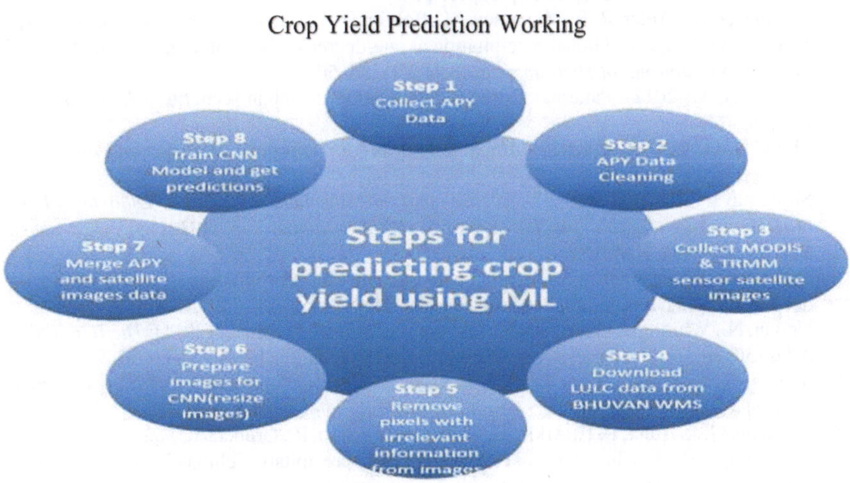

Fig. 3.20 Crop yield prediction working

References

1. Huffman, G. J., Adler, R. F., Bolvin, D. T., Gu, G., Nelkin, E. J., Bowman, K. P., Hong, Y., Stocker, E. F., & Wolff, D. B. (2007). The TRMM multi-satellite precipitation analysis: Quasi- global, multi-year, combined-sensor precipitation estimates at fine scale. *Journal of Hydrometeorology, 8*(1), 38–55.

2. Acharya Vidhyadhar Shukla, Charka Samhita of Agnivesh, 4th ed. (1996). Elaborated by Charka and Dridhbala, Volume—1, with Charka-Chandrika Hindi commentary, Triaishaniya Adhyaya, Chaukhambha Subharati Prakashan, Varanasi, Sutrasthan, 11/54.
3. Zimmet, P. Z. (2017). Diabetes and its drivers: The largest epidemic in human history? *Clinical Diabetes Endocrinology, 3*, 1.
4. Huffman, G. J., Adler, R. F., Rudolph, B., Schneider, U., & Keehn, P. (1995). Global precipitation estimates based on a technique for combining satellite-based estimates, rain gauge analysis, and NWP model precipitation information. *Journal of Climate, 8*, 1284–1295.
5. Kirkhorn, S., & Schenker, M. B. (2001). *Human health effects of agriculture: Physical diseases and illnesses*. National Ag Safety Database.
6. You, J., Li, X., Low, M., Lobell, D., & Ermon, S. (2017, February). Deep gaussian process for crop yield prediction based on remote sensing data. In *Thirty-First AAAI Conference on Artificial Intelligence.*
7. Mamta, S., Sengupta, B., & Pandya, P. (2008). Controlling the microflora in outdoor environment: Effect of Yagya, Indian. *Journal of Air Pollution Control, VIII*(2), 30–36.
8. Brahmavarchas, 1st ed. (2010). *Yagya Chikitsa (Hindi), Chap. 5 Diabetes ki Vishishta Hawan samagri* (p. 103), Published by Shri Vedmata Gayatri Trust (TMD), Shantikunj, Haridwar, Uttarakhand, India.
9. Mamta, S., Kumar, B., & Matharu, S. (2018). Impact of Yagya on particulate matters. *Interdisciplinary Journal of Yagya Research, 1*(1), 01–08.
10. Brahmavarchas. (1994). *Yagya Chikitsa (Hindi)* (1st edn.), Published by Shri Vedmata Gayatri Trust (TMD) 2010 Shantikunj, Haridwar, Uttarakhand, India.
11. Dash, S., Shakyawar, S. K., Sharma, M., & Kaushik, S. (2019). Big data in healthcare: management, analysis and future prospects, 54.
12. Huffman, G. J., Adler, R. F., Morrissey, M., Bolvin, D. T., Curtis, S., Joyce, R., McGavock, B., & Susskind, J. (2001). Global precipitation at one-degree daily resolution from multi-satellite observations. *Journal of Hydrometeorology, 2*(1), 36–50.
13. Kalaiselvi, V. (2012). Patterns of crop diversification in Indian scenario. *Scholars Research Library, 3*(4).
14. Maity, N. G., & Das, S. (2017). *Machine learning for improved diagnosis and prognosis in healthcare*. IEEE.
15. Shrivastava, V., et al. (2016). *A case study-management of type II Diabetes Mellitus (T2DM) through herbal medicinal-smoke (Dhoom-Nasya)* (pp. 103–118). Dev Sanskrit Vishwavidyalaya, Haridwar.
16. University of Cincinnati. (2018). UC business analytics R programming guide, feedforward deep learning models, simple feed forward neural network.
17. Unwin, N., WhitingD, GuariguataL., Ghyoot, G., & Gan, D. (Eds.). (2011). *The IDF diabetes Atlas* (5th ed., pp. 7–12). Brussels, Belgium: International Diabetes Federation.
18. Mamta, S., Sengupta, B., & Pandya, P. (2007). A study of the impact of Yagya on indoor microbial environments, Indian. *Journal of Air Pollution Control, VII*(1), 6–15.
19. Huffman, G. J., Adler, R. F., Arkin, P., Chang, A., Ferraro, R., Gruber, A., Janowiak, J., McNab, A., Rudolph, B., & Schneider, U. (1997). The global precipitation climatology project (GPCP) combined precipitation dataset. *Bulletin of the American Meteorological Society, 78*, 5–20.
20. Schaaf, C., & Wang, Z. (2005). MCD43A4 MODIS/Terra+Aqua BRDF/Albedo Nadir BRDF adjusted ref daily L3 Global—500m V006. NASA EOSDIS Land Processes DAAC.
21. van Evert, F. K., Fountas, S., Jakovetic, D., Crnojevic, V., Travlos, I., Kempenaar, C. (2017). Big data for weed control and crop protection. *Weed Research, 57*.
22. Anjana, R. M., Deepa, M., Pradeepa, R., Mahanta, J., Narain, K., Das, H. K. (2017). Prevalence of diabetes and prediabetes in 15 states of India: Results from the ICMR-INDIAB population-based cross-sectional study. *Lancet Diabetes Endocrinology, 5*, 585–596.
23. Ahamed, A., & Dewar, N. (2009). Predicting agricultural productivity in California using satellite data and machine learning.
24. Brahmavarchas (Ed.). (2012). Yagya – Ek Samagra Upachar Prakriya (Hindi) (Yagya—A holistic therapy), Pandit Shriram Sharma Acharya Samagra Vangamaya—Volume 26, Akhand Jyoti Sansthan, Mathura, Uttar Pradesh, India.

25. Mathers, C. D., & Loncar, D. (2006). Projections of global mortality and burden of disease from 2002 to 2030. *PLOS Medicine, 15.*
26. Didan, K. (2015). MOD13Q1 MODIS/Terra vegetation indices 16-day L3 global 250m SIN grid V006. NASA EOSDIS Land Processes DAAC.
27. Adler, R. F., Huffman, G. J., Chang, A., Ferraro, R., Xie, P., Janowiak, J., Rudolf, B., Schneider, U., Curtis, S., Bolvin, D., Gruber, A., Susskind, J., Arkin, P., & Nelkin, E. J. (2003). The version 2 global precipitation climatology project (GPCP) monthly precipitation analysis (1979-Present). *Journal of Hydrometeorology, 4*(6), 1147–1167.
28. Huffman, G. J. (2012). *Algorithm theoretical basis document (ATBD) version 3.0 for the NASA global precipitation measurement (GPM) integrated multi-satellite retrievals for GPM (I-MERG)* (p. 29). Greenbelt, MD: GPM Project.
29. Huffman, G. J. (1997). Estimates of root-mean-square random error for finite samples of estimated precipitation. *Journal of Applied Meteorology*, 1191–1201.

Chapter 4
Block Chain Application in Automobile Registration: A Novel Approach for Sustainable Smart Cities with Industry 4.0

Rohit Rastogi, Bhuvneshwar Sharma, Pardeep Kumar, and Muskan Gupta

Abstract Block chain Technology replaces centralized applications to distributed computing. Modern economy is estimated by the place of motor transport in the infrastructure of the national economy. An automobile registration system is a unified information system. This information system takes care of every information of an automobile registration. It is administered by a national registry entity and has access by other government and non-government of services that handles automobile information. Cyber Physical System (CPS) is defined as the combination of computation and physical process. It is mainly used in ICT section. It is also focused on resolving the problems related to authors of the data regarding transparency, media and storage problems by technical handling. The presented manuscript uses all above concepts at one place and integrate them to build a useful application. The presented frame allows car manufacturers, owner, repairing companies and insurance agencies to register and add new entries for cars in a simple method. Four different smart contract control blocks are updated in DriveLoop. In addition, database technology has been leveraged to cache intermediate data. It efficiently uses the Industrial IoT and 5G technologies. Many researchers have been called for rules and applications to draw old maps into the blockchain based on distributed applications. New protocols are available in this work for the International Automated Vehicle Management System, called DriveLoop, was proposed and developed.

R. Rastogi (✉) · B. Sharma · M. Gupta
Department of CSE, ABES Engineering College, U.P, Ghaziabad 201009, India
e-mail: rohit.rastogi@abes.ac.in

B. Sharma
e-mail: bhuvneshwar.sharma@abes.ac.in

M. Gupta
e-mail: muskan.18bcs1175@abes.ac.in

P. Kumar
Department of Software Engineering, Quaid-e-Awam University of Engineering, Science and Technology (QUEST), Nawabshah, Pakistan
e-mail: pardeep.kumar@quest.edu.pk

P. Kumar et al. (eds.), *A Fusion of Artificial Intelligence and Internet of Things for Emerging Cyber Systems*, Intelligent Systems Reference Library 210,
https://doi.org/10.1007/978-3-030-76653-5_4

Keywords Driveloop · Blockchain · Peer to Peer (P2P) · Hashing algorithms · Car registration · Overlay network

4.1 Introduction

An individual, whenever bought one's own vehicle or have sold it, or have been a part of an automobile manufacturing or dealing at any stage of the cycle, one would be familiar with the complication that is process of Vehicle Registration. Given the fact that all these vehicles in the market have been sold and resold as they are passed through multiple hands, it becomes a cumbersome task to maintain a legitimate record of the history of each vehicle and make it available when needed.

But before one goes on to talk about the problems of the process of Vehicle Registration, one first needs to understand why Vehicle Registration is such an important aspect of automobile dealing. Car ownership may get changed as many times as you can imagine.

Whether you look at it in the terms of dealing in spare parts or in assembled vehicles, dealings by the middlemen or by the retailer who makes the final sale to a consumer, or in terms of the resale of a second-hand vehicle, there are a number of stakeholders who would very much want to know about all the history of the vehicle they are buying. Not to mention the insurance agencies, the police and other authorities and well, the government too need to keep tabs on the automobiles for various reasons.

The fact of the matter is that, all these stakeholders need information about the vehicles, starting from its manufacturing story, covering its first sale, the accidents, if any, that it has been in, and any all repairs and maintenances. This is crucial not for just to maintain a track record of the vehicle in question to determine its market value, but also for legal and insurance purposes.

Vehicle Registration is a way to facilitate this record keeping by maintaining a link between the vehicle and its owner. It might be or not be compulsory, depending on the law of the land. This helps the authorities with regards to taxation, insurance, or crime detection purposes. Also, it is a way for the automobile dealer to keep a track of their vehicles [1, 6, 15].

4.1.1 Concept of Smart Cities

The urban development has resulted in a change of archetype in twenty-first century. Research activities for smarter cities have become priority task. The life had been improved in the last century in terms of technologies and services. Smart City is the demanding solution of sustainability and urbanization. Smart Cities may lead to a dystopian world that is regulated by technocratic governments which propel citizens to subaltern roles. However, the massive industrialization and the increasing population in the big cities has been a big challenging for urban planner, architects, and administrators.

The service platforms of smart cities are Internet of Things (IoT), big data systems, and mobility. Connected automobile with its advanced technology reduces the chances of accident and help drivers to save time and gasoline in their limits. Increase in population in urban areas, often leads to the problem of parking spaces. Smart parking is one the most important parts of smart city. Sensors are placed in smart cities with good internet connectivity. More urban our planet becomes, smarter the cities have to be. The cities of tomorrow will be more prone to transformation embellishment than the cities of yesterday [12].

4.1.2 *Problem of Car Registration and Motivation*

The process of registering a car has always been difficult. This is a lengthy process involving several parties, and there is also the risk of manipulating information, replicating data and various errors. In this case, critical information can be very vulnerable to fraud or data falsification, or even available for tracking.

By bringing the power of Distributed Ledger Technology called Blockchain into the picture and moving the entire process of registering a car on to Blockchain, a lot of these vulnerabilities can easily be resolved [16].

4.1.2.1 Research Objectives

Blockchain comes to the rescue by reducing the average response time. The Blockchain will allow parties to send data in the form of an intellectual contract or chain code, which will eventually become the single source of unchanged data for all parties. In addition, the Blockchain in the vehicle registration ecosystem will help reduce the risk of fraud and aggression, since only authorized personnel can use the data when updating the private key in province.

In fact, any attempt to track fake data can be easily done on the Blockchain. The best part is that Blockchain provides one single idea of the lifecycle of the car in one book, which is not currently available [2].

4.1.2.2 Scope of the Research Work

This research experiment is a generalized project implemented using open source technologies developed by Linux Foundation called hyper ledger Fabric in a permission model. Anyone can use this project by taking the authorization and adding their stakeholders into the system.

4.1.3 5-G Technology and Its Implications

With an advanced access technology and with an increase in the demand of the users, 4G will now be easily replaced with 5G. There are several reasons to switch to 5G: have higher capacity, increase data rate, lower end-to-end interruption, massive device connectivity, reduced cost and consistent quality of experience [18].

5G consists microcells, small cells, and relays and hence heterogeneous. Device to Device communicative (D2D) and Internet of Things (IoT) are major concerns. 5G provide a good policy for future 5G standardization network MBB mobile broadband. 5G will allow wireless networks to matter data rates and use case that are currently handled by fiber access. One of the widely used technology in today's era is IoT. IoT further consists two technologies. These technologies used to describe a key focus area for the ICT sector [4, 5].

a. Cyber Physical System (CPS)—This system is used to describe a key focus area for the ICT section. It is basically defined as the unification of computation and physical processes.
b. Machine to Machine (M2M)—It represents the way in which machine can communicate between themselves.

5G validate IoT for new use cases and economic sectors. Objective of 5G is to meet projected mobile traffic demand and to heuristically address the communications needs most sectors of the economy. Also, the aim of group is to promote the development of 5G technologies in China. South Korea's 5G forum is also a public private partnership program that is formed in May, 2013.

4.1.4 IoT and Its Applications in Transportation

Application in Automobile.
If you have ever bought your own vehicle or have sold one, or have been a part of an automobile manufacturing or dealing at any stage of the cycle, you would be familiar with the complication that is Vehicle Registration.

Given the fact that all vehicles in the market have been sold and resold as they are passed through multiple hands, it becomes a cumbersome task to maintain a legitimate record of the history of each vehicle and make it available when needed. By applying Blockchain and IoT technologies and the whole process of registering of vehicles in Blockchain, many of these problems can be easily solved.

4.1.5 Usage of AI, ML in IoT and Blockchain

A good working model could be IoT generating data from a multitude of sensors and analytics, Blockchain storing data and, AI/ML drawing intelligence from the same data. An example of the above is in a supply chain, where IoT can measure a lot of different metrics from environment to trip record to motion sensing, use Blockchain to store that data and then use AI on that data to make human-like decisions. The purpose of Blockchain in this solution is to provide transparency across organization and immutability of data as well as executing smart contracts.

This is not just true for supply chain but is possible in many sectors such as healthcare manufacturing, identity, and security applications and even finance industries. For example, a bank offering line of credit to SMEs may depend on these technologies to make faster, accurate and error-free assessment by using IoT to measure goods, raw materials, finished products, assets, etc. of an SME, store these in Blockchain for audit and other decision-making purposes and employ AI to make recommendations [4, 6, 7].

Each technology in itself is capable of transformation. They don't need one another to be useful. But together, they are even more powerful catalysts to solve problems that are difficult to handle otherwise. Take an example of healthcare. Healthcare issues such as surgical infections, hygiene, negligence etc. can have a bad impact on the patient as well as the hospital in itself. The combination of IoT, Blockchain and AI can be used effectively to bring accountability, efficiency and better and faster patient recovery [19].

4.2 Related Work

Blockchain is not a new technology. It is a set of existing methods, which are organized in a new specific order to solve problems related to different strengths, security and sharing. Many applications are suggested to move from a normal or normal operation to a Blockchain. In addition, many surveys were written to obtain information about applications. The following are some of the previous works related to Driveloop. Two important Blockchain systems for this application are CarChain and Fabcar IBM Blockchain [20].

4.2.1 Carchain

The Carchain is a distributed and decentralized system that connects the car owner and tenant, securely leases and secures financial exchange based on the time spent. The system operates in the open network Blockchain—Ethereum and can be moved to a private Blockchain—Hyper ledger.

It consists of an intellectual agreement that integrates systems and applications into the system (for web application owners, for the user's mobile phone), to manage the system, to send information to the Blockchain and to make changes to the system. It uses an electronic signature method that allows you to unlock the car on arrival.

4.2.2 Fabcar IBM Blockchain

This code demonstrates network configuration on the standard IBM blockchain platform and the implementation of the Fabcar smart contract on the network. We then configure our application to interact with the network, including identity, to send transactions in a smart contract. The application is configured with Node.js using the Fabric Node SDK to handle network requests and the Angular client to open the web interface [21].

Nowadays, career opportunities are rising rapidly. To achieve success, every field needs lots of dedication and hard work. Automobile Engineering career is one of the best careers that is very creative and fast paced. It mainly deals with construction, manufacturing and design of automobile. Due to rapid growth of auto component in automobile sector because of an advanced technology, the jobs in automobile engineering is increasing everyday and the reason behind it are automobile engineers.

4.2.3 Blockchain and Future of Automobiles

The Authors Pham and team explained the future Scope and limitations as below-

As future perspective, it can be said that nowadays, career opportunities are rising rapidly. Any field requires lots of dedication and hard work to learn any profession and achieve the success.

Basically, in this research, authors have presented a write-up for an automobile registration or automobile parking using Blockchain. Here, scientists are using an automobile which is designed for passenger and is run by an internal combustion engine with the help of volatile fuel. In today's world, people prefer vehicle to go anywhere whether it is miles away or it is near to the location. It is the daily need of the person as they have to go for their work or to fulfil their needs. The smoothing lubrication of an automobile helps to move vehicles fast and easy which make our life so simple.

As it's known that nowadays people move to the big cities for better jobs, excellent education and of their bright future. This migration often leads to the increase in population which further leads to the problem of parking spaces. Mostly, many people cannot find safe parking spaces in a crowded area. So, this is insecure solution of centralised based car parking system. An automobile registration system is a unified information system. These information system controls of every information related to an automobile registration. Blockchain is being used nowadays as one of the most emerging domains [14, 17].

The authors' team have applied the methodology for the help of assigned unique ids and without disclosing their personal information, vehicles can communicate with deployed parking lots. Then register vehicle book parking by requesting the controller. Then the controller check for parking space around their establishment when receive a request from the ordinary. Then the complete information is sent to the ordinary node and then the ordinary node reserves the parking and pays for it.

In limitations, one can see that the study was a good learning process and was a very satisfying experience. Yet there are several factors that limited this researches plan to study as every researcher desired limitations are as follows.

(a) **Access to Documentation and information**

- Required data was not readily available. The process of documentation during design and development is not a regular practise. Due to confidentiality of the companies, an R&D and Design activity, the information shared was limited about the processes that are followed for a particular product category.

(b) **Automobile Industry**

- The R&D and Design executives in the automotive industry are tied up because of many rules and policies.
- Data sharing is very limited. It is not the general practise in the corporation culture to openly and willingly share the information.

In concluding remarks, they explained that they implemented the blockchain technology to maintain trust, security, and clarity in the system. We use many technologies and one of the technologies is IoT, Ethereum.

They tested proposed idea on the basis of latency of blockchain, the throughput of blockchain, the accuracy of transactions, latency upon TAIVs and throughput upon TAIVs [13].

4.2.4 *Significance of 5-G Technology*

One of the widely used technologies in today's era is IoT. IoT further consists two technologies. These technologies used to describe a key focus area for the ICT sector. (b) Machine to Machine (M2M)—It defines the way of communication of machine between them.

The purpose of 5G is as follows,

(a) To meet projected mobile traffic demand [24, 25].
(b) To address the communications that is mostly needed by the economy sectors.

4.3 Presented Methodology

With Blockchain, Stakeholders, such as automotive vehicle manufacturers, agents, customers and agencies, can easily participate in accessing and updating vehicle data based on their access to security. The solution also ensures that the most secure and complete information is stored and shared securely and economically [8].

To further explain it, let's first look at the roles of the various stakeholders involved in the vehicle registration process. We also looked at some basic workflows and understood how they were simplified with Blockchain.

(a0 **Manufacturer**: Push the vehicle towards blockchain by adding details including make, model, version, chassis number, engine number, etc. and the selling date of vehicles.
(b) **Dealer**: Car sales are applied to end customers.
(c) **Insurance Agency**: Checks customer and car information and provides insurance.
(d) **Registration Authority**: The RTO will be responsible for approving registries and providing registration numbers, sending vehicle transfers and resetting vehicles.
(e) **Police**: It issues vehicle licenses and transfer certificates, as well as traffic invoices.
(f) **Service Center**: Parts of the service are included as work cards and replacement parts.
(g) **Customer/ Car owners**: Allow the exchange of confidential information as PII [27].

4.4 Software Requirement Specification

The Following Software Requirements have to be fulfilled.

4.4.1 Product Perspective

This idea is not totally implemented anywhere in this world. There exists an app named "Carchain" which provides a way to connect the car owner and tenant securely leases and secures financial exchange based on the time spent.

4.4.1.1 Similarities Between Carchain and Our Application

(a) Both Carchain and this application are service based applications.
(b) In both the applications there are customers who want to avail the services and the professionals who want to provide those services.
(c) One can join as a service provider in both applications.
(d) Feedback can be provided for both the applications.

4.4.1.2 Differences Between Carchain and Our Application

(a) Our application provides an automated way of purchasing a car right from the first step to the last step. Carchain doesn't involve selling cars.
(b) Carchain uses the Ethereum network to implement the blockchain but our application uses hyper ledger Fabric—a private network.

4.4.2 System Interfaces

(a) HTML5, JavaScript, CSS3 and Bootstrap are used for the front end portion of the application.
(b) Node JS is used to write the chain codes for the backend.
(c) Docker is used as a service product that uses OS-level virtualization to deliver software in packages called containers. The containers are isolated and group their own software, libraries and configuration files, they can communicate through clearly defined channels.
(d) Hyper ledger Fabric is used as a platform to operate the application.
(e) Two databases are used—LevelDB for storing the transaction data and CouchDB for storing the asset data.
(f) Visual studio code is used as a source code editor.
(g) Postman is used to create, share, test and document APIs.

4.4.3 Interfaces (Hardware and Software and Communication)

We use many interfaces like.

(a) **Login/Signup**—This interface lets a customer enter the application and avail services and if someone is not a customer to this application, it also helps them to become a registered customer.
(b) **Main Page**—This interface consists of all the services available also it is a connecting medium to all interfaces.
(c) **Contact us**—This interface lets any customer with any issue to contact us.
(d) **Manufacturer**—This interface lets the manufacturer push the vehicle towards blockchain by adding details including make, model, version, chassis number, engine number, etc. and whenever one sells vehicles.
(e) **Dealer**—Car sales are applied to end customers.
(f) **Registration Authority**—The RTO will be responsible for approving registries and providing registration numbers, sending vehicle transfers and resetting vehicles.
(g) **Police**—It issues the vehicle license and transfer certificate.
(h) **Customer**—Allows the exchange of confidential information as PII.

4.4.3.1 Hardware Interfaces

(a) Processor: Intel i5-6200U / Intel Core or better.
(b) GPU:2.30Ghz
(c) Ram: 8 GB or more.
(d) Hard Disk: 20 GB or more.
(e) Operating System: Linux/Mac.
(f) Input Device: Standard Keyboard, Mouse and USB.
(g) A browser which supports HTML and Java script.
(h) Internet Connection.

4.4.3.2 Software Interfaces

(a) **Ubuntu 20.04**—Team has chosen Linux operating system for its best support and user friendliness for this project.
(b) **Hyper ledger Fabric v0.20**—It is used as a modular blockchain structure, which serves as the basis for the development of blockchain-based products, solutions and applications using plug-and-play components intended for use in private companies.
(c) **Node Js v12.16.0-×64**—It is been used to write down the backend logic i.e. Chain code for the automation of the transactions.
(d) **Docker 19.03.8**—It is used as a service product that uses OS-level virtualization to deliver software in packages called containers.
(e) **Postman 7.24.0**—It is used to create, share, test and document APIs. This is achieved because users can create and save simple and complex HTTP/s requests and their responses. This results in more effective and less tiring work.

4.4.3.3 Communications Interfaces

This project supports all types of web browsers. The team is using simple forms for the registration forms, feedback, availing the services etc.

Memory Constraints
Primary Memory: 8 GB or above.
Secondary Memory: 20 GB or above.

4.4.4 Operations (Product Functions, User Characteristics)

Following operations will be performed by our software.

4.4.4.1 Product-Functions

(a) It allows people to register onto the application who want to use its services.
(b) The Manufacturer can add a new car into the blockchain for the sale purpose.
(c) The Dealer can sell a car and can change the ownership of the car after some validations.
(d) The Registration authority can validate a car for changing its ownership from one person to another.
(e) The Customer can check all the steps involved in a registration process directly from a single dashboard.
(f) Any change done anywhere is reflected everywhere in the network.

4.4.4.2 User Characteristics

(a) Only 18 + adults can register or can provide service to other needed people.
(b) Basic technical knowledge of using the computer system is required.
(c) 2-week hands-on training is enough for using the software.

4.4.5 *Use Case, Sequence Diagram*

4.4.5.1 Use Case

The following are the various Use case diagrams of the various Actors involved in the project.

Manufacturer—The Fig. 4.1 depicts the relationship of manufacturer and the various use cases.

Dealer—The Fig. 4.2 depicts the relationship of Dealer and the various use cases.

Registration Authority—The Fig. 4.3 depicts the relationship of the Registration Authority and the various use cases.

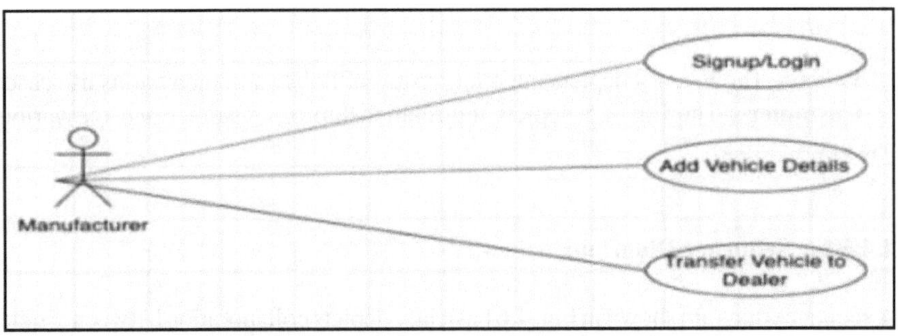

Fig. 4.1 Use case of manufacturer

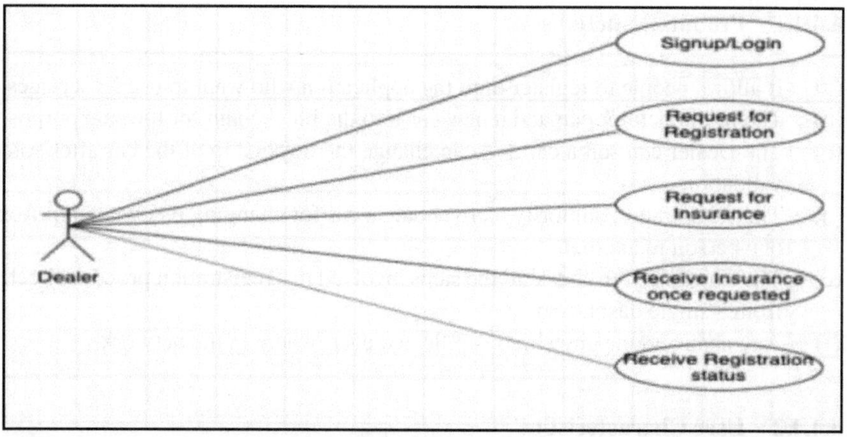

Fig. 4.2 Use case of dealer

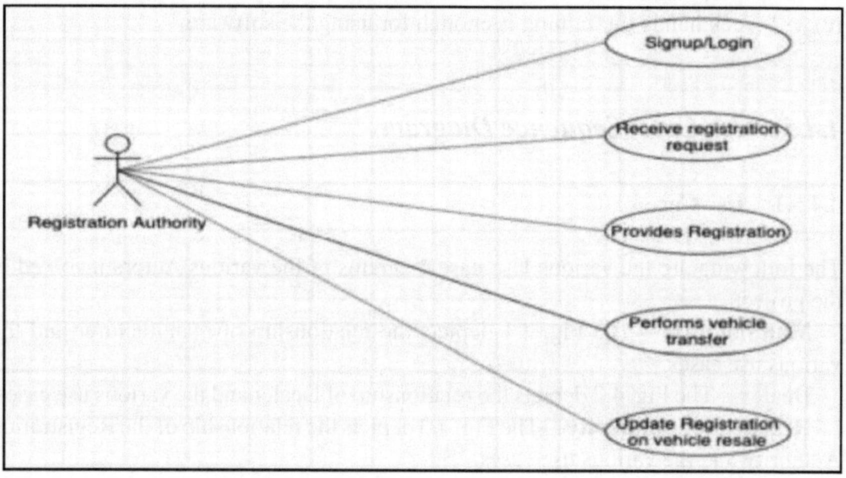

Fig. 4.3 Use Case of Registration Authority

Police—The Fig. 4.4 depicts the relationship of Police and the various use cases.

Customer—The Fig. 4.5 depicts the relationship of Customer and the various use cases.

4.4.5.2 Sequence Diagrams

A sequence diagram shown in Fig. 4.6 basically depicts collaboration between articles in a sequential order. This diagram shows how the client enters into the network and a new block of transactions is created and finally added into the block chain network.

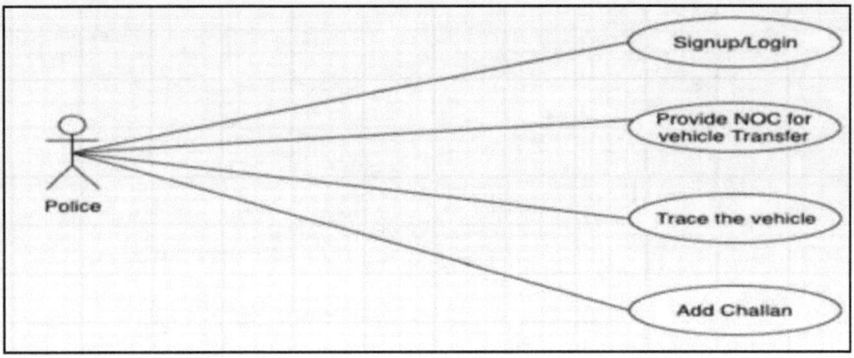

Fig. 4.4 Use case of police

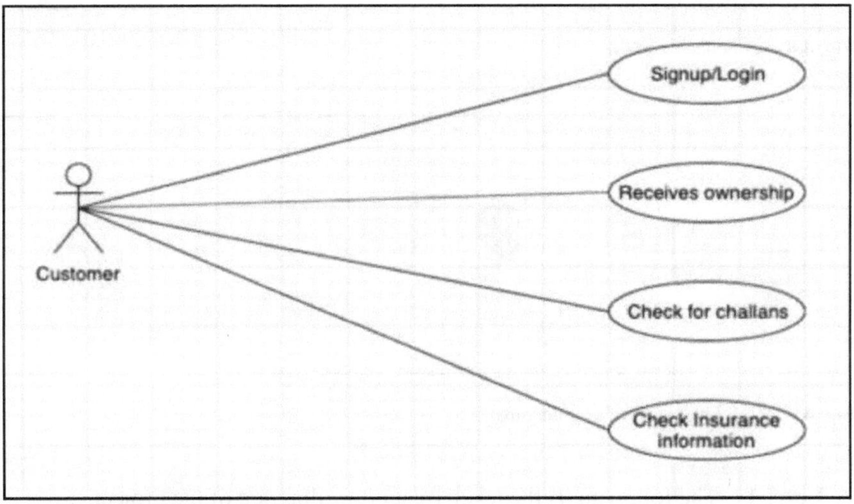

Fig. 4.5 Use case of customer

4.4.6 System Design

System design is the way towards defining the engineering, modules, interfaces, and information for a system to fulfill indicated prerequisites.

Architecture Diagrams, Data Flow Diagrams, Activity Diagram, ER Diagram, Database schema Diagrams (as per Figs. 4.7, 4.8, 4.9, 4.10, 4.11, 4.12 and Fig. 4.13).

Architecture Diagrams
The following is the system architecture design for the project.

Data Flow Diagram
Level 0—The following is the level-0 data flow diagram of the project.

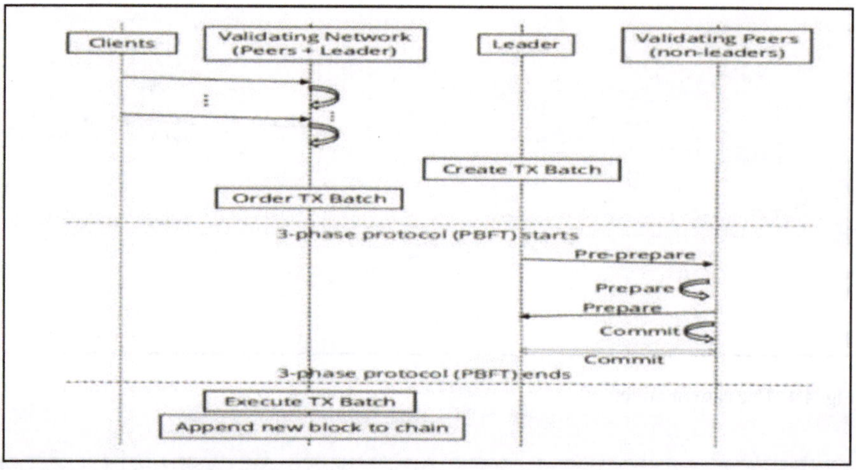

Fig. 4.6 Sequence diagram

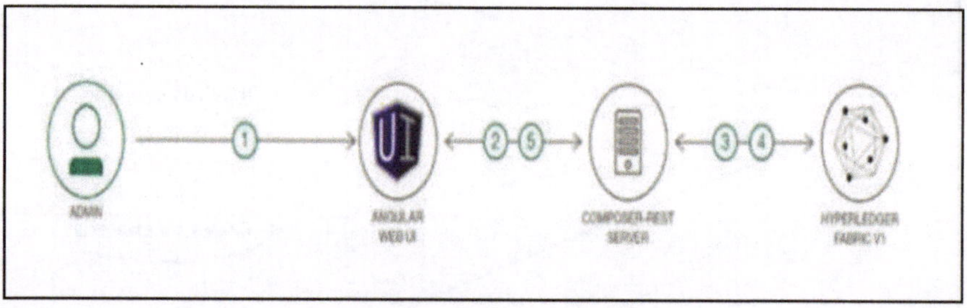

Fig. 4.7 High level view of architecture

Level 1—The following is the level-1 data flow diagram of the project.
Level 2—The following is the level-2 data flow diagram of the project.

Activity Diagram
The following is the Activity diagram showing the Login of the customer into the system.

ER Diagram
The following is the Entity-Relationship Diagram for the system.

Database Schema Diagrams
Hyper ledger Fabric supports two types of peer databases: LevelDB is the default state database embedded in the peer node and stores chain code data as simple key-value pairs; and CouchDB is an alternate state database that supports advanced queries when modeling chain code data values as JSON (as per Figs. 4.14, 4.15, 4.16, 4.17, 4.18 and 4.19).

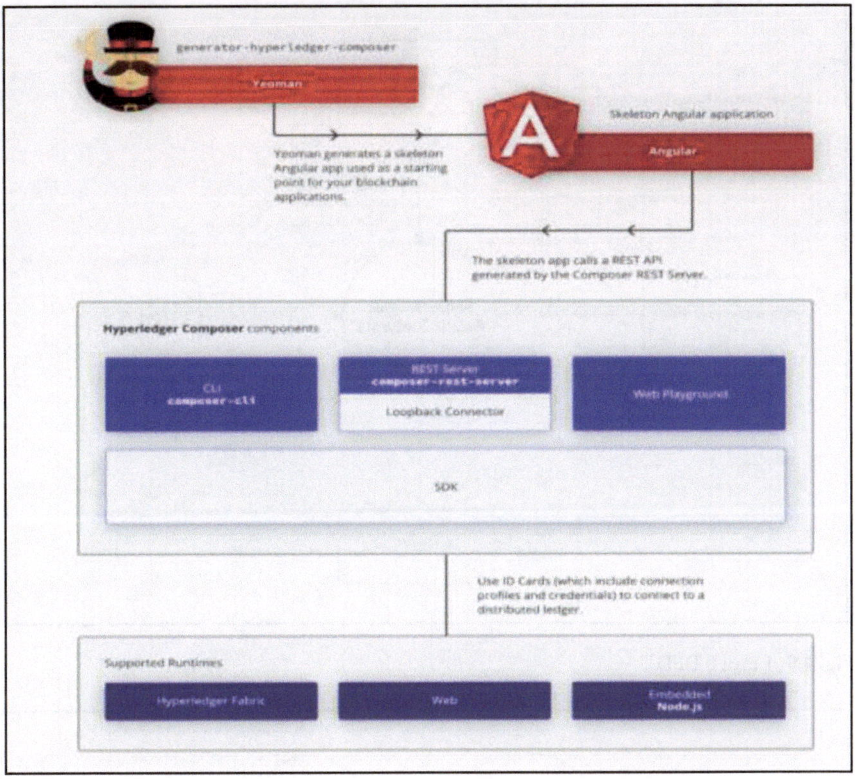

Fig. 4.8 Detailed view of architecture

1. **Assets**
   The following is the schema for the asset data stored in the system.

2. **Manufacturer**
   The following is the schema for the manufacturer data stored in the system.

3. **RTO**
   The following is the schema for the RTO data stored in the system.

4. **Dealer**
   The following is the schema for the dealer data stored in the system.

5. **Police**
   The following is the schema for the police data stored in the system.

6. Customer
   The following is the schema for the customer data stored in the system.

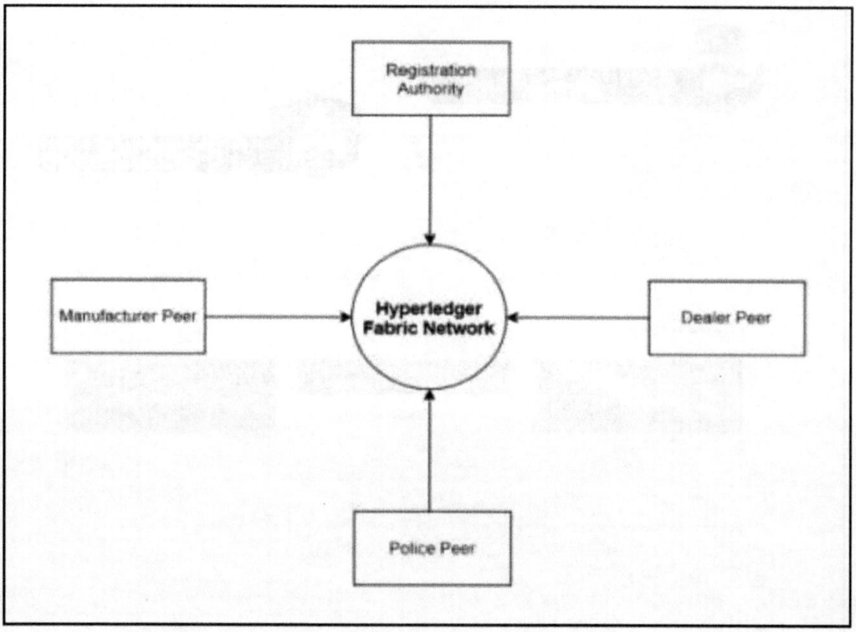

Fig. 4.9 Level 0 DFD

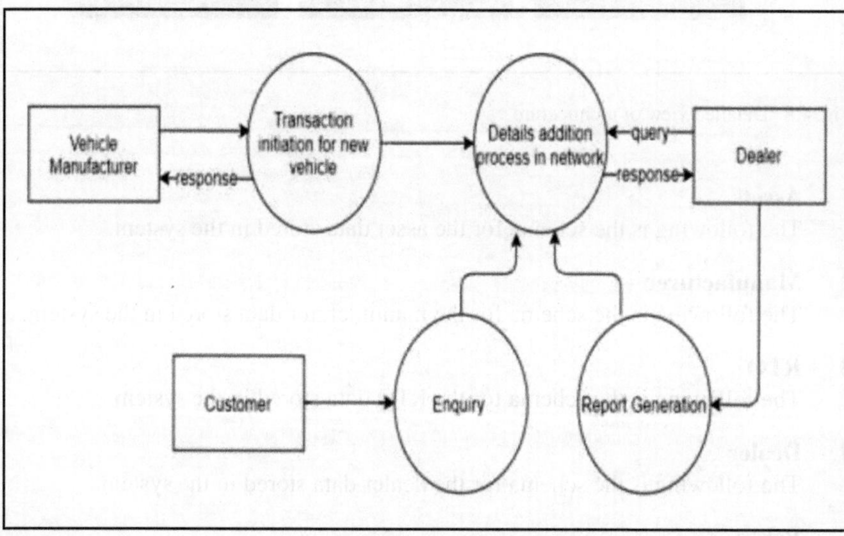

Fig. 4.10 Level 1 DFD

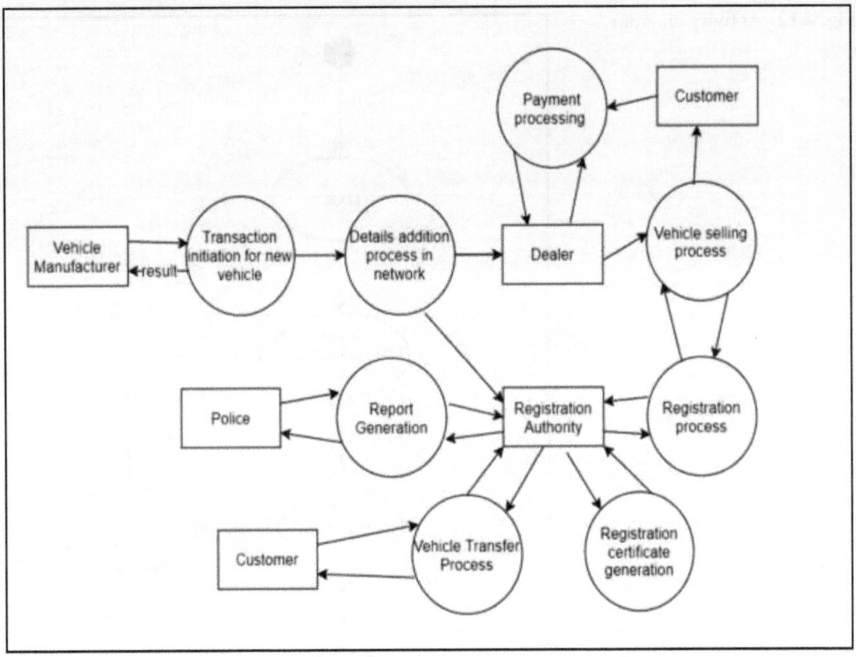

Fig. 4.11 Level 2 DFD

4.5 Software and Hardware Requirements

4.5.1 Software Requirements

(a) **Ubuntu 20.04**: Researchers' Team has chosen Linux operating system for its best support and user friendliness for this project.

(b) **Hyper ledger Fabric v0.20**: It is used as a modular blockchain structure, which serves as the basis for the development of blockchain-based products, solutions and applications using plug-and-play components intended for use in private companies.

(c) **Node JS v12.16.0-x 64**: It is been used to write down the backend logic i.e. Chain code for the automation of the transactions.

(d) **Docker 19.03.8**: It is used as a service product that uses OS-level virtualization to deliver software in packages called containers.

(e) **Postman 7.24.0**: It is used to create, share, test and document APIs. This is achieved because users can create and save simple and complex HTTP/s requests and their responses. This results in more effective and less tiring work.

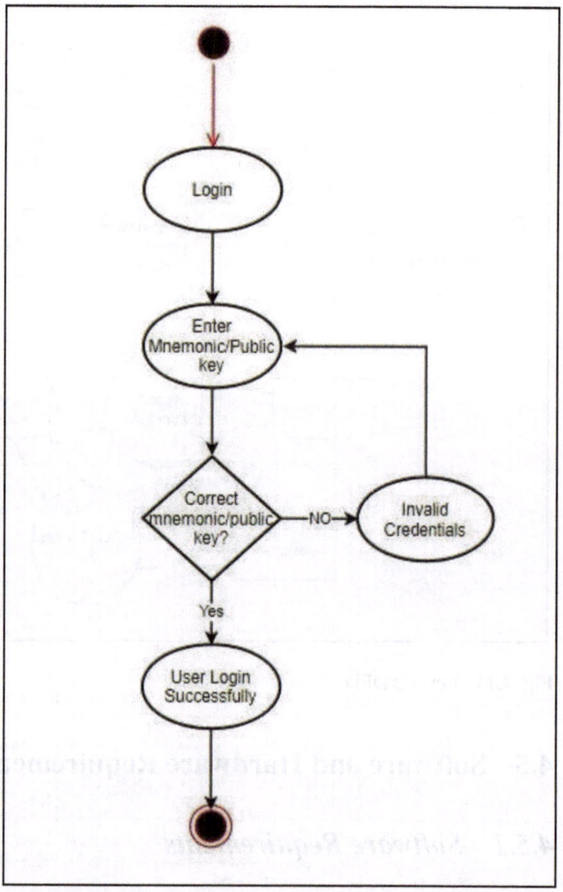

Fig. 4.12 Activity diagram

4.5.2 Hardware Requirements

(a) **Processor**: Intel i5-6200U/Intel Core or better.
(b) **GPU**: 2.30Ghz
(c) **Ram**: 8 GB or more.
(d) **Hard Disk**: 20 GB or more.
(e) **Input Device**: Standard Keyboard, Mouse and USB.

4.6 Implementation Details

Snapshots of Interfaces (As per Figs. 4.20, 4.21, 4.22, 4.23, 4.24, 4.25, 4.26 and 4.27).

There are few snapshots from are project.

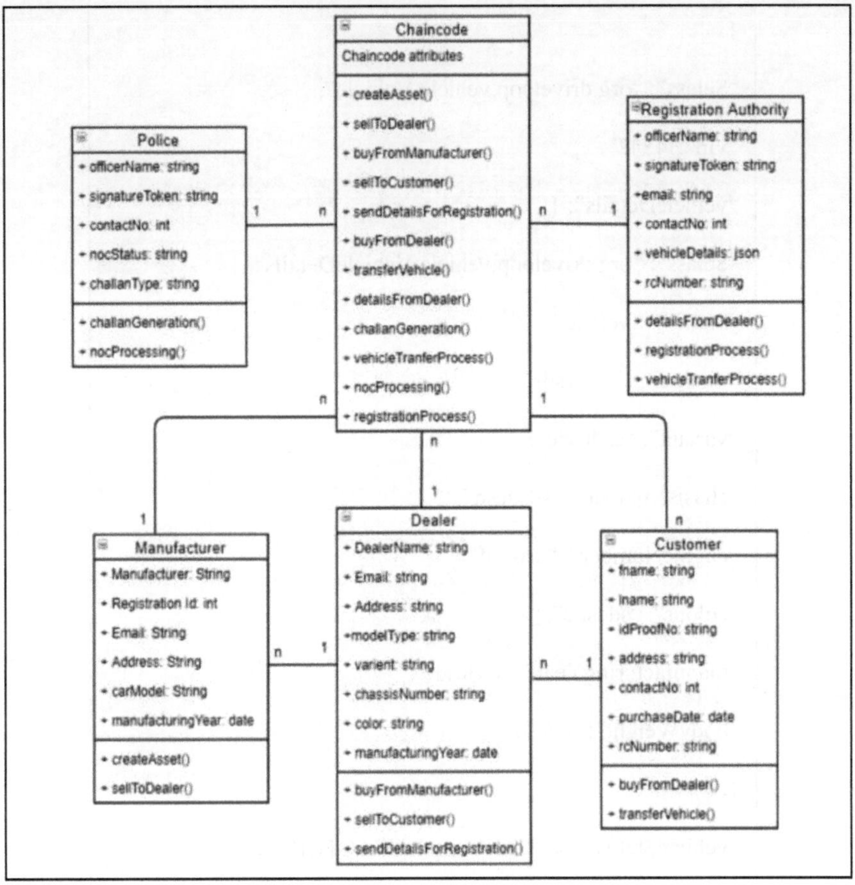

Fig. 4.13 Entity relationship diagram

The above Fig. 4.20 demonstrates the front page where the user can easily go to the platform and using the drive loop and different blockchain techniques, it felicitates the user for automobile registration using authentic and transparent manner.

Figure 4.21 depicts the main page on which user can upload the documents of automobile and can see the all parties which were involved earlier in whole transaction before the registration of this particular vehicle. It will be able to show the whole history of automobile.

Asset can be created by above Fig. 4.22 where the blocks will contain the records for automobile and whole details of all possible transactions will be stored for future purposes. This Asset will be base information and will be authenticated by all parties for transparency.

Above form in Fig. 4.23 is to show the model for manufacturer and displays the process of entering the unique and basic details of automobile by the manufacturer

{

"$class": "org.driveloop.vehicle.Vehicle",

"vin":"4242",

"vehicleDetails": {

"$class": "org.driveloop.vehicle.VehicleDetails",

"make": "fvefvsd",

"modeType": "svsds",

"variant": "csdcsdc",

"chasisNumber": "sdsdcsd",

"engineNumber": "csdcsd"

"colour": "sdcsdc",

"manufacturingYear": "csdcsdc",

"bodyWeight": "cdcsd"

},

"vehicleStatus": "UNDER_MANUFACTURER"

}

Fig. 4.14 Schema for assets

which will help to maintain the transparency and ease in smart contract and future transactions.

The Fig. 4.24 code in hyper ledger shows the transaction history along with all necessary details for a automobile.

Figure 4.25 runs the possible test cases and checks the right functioning of the software.

Test Cases

To run the tests locally, we use a Docker file that builds our environment. The Docker file would be something like this:

```
{
"$class" : "org.driveloop.participant.Manufacturer",
"make": {
"$class": "org.driveloop.partcipant.Make",
"name": "BMW",
"registrationId": "asj5w67dw87wgx87x8vw"
},
"partcioantId": "9192"
}
```

Fig. 4.15 Schema for manufacturer

```
{
"$class": "org.driveloop.partcipant.RegistrationAuthority",
"officerName": "Mr. Joe",
"signatureToken": "43546756545",
"participantId": "8753"
}
```

Fig. 4.16 Schema for RTO

We'll get a response like this if everything occurred nicely in Fig. 4.27 after running all the test cases. The sample snippets are presented for the understanding of the readers.

4.7 Results and Discussions

Driveloop enables all information to be accumulated into one place so that it can be easily accessed and managed. Vehicle Registration, Citations, Insurance Details and

Fig. 4.17 Schema for dealer


```
{
“$class”: “org.driveloop.partcipant.Dealer”,
“dealerName”: “Man Sales”,
“contact” :{
  “$class”: “org.driveloop.partcipant.Contact”,
“email”: “mansales@gmail.com”,
“address”: “south district”
},
“participantId”: “3888”
}
```


Fig. 4.18 Schema for police


```
{
“$class”: “org.driveloop.partcipant.Police”,
“officerName”: “Mr. Joe”,
“signatureToken”: “43546756545”,
“participantId”: “8753”
}
```


everything else accruing to the vehicle in question is integrated on this platform. So, when you log in to find out something about one particular vehicle, what you will find is everything there is to know about it. A comprehensive, all-encompassing history is obtained. Anybody who is even remotely aware of the Blockchain technology will tell you how authentic it is.

It is structured in such a way that only authorized personnel can make entries or change records. Hence, there is no need to worry about any kind of tempering with the data or falsification of information.

```
{
“$class”: “org.driveloop.partcipant.Customer”,
“fName”: “Anuranjan”,
“lname” : “Singh”,
“contact”: {
  “$class”: “org.driveloop.partcipant.Contact”,
“email”: “anuranjansingh@gmail.com”,
 “address”: “ballia”
},
“participantId”: “9317”
}
```

Fig. 4.19 Schema for customer

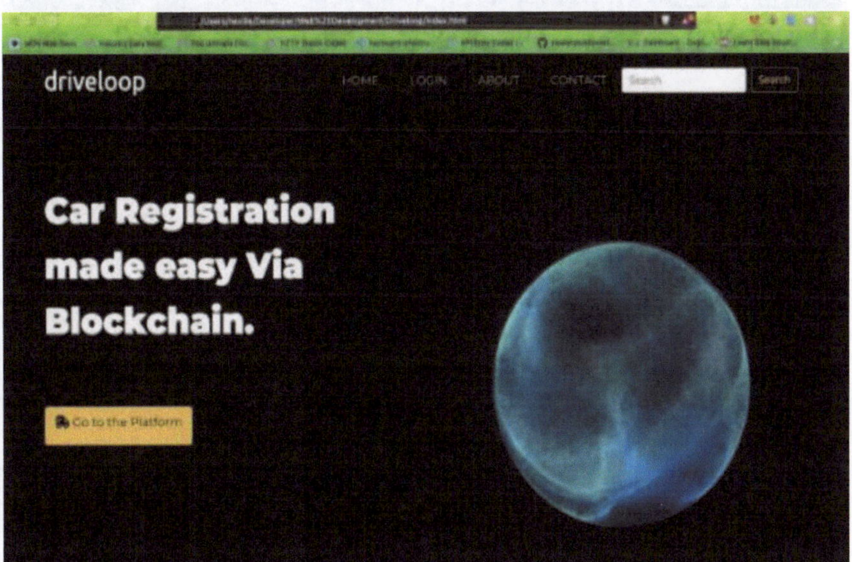

Fig. 4.20 Front page

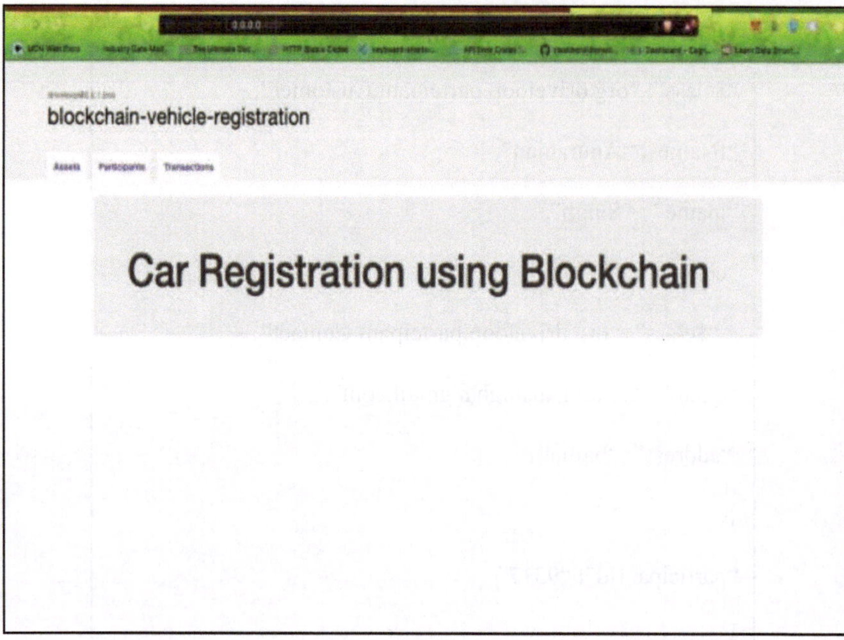

Fig. 4.21 Main page

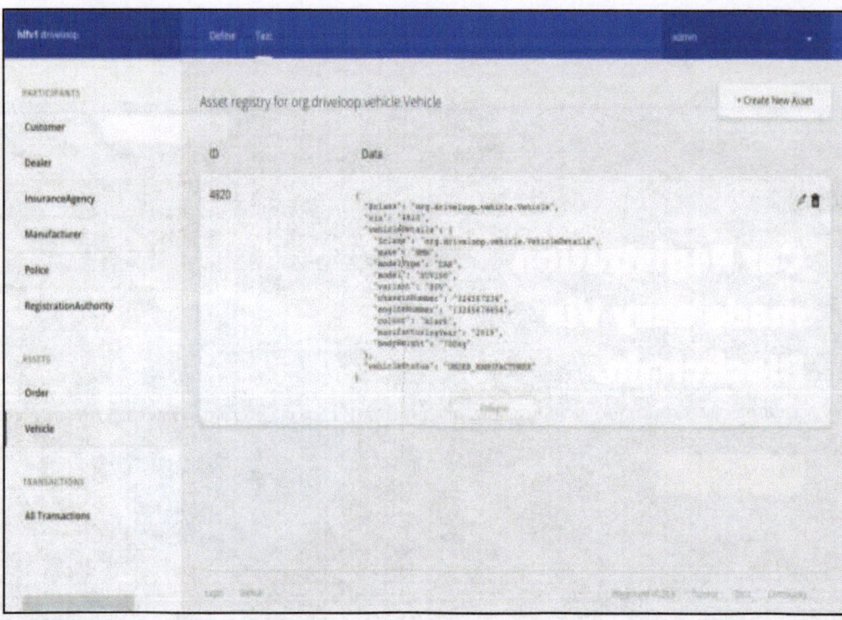

Fig. 4.22 Asset creation

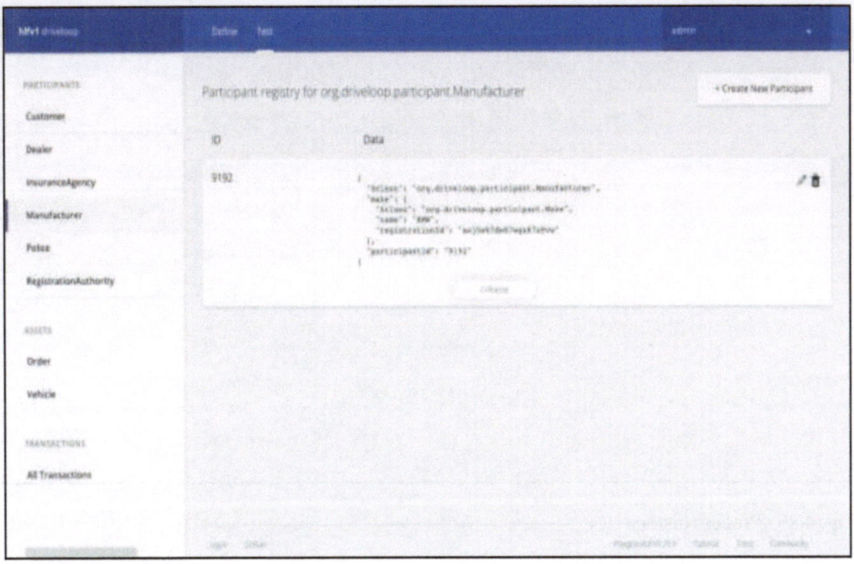

Fig. 4.23 Model testing for manufacturer

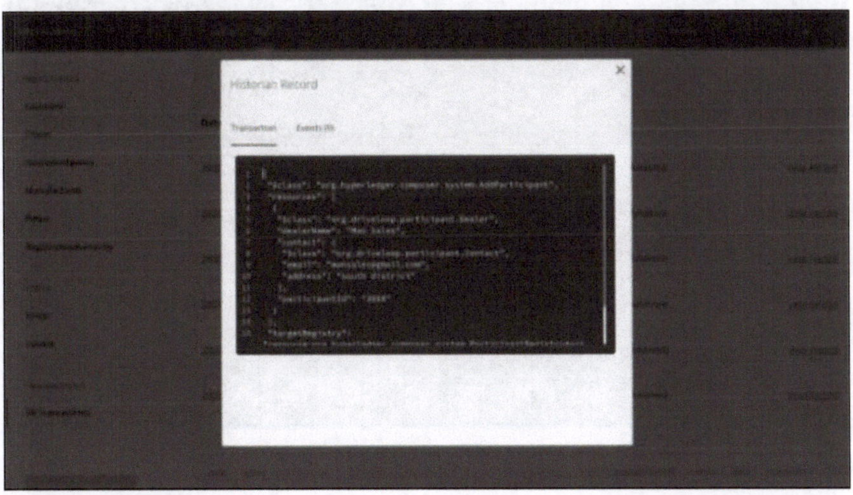

Fig. 4.24 Transaction history

From our research, we implemented the blockchain technology to maintain trust, security, and clarity in the system. We used many technologies and one of the technologies is IoT, Ethereum. On the basis of latency of blockchain, the throughput of blockchain, the accuracy of transactions, we tested our proposed idea.

Comparison with existing state-of-the-art technologies.

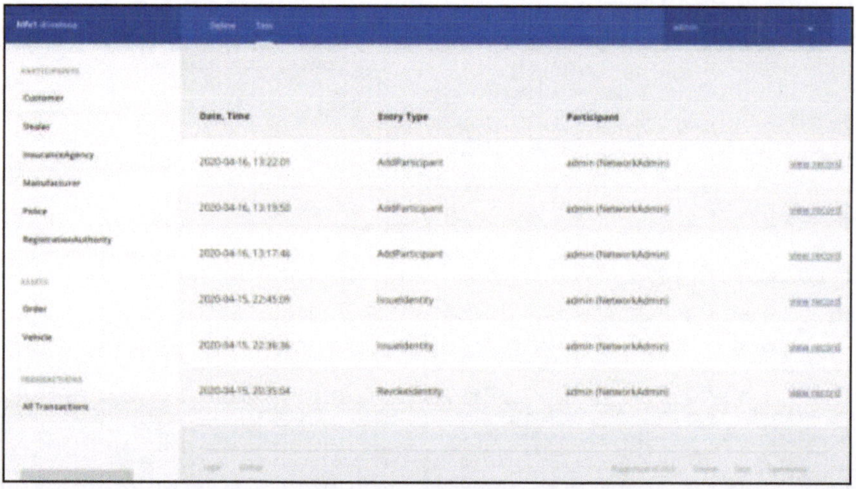

Fig. 4.25 Process history

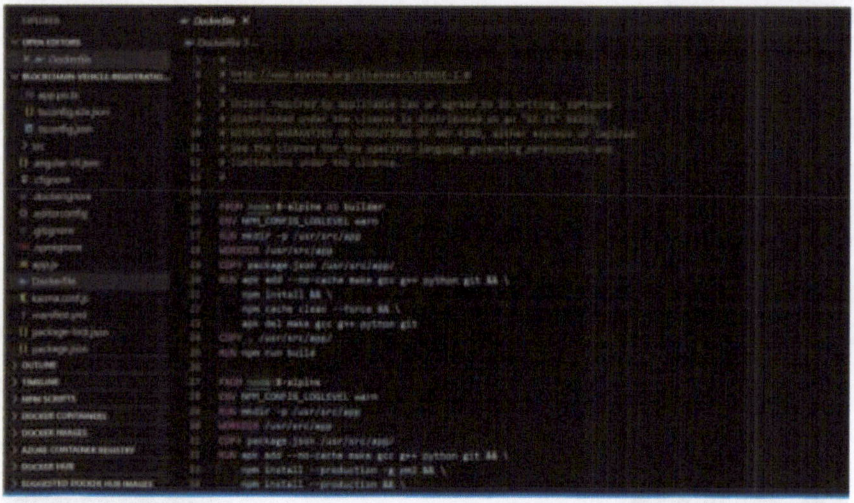

Fig. 4.26 Docker file

Carchain	Driveloop (our application)
1. It is used to maintain the data for Rental cars	1. It is an automated process for buying and selling of cars
2. It uses the Ethereum platform	2. It uses Hyper ledger Fabric platform
3. It is a public blockchain	3. It is a private blockchain

Fig. 4.27 Test cases

4.8 Novelty and Recommendations

Before we go on to talk about the problems of the process of Vehicle Registration, we first need to understand why Vehicle Registration is such an important aspect of automobile dealing. Car ownership of changes as many times as you can imagine. Whether you look at it in the terms of dealing in spare parts or in assembled vehicles, dealings by the middlemen or by the retailer who makes the final sale to a consumer, or in terms of the resale of a second-hand vehicle, there are a number of stakeholders who would very much want to know about all the history of the vehicle they are buying. Not to mention the insurance agencies, the police and other authorities and well, the government too needs to keep tabs on the automobiles for various reasons [9, 10].

The fact of the matter is that all these stakeholders need information about the vehicles, starting from its manufacturing story, covering its first sale, the accidents, if any, that it has been in, and any and all repairs and maintenance. This is crucial not just to maintain a track record of the vehicle in question to determine its market value, but also for legal and insurance purposes [26].

The word automobile is derived from the Greek word auto which means "self" and the French word mobile which means 'moving'. The significance of automobile are as follows,

(a) The increase in the demand for automobiles such as cars and other vehicles increase the income of driver of the automobile industry.
(b) In this foster age, people need to reach destinations rapidly. So, automobiles help one over here and it became popular. With the help of automobile, people from all over the world can travel anywhere. Automobiles play a vital role in the country's socio-economic development.

(c) There is also a worldwide sharing in automotive industry of cars, vehicles, parts and accessories that ranges from 15 to 40% in US, South Korea etc.
(d) The automotive industry provides development of the taxable base and revenues of the state budget.
(e) It also influences scientific and technical progress [11].

4.9 Future Research Directions

There is a huge transformation in the urban development in twenty-first century because of the advanced technologies and various services. Nowadays, research activities become common for growing smarter cities. Smart City is the demanding solution of sustainability and urbanization. Nowadays, corruption is common and mostly it is paid by the poor. It is like a cancer that eats away at a citizen's faith in the government. For example—smart cities may lead to injustice in the world where citizens or people are pushed to subaltern roles and it is regulated by technocratic governments. Increase in population in urban areas often leads to the problem of parking spaces and has been a big challenge for urban planners, architects, and administrators. So keeping all such views, an innovative solution is much more awaited and demanded [3].

There are few future remedies that can be carried out in this project.

(a) We intend to add certain features like location detection through GPS and addition of some more services according to user requirements afterwards.
(b) We also intend to add an Insurance party into our project.
(c) We also intend to increase the scalability of this project worldwide i.e. beyond our country [22, 23].

4.10 Limitations

The study was a good learning process and was a very satisfying experience. Yet there are several factors that limited this researchers plan to study as every researcher desired limitations. Some are as follows:

(a) Access to documentation and information—The required data was not readily available. The process of documentation is not a continual practice.
(b) Automobile Industry—Because of strict rules and many policies, the R&D and design executives are bound up in the automotive industry. Data sharing is very limited. To share the information openly and willingly is not considered good practice in the corporations.

4.11 Conclusions

The interesting parts, like fabricators, conventions, clients and automobile agencies, can easily be facilitated for accrediting and updating the information of the vehicle in its secure function. The solution also guarantees that the information is more precise and completely sealed and transmits secure and economical form.

Performance Evaluation.

(a) The performance of the service providers is based on ratings given to them by service users.
(b) The performance of the service users is based on ratings given to them by service providers.
(c) The performance of the overall website is based on feedback given to us by the users of the website.
(d) The reviews for the website will be taken from mentors, coordinators and peers' students.

Internet of Things (IoT), big data systems and mobility are some of the services programmers of smart cities. Smart parking is the crucial parts of smart city. Connected automobile with its advanced technology reduces the chances of accident and help drivers save time and gasoline in their limits. More urban our planet becomes, smarter the cities have to be. In the coming days, due to the advanced technology, the smart cities would prone to the smarter cities.

References

1. Agarwal, A., Goel, D., Tyagi, A., Aggarwal, A., & Rastogi, R. (2018, February). A Smarter Approach for Better Lifestyle in Indian Societies, In Saeed, K., Chaki, N., Pati, B., Bakshi, S., & Mohapatra, D. (Eds.), *Progress in advanced computing and intelligent engineering. Advances in intelligent systems and computing* (Vol. 563, pp. 355–362). Springer. Print ISBN 978-981-10-6871-3, Online ISBN 978-981-10-6872-0, First Online09. https://doi.org/10.1007/978-981-10-6872-0_33.
2. Aggarwal, S., Rastogi, R., & Mittal, S. (2015, March 11th–13th). A novel approach for communication among blind deaf and dumb people. In *Proceedings of 09th INDIACom; 2nd International Conference on Computing for Sustainable Global Development* (pp. 605–610). ISSN 0973–7529 and ISBN 978-93-80544-14-4 serials. https://ieeexplore.ieee.org/abstract/document/7100321
3. Arora, S., R., Goel, P., Maini, J., & Mallick, P. (2016, March 16th–18th). Increasing efficiency in online studies through web socket. In *Proceedings of 3rd International Conference on Computing for Sustainable Global Development* (pp. 509–512). ISSN 0973–7529 and ISBN978-93-80544-14-4 serials. https://ieeexplore.ieee.org/document/7724316.
4. Goel, D., Agarwal, A., & Rastogi, R. (2016). A novel approach for residential society maintenance problem for better human life. In Rajesh, R., & Mathivanan, B. (Eds.), *Communication and power engineering* (pp. 177–185).
5. Goel, D., Agarwal, A., & Rastogi, R. (2016, September). A novel approach for residential society maintenance problem for better human life. *International Journal of Urban Design for Ubiquitous Computing (IJUDUC) 4*, 1–8. (Affiliated to the National Library of Australia,

Global Vision School Publication, Sandy Bay, Tasmania, Australia). https://doi.org/10.21742/ijuduc.2016.4.2.01.

6. Gupta, R., Rastogi, R., & Singh, A. (2015, March 11th–13th). A novel approach for vehicle tracking system for traffic jam problem. In *Proceedings of 9th INDIACom; 2nd International Conference on Computing for Sustainable Global Development* (pp. 69–174). ISSN 0973–7529 and ISBN 978-93-80544-14-4 serials. https://ieeexplore.ieee.org/document/7100240.
7. Gupta, R., Rastogi, R., Mondal, P., & Aggarwal, K. (2014, January–June). GA based clustering of mixed data type of attributes (numeric, categorical, ordinal, binary, ratio-scaled). *BVICA M's International Journal of Information Technology*, *7*(2), 861–866. ISSN 0973 – 5658. Indexed in DBLP, published by Springer. http://bvicam.ac.in/bjit/issues.asp?issue=14.
8. Kaul, U., Rastogi R., Agarwal S., Sharma P., & Jain S. (2015). A novel D&C approach for efficient fuzzy unsupervised classification for mixed variety of data. In: Satapathy, S., Govardhan, A., Raju, K., & Mandal J. (Eds.), *Emerging ICT for Bridging the Future—Proceedings of the 49th Annual Convention of the Computer Society of India, CSI Volume 2. Advances in Intelligent Systems and Computing* (Vol. 338, pp. 553–563), Springer, Cham. Print ISBN 978-3-319-13730-8 Online ISBN 978-3-319-13731-5. https://doi.org/10.1007/978-3-319-13731-5. https://doi.org/10.1007/978-3-319-13731-5_60 . https://link.springer.com/chapter/https://doi.org/10.1007/978-3-319-13731-5_60
9. Khanna, A., & Anand, R. (2016). IoT based smart parking system. In: *International Conference on Internet of Things and Applications (IOTA)*. IEEE 266–270.
10. Mejri, N., Ayari, M., Langar, R., Kamoun, F., Pujolle, G., & Saidane, L. (2014). Cooperation versus competition towards an efficient parking assignment solution. In *IEEE International Conference on Communications (ICC)* (pp. 2915–2920).
11. Nigam, A., Rastogi, R., Mishra, R., Arya, P., & Sharma, S. (2015, June). Security of data transmission using logic gates and crypt analysis. *CSI Communication*, *17*. http://www.csi-india.org/communications/CSI%20June15%20Combine.pdf.
12. Nigam, A., Rastogi, R., Mishra, R., Arya, P., & Sharma, S. (2015, March 11th–13th). Security of data transmission using logic gates and crypt analysis. In *Proceedings of 9th INDIACom; 2015 2nd International Conference on Computing for Sustainable Global Development* (pp. 101–105). ISSN 0973–7529 and ISBN 978-93-80544-14-4 serials. https://ieeexplore.ieee.org/document/7100229.
13. Pham, T. N., Tsai, M.-F., Nguyen, D.B., Dow, C.-R., Deng, D.-J. (2015). A cloud-based smart-parking system based on Internet-of-Things technologies. *IEEE Access*, *3*, 1581–1591
14. Pilkington, M. (2016). Blockchain technology: Principles and applications. *Research Handbook on Digital Transformations, 225*.
15. Rastogi, R., Aggarwal, K., & Mondal, P. (2015). An exhaustive review for infix to postfix conversion with applications and benefits. In *Proceedings of 9th INDIACom; 2nd International Conference on Computing for Sustainable Global Development*, (11th–13th March, 2015, 95–100). ISSN 0973–7529 and ISBN 978-93-80544-14-4 serials. https://ieeexplore.ieee.org/document/7100228.
16. Rastogi, R. Agarwal , S., Sharma, P., Kaul, U., & Jain, S. (2014). Unsupervised classification of mixed data type of attributes using genetic algorithm (numeric, categorical, ordinal, binary, ratio-scaled). In: Pant M., Deep K., Nagar A., & Bansal J. (Eds.), *Proceedings of the Third International Conference on Soft Computing for Problem Solving. Advances in Intelligent Systems and Computing* (Vol. 258, pp. 121–131). Springer. First Online04. Print ISBN 978-81-322-1770-1. Online ISBN 978-81-322-1771-8. https://doi.org/10.1007/978-81-322-1771-8_11.
17. Rhodes, C., Blewitt, W., Sharp, C., Ushaw, G., & Morgan, G. (2014.) Smart routing: A novel application of collaborative pathfinding to smart parking systems. In *CBI* (Vol. 1, pp. 119–126).
18. Rungta, S., Srivastava, S., Yadav, U.S., & Rastogi, R. (2014). A comparative analysis of new approach with an existing algorithm to detect cycles in a directed graph. In: Satapathy, S., Avadhani, P., Udgata, S., & Lakshminarayana S. (Eds.), *ICT and Critical Infrastructure: Proceedings of the 48th Annual Convention of Computer Society of India* (Vol. II., Vol. 249, pp. 37–47), Advances in Intelligent Systems and Computing, Springer, Cham. Print ISBN

978-3-319-03094-4. Online ISBN978-3-319-03095-1. https://doi.org/10.1007/978-3-319-03095-1_5.
19. Sharma, R., Jain, A., & Rastogi, R. (2013, August 8–10). A new face to photo security of Facebook. In *Proceedings of the Sixth International Conference on Contemporary Computing, Jointly Organized by Jaypee Institute of Information Technology & University of Florida, IC3* (pp. 415–420). https://doi.org/10.1109/IC3.2013.6612231, https://ieeexplore.ieee.org/document/6612231, 3 Papers citations yet.
20. Sharma, S., Rastogi, R., & Kumar, S. (2015, March 11th–13th). A revolutionary technology to help the differently abled person. In *Proceedings of 9th INDIACom; 2015, 2nd International Conference on Computing for Sustainable Global Development* (pp. 622–624). ISSN 0973–7529 and ISBN 978-93–80544-14-4 serials. https://ieeexplore.ieee.org/document/7100324.
21. Sharma, P., Rastogi, R., Aggarwal, S., Kaul, U., & Jain, S. (2014, January–June). Business analysis and decision making through unsupervised classification of mixed data type of attributes through genetic algorithm. *BVICAM's International Journal of Information Technology, 6*(1), 683–689. ISSN 0973 – 5658. Special Issue of BIJIT on Fuzzy Logic. http://bvicam.ac.in/bjit/issues.asp?issue=11. Indexed in DBLP, published by Springer.
22. Shekhar, S., Rastogi, R., & Mittal, S. (2015, 11th–13th March). Linear algorithm for Imbricate Cryptography using Pseudo Random Number Generator, In *Proceedings of 09th INDIACom, 2nd International Conference on Computing for Sustainable Global Development* (pp. 89–94). ISSN 0973–7529 and ISBN 978–93–80544–14–4 serials. https://ieeexplore.ieee.org/document/7100227.
23. Srivatava, S., Rastogi, R., Rungta, S., & Yadav, U. (2014, July–December) A methodology to find the cycle in a directed graph using linked list. *BVICA M's International Journal of Information Technology, 6*(2), 743–749. ISSN 0973 – 5658. http://bvicam.ac.in/bjit/issues.asp?issue=12. Indexed in DBLP, published by Springer.
24. Wood, G. (2014). Ethereum: A secure decentralized generalized transaction ledger. *Ethereum Project Yellow Paper, 151*, 1–32.
25. Wu, H., Li, Z., King, B., Miled, Z. B., Wassick, J., & Tazelaar, J. (2017). A distributed ledger for supply chain physical distribution visibility. *Information, 8*(4), 137.
26. Zheng, Z., Xie, S., Dai, H., & Wang, H. (2016). Blockchain challengesand opportunities: A survey. *International Journal of Web and Grid Services*, 23–29 (2016).
27. Zyskind, G., Nathan, O., et al. (2015). Decentralizing privacy: using blockchain to protect personal data. In *Security and Privacy Workshops (SPW)*, IEEE, pp. 180–184.

Chapter 5
Nonparametric Test for Change-Point Detection of IoT Time-Series Data

Dmitriy Klyushin and Andrii Urazovskyi

Abstract The Internet of Things is a concept of computer networks consisting of devices interacting with each other without human intervention. The complete or partial absence of human participation in the operation of such a network makes it necessary to solve the problem of automatic recognition of a significant deviation of the current state of the network from its normal state. The concept of "significance" is vague until it is formalized with the help of rigorous statistical concepts, in particular, the concept of confidence intervals with a given level of significance. In this chapter, we offer a new effective online algorithm for detection of change-points in data generated by IoT devices (tracking data, health rate data etc.). This allows detecting failures, cyber attacks and other deviations in the data. We describe a non-parametric test for evaluation of the statistical hypothesis that data in two adjacent time intervals of a time series have the same distribution. When this hypothesis is true, we detect a change-point. The significance level for the test is less than 0.05. The test is universal in two aspects: it permits ties in a sample and it save the sensitivity when the distributions are greatly overlapped. We demonstrate the prevalence of the proposed test over widely used the Kolmogorov–Smirnov test and the Wilcoxon signed-rank tests using both artificial samples with various distributions and properties (disjoint, slightly overlapped, and strongly overlapped) and real examples. We show that our approach provides robust, sensitive and accurate results.

Keywords IoT · Time series · Change-point problem · Tracking data · Health rate control · Nonparametric test · Homogeneity measure

5.1 Introduction

The widespread use of IoT technology in various spheres of life (from everyday life to high-tech enterprises) creates new problems associated with the analysis of data coming from devices. The high rate of data generation and the large amount

---

D. Klyushin (✉) · A. Urazovskyi
Taras Shevchenko National University of Kyiv, Kyiv, Ukraine

P. Kumar et al. (eds.), *A Fusion of Artificial Intelligence and Internet of Things for Emerging Cyber Systems*, Intelligent Systems Reference Library 210,
https://doi.org/10.1007/978-3-030-76653-5_5

of memory required to store it highlight the challenges of online data processing in small chunks. One of the important and rather difficult tasks is to detect change-points in data streams. A change-point is a point in time before and after which the data has a different distribution.

The algorithm for change-point detection may be categorized as online and offline. Online algorithms process data "in fly" and does not require to store huge amount of data. Offline algorithm use entire information on data stream and demands to store all the data collected during given period. Obviously, offline algorithms are most accurate since they use complete datasets rather than chunks of information. However, their practice applicability depends on availability of high-performance computers and storage for big data. Thus, online algorithms are useful when we need fast and robust estimation of data stream features. Specifically, these algorithms are necessary for IoT devices that cannot have huge memory.

The statistical tests for change-point detection on random data stream are classified as parametric and nonparametric. Parametric tests suppose some functional form of data stream (see for example [1, 2] etc.). This allows discarding of training samples after learning, but narrows the applicability of parametric tests. Nonparametric tests do not use any assumptions on functional form of data distribution excepting very general ones, e.g. absolute continuity. Some authors state that nonparametric is more effective for large datasets [3].

If samples are drawn from the same populations they are called homogeneous. The two-sample tests for homogeneity are divided on purely nonparametric (Smirnov [4, 5], Dickson [6], Wald and Wolfowitz [7], Mathisen [8], Wilcoxon [9], Mann–Whitney [10], Wilks [11] etc.) and conditionally nonparametric (Pitman [12], Lehmann [13], Rosenblatt [14], Dwass [15], Fisz [16], Barnard [17], Birnbaum [18], Jockel [19], Allen [20], Efron and Tibshirani [21], Dufour and Farhat [22] etc.).

The chapter describes a new effective online algorithm for detection of change-points in data generated by IoT devices based on a non-parametric approach. The further content is organized as follows. Section 2 introduces the Klyushin–Petunin tests for heterogeneity of two samples following absolutely continuous distribution functions. Section 3 describes numerical experiments. Section 4 presents results of the numerical experiments and their discussion. Section 5 concludes the chapter and poses some open problems.

5.2 Nonparametric Test for Absolutely Continuous Distribution Functions

Suppose that $x = (x_1, x_2, ..., x_n)$ and $y = (y_1, y_2, ..., y_n)$ are samples from populations G_1 and G_2 following absolutely continuous distribution functions F_1 and F_2. The null hypothesis states that these distributions are identical and the alternative hypothesis states the opposite. Consider the purely non-parametric Klyushin–Petunin test [23] based on the Hill's assumption $A_{(n)}$ [24] that states that if random values

$x_1, x_2, ..., x_n \in G$ are exchangeable and belong to absolutely continuous distribution then

$$P\big(x \in \big(x_{(i)}, x_{(j)}\big)\big) = \frac{j-i}{n+1},\ \ j < i, \tag{5.1}$$

where x is a sample value from G, and $x_{(i)}$ and $x_{(j)}$ are the ith and jth order statistics. This assumption was proved for independent identically distributed random values in [25] and for exchangeable identically distributed random values in [26]. Finding the relative frequency h_{ij} of the event $y_m \in \big(x_{(i)}, x_{(j)}\big)$, we can estimate how h_{ij} differs from (1) using a confidence interval for binomial proportion $I_{ij}^{(n)} = \Big(p_{ij}^{(1)}, p_{ij}^{(2)}\Big)$ (for example, the Wilsoninterval as in the following numerical experiments). The similarity measure of samples x and y (p-statistics) is defined by the equation

$$h = \frac{2\#\Big\{p_{ij} = \frac{j-i}{n+1} \in I_{ij}^{(n)}\Big\}}{(n-1)n}. \tag{5.2}$$

As far as the p-statistics (2) is the relative frequency of the event $\Big\{\frac{j-i}{n+1} \in I_{ij}^{(n)}\Big\}$, we may construct the confidence interval I_n for the p-statistics (for example, the Wilsoninterval) an formulate a decision rule of the test: if I_n contains 0.95, the null hypothesis is accepted, else the null hypothesis is rejected. If the null hypothesis holds, and $\lim\limits_{n\to\infty} \frac{j-i}{n+1} \in (0, 1)$, and $\lim\limits_{n\to\infty} \frac{i}{n+1} \in (0, 1)$, then the asymptotic significance level β of a sequence of confidence intervals $I_{ij}^{(n)}$ is less than 0.05 [23].

5.3 Numerical Experiments

To test proposed method we carried out numerical experiments with stationary time series having one change-point. Time-series of such kind were studied, for example, in [27–29]. We divided the stationary time series in segments of size n and considered two adjacent non-overlapping sliding windows of length n, where $n = 10, 40, 70$, and 100. The lengths of the entire time series were equal 1000 and 1500. We used two kinds of time series: (1) without a jump, and (2) with a jump (Fig. 5.1). The first segment consisted of the real numbers drawn from the Gaussian distribution $N(a_1, \sigma_1)$, and the second segment consisted of the numbers drawn from the Gaussian distribution $N(a_2, \sigma_2)$, where either $a_1 = a_2$ and $\sigma_1 < \sigma_2$ or $a_1 < a_2$ and $\sigma_1 = \sigma_2$. Since the tests can detect any number of change-points (true or false) or do not detect change-points at all, we estimated two kinds of errors: the mean error of number of change-points and the mean error of coordinates of change-points. Averaging was made over 100 trials with sliding windows of given size. If the samples $x_1, x_2, ..., x_n$ and $x_{n+1}, x_2, ..., x_{2n}$ were heterogeneous, i.e. they had different distributions, then

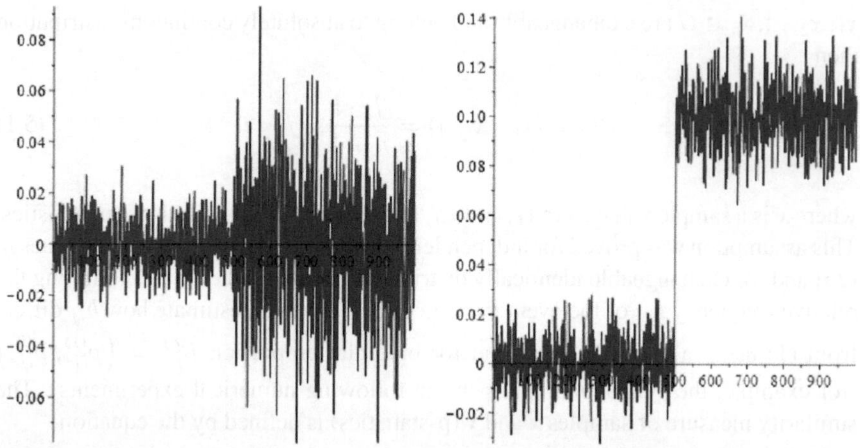

Fig. 5.1 Stationary time series without jump (left) and with a jump (right)

the point x_{n+1} was considered as a change-point. Then we move the sliding window ahead by n elements and repeat the procedure. As alternative tests for change-point detection, we used the Kolmogorov–Smirnov test and the Wilcoxon test. These tests are widely used tools for change-point detection [30–33].

5.4 Results and Discussion

Consider the case with two segments of difference sample variances (Fig. 5.1, left). Let us see how the change of the variance of the segments effects on the accuracy of the tests. The results obtained for the case (Fig. 5.2) when $a_1 = a_2 = 0$, $\sigma_1 = 0.01$, and $\sigma_1 = 0.025$ are provided in Table 5.1. As we see, the p-statistics produced the best relative mean for number of change-points, averaged by windows widths, the Kolmogorov–Smirnov statistics had modest error and the Wilcoxon test was the worst one. The p-statistics and the Kolmogorov–Smirnov statistics produced almost the same error. Note, that the accuracy of the p-statistics and the Kolmogorov–Smirnov statistics increased when the window width increased, but for the Wilcoxon test failed in every experiment (N/A means that the test did not a change-point in any trial at all).

The results obtained for the case (Fig. 5.3) when $a_1 = a_2 = 0$, $\sigma_1 = 0.01$, and $\sigma_2 = 0.05$ are provided in Table 5.2. Here, the p-statistics produced the best relative mean for number of change-points, the Kolmogorov–Smirnov statistics had modest error and the Wilcoxon test was the worst one. The p-statistics and the Kolmogorov–Smirnov statistics produced almost the same error, while the Wilcoxon test failed in every experiment.

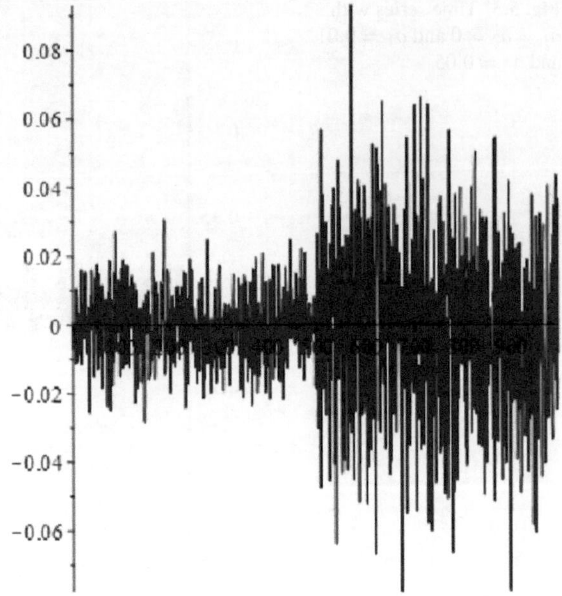

Fig. 5.2 Time series with $a_1 = a_2 = 0$ and $\sigma_1 = 0.01$, and $\sigma_2 = 0.025$

Table 5.1 The results of the experiment with $a_1 = a_2 = 0$ and $\sigma_1 = 0.01$, and $\sigma_2 = 0.025$

Window width, points	Relative mean error for number of change-points			Absolute mean error of coordinates of change-point		
	P	KS	W	*P*	KS	W
10	1.00	N/A	N/A	N/A	N/A	N/A
40	0.13	0.96	N/A	22.11	2.37	N/A
70	0.04	0.50	N/A	25.82	23.66	N/A
100	0.03	0.14	N/A	29.17	51.31	N/A
Average	0.30	0.65	N/A	25.70	27.78	N/A

The results obtained for the case (Fig. 5.4) when $a_1 = a_2 = 0$, $\sigma_1 = 0.01$, and $\sigma_1 = 0.075$ are provided in Table 5.3. Here, the p-statistics produced the best relative mean error for number of change-points, the Kolmogorov–Smirnov statistics had modest error and the Wilcoxon test was the worst one. The p-statistics and the Kolmogorov–Smirnov statistics produced almost the same error. When the sliding window width increases the accuracy of the Wilcoxon test increases also.

The results obtained for the case (Fig. 5.5) when $a_1 = a_2 = 0$ and $\sigma_1 = 0.01$, and $\sigma_2 = 0.1$ are provided in Table 5.4. Here, again the p-statistics produced the best relative mean error for number of change-points, the Kolmogorov–Smirnov statistics had modest error and the Wilcoxon test was the worst one. The p-statistics and the Kolmogorov–Smirnov statistics produced rather the same error in finding of coordinates of the change-point, and the Wilcoxon test failed.

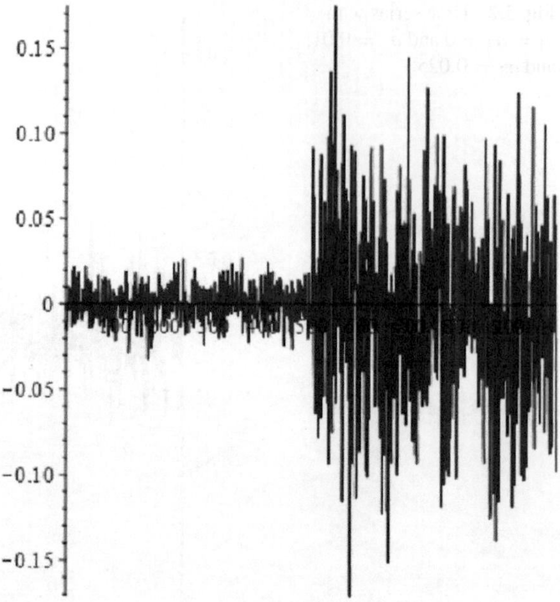

Fig. 5.3 Time series with $a_1 = a_2 = 0$ and $\sigma_1 = 0.01$, and $\sigma_2 = 0.05$

Table 5.2 The results of the experiment with $a_1 = a_2 = 0$ and $\sigma_1 = 0.01$, and $\sigma_2 = 0.05$

Window width, points	Relative mean error for number of change-points			Absolute mean error of coordinates of change-point		
	P	KS	W	*P*	KS	W
10	0.92	N/A	N/A	1.11	N/A	N/A
40	0.05	0.04	N/A	14.29	22.86	N/A
70	0.03	0.00	N/A	16.41	32.13	N/A
100	0.05	0.00	N/A	23.30	42.39	N/A
Average	0.26	0.26	N/A	13.78	32.46	N/A

The results obtained for the case (Fig. 5.6) when $a_1 = a_2 = 0$ and $\sigma_1 = 0.01$, and $\sigma_2 = 0.125$ are provided in Table 5.5. These results are very similar for the results from Table 5.4. This is evidence that the increasing of σ_2 leads to improving of the accuracy of the p-statistics and the Kolmogorov–Smirnov statistics in change-point detection, but the Wilcoxon test failed in every experiment.

The results obtained for the case (Fig. 5.7) when $a_1 = a_2 = 0$ and $\sigma_1 = 0.01$, and $\sigma_2 = 0.15$ are provided in Table 5.6. These results are very similar for the results from Table 5.4. This is evidence that the increasing of σ_2 leads to improving of the accuracy of the p-statistics and the Kolmogorov–Smirnov statistics in change-point detection, but the Wilcoxon test failed in every experiment.

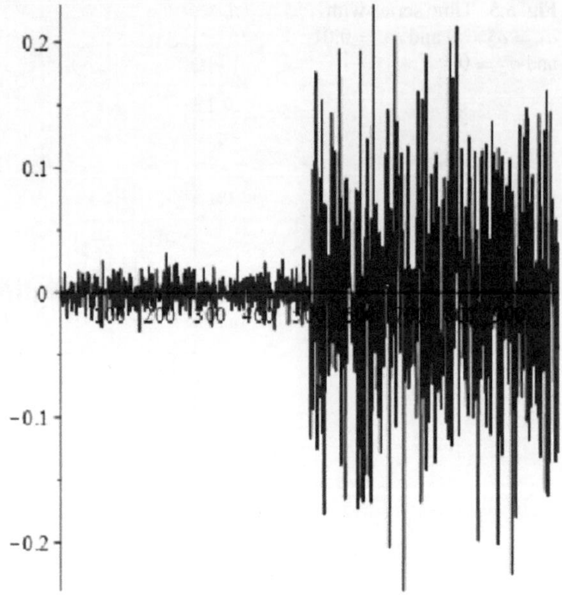

Fig. 5.4 Time series with $a_1 = a_2 = 0$ and $\sigma_1 = 0.01$, and $\sigma_2 = 0.075$

Table 5.3 The results of the experiment with $a_1 = a_2 = 0$ and $\sigma_1 = 0.01$, and $\sigma_2 = 0.075$

Window width, points	Relative mean error for number of change-points			Absolute mean error of coordinates of change-point		
	P	KS	W	*P*	KS	W
10	0.72	N/A	N/A	6.64	N/A	N/A
40	0.06	0.00	N/A	13.51	20.10	N/A
70	0.07	0.01	0.99	24.30	29.56	3.13
100	0.07	0.00	0.99	29.02	37.98	3.30
Average by width	0.23	0.25	0.99	18.37	29.21	3.22

The results obtained for the case (Fig. 5.8) when $a_1 = a_2 = 0$ and $\sigma_1 = 0.01$, and $\sigma_2 = 0.15$ are provided in Table 5.7. These results are very similar for the results from Table 5.6 because the situation is the same.

The results obtained for the case (Fig. 5.9) when $a_1 = a_2 = 0$ and $\sigma_1 = 0.01$, and $\sigma_2 = 0.2$ are provided in Table 5.8. These results are very similar for the results from Table 5.7 because the situation is the same.

Resuming, we can say that the increasing on the window width improves the accuracy of the p-statistics and the Kolmogorov–Smirnov test, but does not impact on the results of the Wilcoxon test.

Now, we pass to the case of two segments with different sample means (Fig. 5.2, right). Let us see how the change of the difference between the means of the segments effects on the accuracy of the tests.

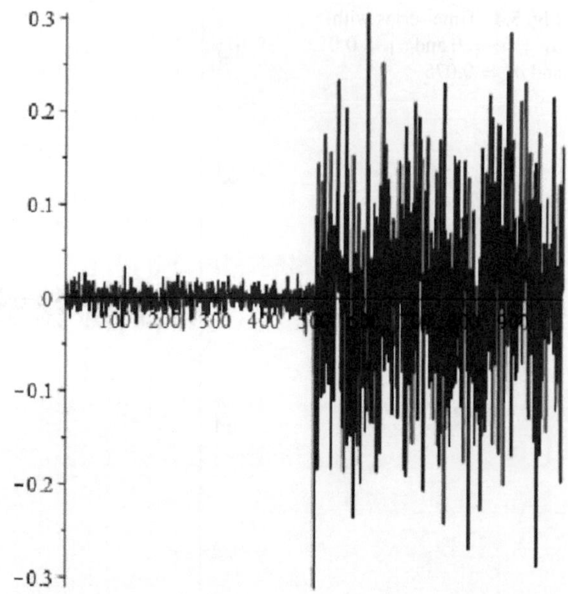

Fig. 5.5 Time series with $a_1 = a_2 = 0$ and $\sigma_1 = 0.01$, and $\sigma_2 = 0.1$

Table 5.4 The results of the experiment with $a_1 = a_2 = 0$ and $\sigma_1 = 0.01$, and $\sigma_2 = 0.1$

Window width, points	Relative mean error for number of change-points			Absolute mean error of coordinates of change-point		
	P	KS	W	*P*	KS	W
10	0.38	1.00	N/A	11.33	N/A	N/A
40	0.07	0.00	N/A	12.27	19.57	N/A
70	0.04	0.01	N/A	14.55	27.18	N/A
100	0.06	0.00	N/A	20.72	35.79	N/A
Average by width	0.14	0.26	N/A	14.72	20.64	N/A

Consider the results for the case (Fig. 5.10) when $a_1 = 0$, $a_2 = 0.1$, $\sigma_1 = \sigma_2 = 0.1$ (Table 5.9). As we see, the p-statistics, the Kolmogorov–Smirnov test and the Wilcoxon test produced comparable results at every width of the sliding window and in average. This is an expected effect all these tests are very sensitive to the location shift.

The results for the case (Fig. 5.11) when $a_1 = 0$, $a_2 = 0.2$, $\sigma_1 = \sigma_2 = 0.1$, are provided in Table 5.10. As in the previous experiment see, the Kolmogorov–Smirnov statistics and the Wilcoxon test produced better, but comparable results at every width of the sliding window and in average.

Consider the results for the case (Fig. 5.12) when $a_1 = 0, a_2 = 0.3, \sigma_1 = \sigma_2 = 0.1$ (Table 5.11). The results show that in this case accuracy of the Kolmogorov–Smirnov tests and Wilcoxon tests is higher, but comparable with the accuracy of p-statistics.

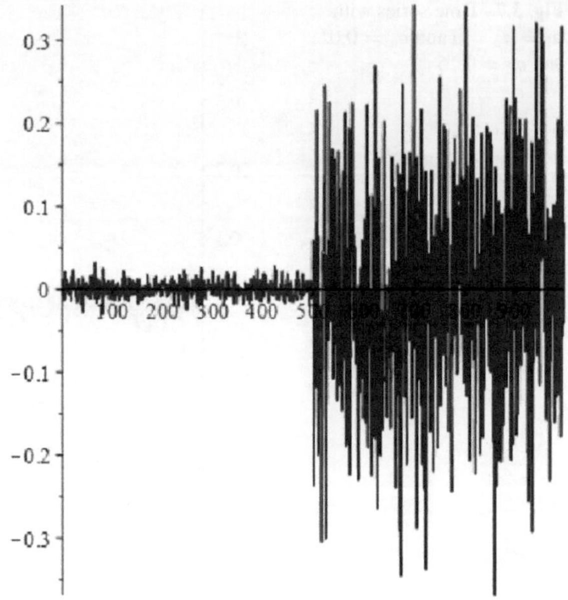

Fig. 5.6 Time series with $a_1 = a_2 = 0$ and $\sigma_1 = 0.01$, and $\sigma_2 = 0.125$

Table 5.5 The results of the experiment with $a_1 = a_2 = 0$ and $\sigma_1 = 0.01$, and $\sigma_2 = 0.125$

Window width, points	Relative mean error for number of change-points			Absolute mean error of coordinates of change-point		
	P	KS	W	*P*	KS	W
10	0.21	N/A	N/A	12.67	N/A	N/A
40	0.05	0.00	N/A	11.41	18.82	N/A
70	0.10	0.00	N/A	18.75	24.40	N/A
100	0.06	0.00	N/A	26.30	32.11	N/A
Average by width	0.14	0.00	N/A	17.28	18.83	N/A

Consider the results for the case (Fig. 5.13) when $a_1 = 0, a_2 = 0.4, \sigma_1 = \sigma_2 = 0.1$ (Table 5.12). The results shows that accuracy of the Kolmogorov–Smirnov tests and Wilcoxon tests is higher, but comparable that the accuracy of p-statistics.

Consider the results for the case (Fig. 5.14) when $a_1 = 0, a_2 = 0.5, \sigma_1 = \sigma_2 = 0.1$ (Table 5.13). The results show that accuracy of the Kolmogorov–Smirnov tests and Wilcoxon tests is higher, but comparable with the accuracy of p-statistics.

Consider the results for the case (Fig. 5.15) when $a_1 = 0, a_2 = 0.6, \sigma_1 = \sigma_2 = 0.1$ (Table 5.14). The results show that accuracy of the Kolmogorov–Smirnov tests and Wilcoxon tests is higher, but comparable with the accuracy of p-statistics.

Consider the results for the case (Fig. 5.16) when $a_1 = 0, a_2 = 0.7, \sigma_1 = \sigma_2 = 0.1$ (Table 5.15). The results show that accuracy of the Kolmogorov–Smirnov tests and Wilcoxon tests is higher, but comparable with the accuracy of p-statistics.

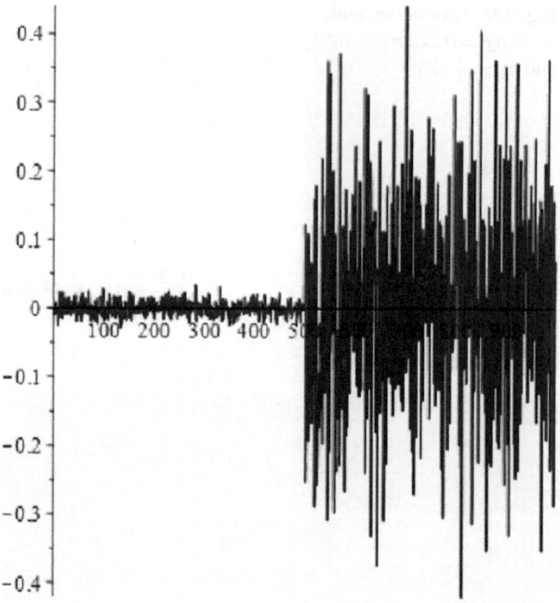

Fig. 5.7 Time series with $a_1 = a_2 = 0$ and $\sigma_1 = 0.01$, and $\sigma_2 = 0.15$

Table 5.6 The results of the experiment with $a_1 = a_2 = 0$ and $\sigma_1 = 0.01$, and $\sigma_2 = 0.15$

Window width, points	Relative mean error for number of change-points			Absolute mean error of coordinates of change-point		
	P	KS	W	*P*	KS	W
10	0.10	N/A	N/A	12.33	N/A	N/A
40	0.05	0.00	N/A	15.44	18.55	N/A
70	0.10	0.02	N/A	22.81	28.41	N/A
100	0.07	0.00	N/A	22.60	33.67	N/A
Average	0.08	0.26	N/A	18.30	20.16	N/A

Consider the results for the case (Fig. 5.17) when $a_1 = 0, a_2 = 0.8, \sigma_1 = \sigma_2 = 0.1$ (Table 5.16). The results show that accuracy of the Kolmogorov–Smirnov tests and Wilcoxon tests is higher, but comparable with the accuracy of p-statistics.

Consider the results for the case (Fig. 5.18) when $a_1 = 0, a_2 = 0.9, \sigma_1 = \sigma_2 = 0.1$ (Table 5.17). The results show that accuracy of the Kolmogorov–Smirnov tests and Wilcoxon tests is higher, but comparable with the accuracy of p-statistics.

Consider the results for the case (Fig. 5.19) when $a_1 = 0, a_2 = 1.0, \sigma_1 = \sigma_2 = 0.1$ (Table 5.18). The results shows that accuracy of the Kolmogorov–Smirnov tests and Wilcoxon tests is higher, but comparable with the accuracy of p-statistics.

Thus, as it was expected, when the means of distributions are different and the variance is the same, all the tests have comparable accuracy.

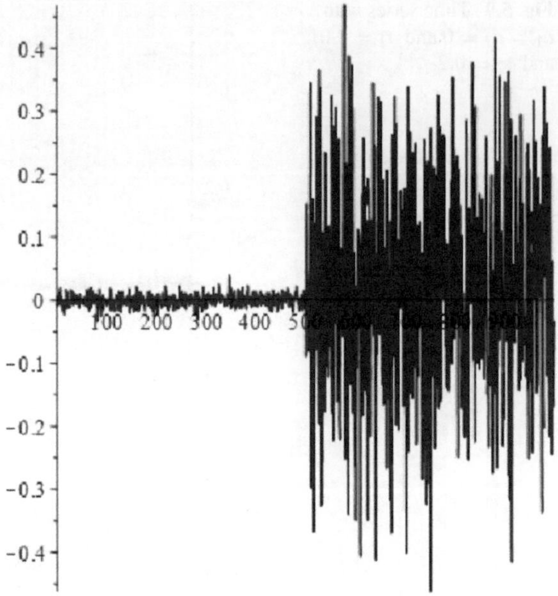

Fig. 5.8 Time series with $a_1 = a_2 = 0$ and $\sigma_1 = 0.01$, and $\sigma_2 = 0.175$

Table 5.7 The results of the experiment with $a_1 = a_2 = 0$ and $\sigma_1 = 0.01$, and $\sigma_2 = 0.175$

Window width, points	Relative mean error for number of change-points			Absolute mean error of coordinates of change-point		
	P	KS	W	*P*	KS	W
10	0.06	1.00	N/A	9.25	0.00	N/A
40	0.03	0.00	N/A	12.73	17.79	N/A
70	0.03	0.00	N/A	12.41	23.92	N/A
100	0.06	0.01	N/A	23.16	32.72	N/A
Average	0.05	0.26	N/A	14.39	18.71	N/A

5.5 Conclusion

The proposed algorithm for change-point detection has the comparable accuracy to the Kolmogorov–Smirnov and Wilcoxon tests if distributions of segments of data sequence has different mathematical expectations and the same variance and it is more accurate for the distributions with the same mathematical expectations and different variance. This algorithm does not require storing big amounts of data and allows easy implementation.

There are several directions of future work on this algorithm: (1) developing of multivariate version, (2) extension on different types of anomalies, and (3) extension on the cases with smooth transition segments.

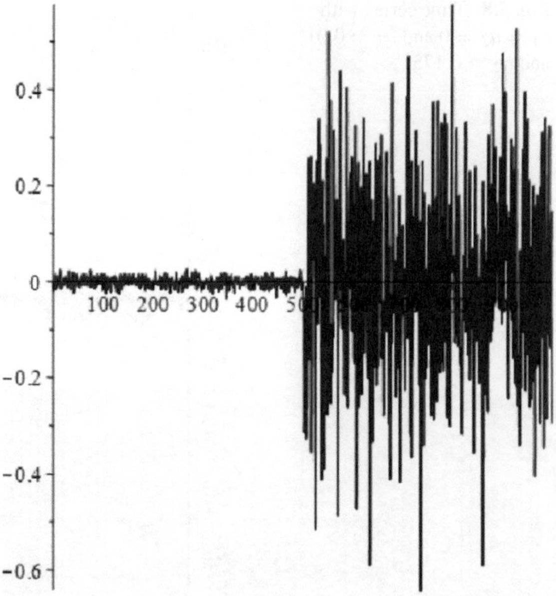

Fig. 5.9 Time series with $a_1 = a_2 = 0$ and $\sigma_1 = 0.01$, and $\sigma_2 = 0.2$

Table 5.8 The results of the experiment with $a_1 = a_2 = 0$ and $\sigma_1 = 0.01$, and $\sigma_2 = 0.2$

Window width, points	Relative mean error for number of change-points			Absolute mean error of coordinates of change-point		
	P	KS	W	*P*	KS	W
10	0.09	N/A	N/A	9.26	N/A	N/A
40	0.05	0.00	N/A	13.65	17.23	N/A
70	0.05	0.00	N/A	15.80	23.35	N/A
100	0.05	0.01	N/A	19.96	29.68	N/A
Average	0.06	0.25	N/A	14.67	17.57	N/A

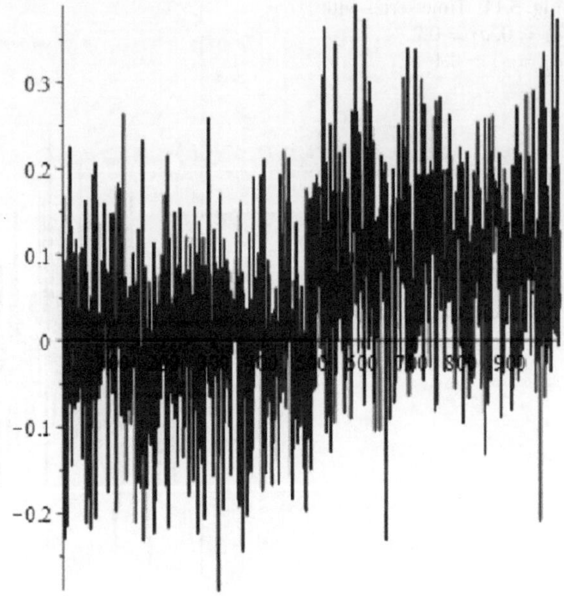

Fig. 5.10 Time series with $a_1 = 0, a_2 = 0.1,$ $\sigma_1 = \sigma_2 = 0.1$

Table 5.9 The results of the experiment with $a_1 = 0, a_2 = 0.1, \sigma_1 = \sigma_2 = 0.1$

Window width, points	Relative mean error for number of change-points			Absolute mean error of coordinates of change-point		
	P	KS	W	*P*	KS	W
10	0.99	0.99	0.99	0.05	0.55	0.010
40	0.33	0.19	0.14	20.37	19.88	17.93
70	0.04	0.01	0.00	28.81	27.70	27.15
100	0.03	0.00	0.00	39.15	35.93	32.22
Average	0.35	0.30	0.28	22.10	21.02	19.33

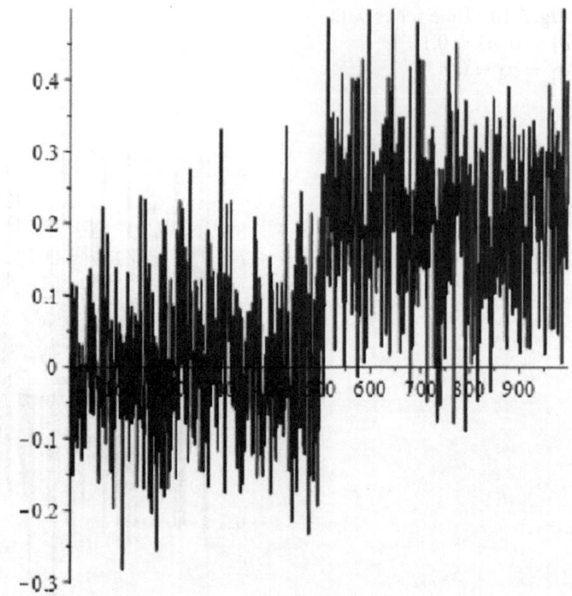

Fig. 5.11 Time series with $a_1 = 0, a_2 = 0.2$, $\sigma_1 = \sigma_2 = 0.1$

Table 5.10 The results of the experiment with $a_1 = 0, a_2 = 0.2, \sigma_1 = \sigma_2 = 0.1$

Window width, points	Relative mean error for number of change-points			Absolute mean error of coordinates of change-point		
	P	KS	W	*P*	KS	W
10	0.74	0.26	0.38	4.86	7.70	6.31
40	0.04	0.00	0.01	12.50	10.23	10.85
70	0.04	0.00	0.00	19.34	14.98	14.41
100	0.05	0.01	0.01	23.74	21.21	18.83
Average	0.22	0.07	0.01	15.11	13.53	12.6

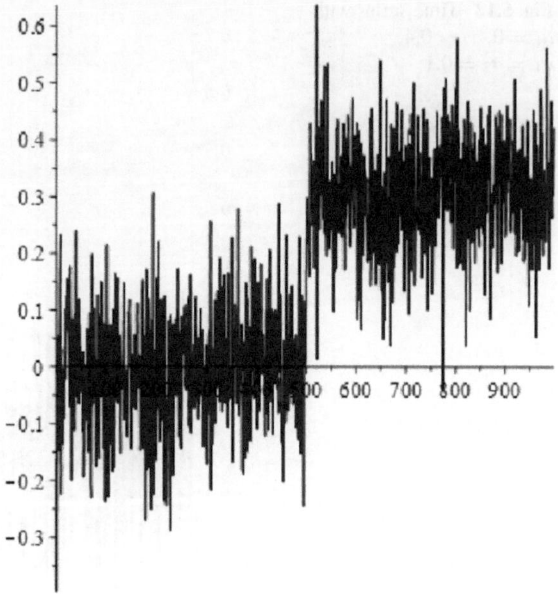

Fig. 5.12 Time series with $a_1 = 0, a_2 = 0.3$, $\sigma_1 = \sigma_2 = 0.1$

Table 5.11 The results of the experiment with $a_1 = 0, a_2 = 0.3, \sigma_1 = \sigma_2 = 0.1$

Window width, points	Relative mean error for number of change-points			Absolute mean error of coordinates of change-point		
	P	KS	W	*P*	KS	W
10	0.11	0.00	0.00	6.33	3.72	3.62
40	0.05	0.00	0.00	10.09	8.22	7.19
70	0.07	0.00	0.02	25.53	15.27	14.6
100	0.04	0.00	0.01	20.29	15.58	13.3
Average	0.07	0.00	0.01	15.56	10.70	9.67

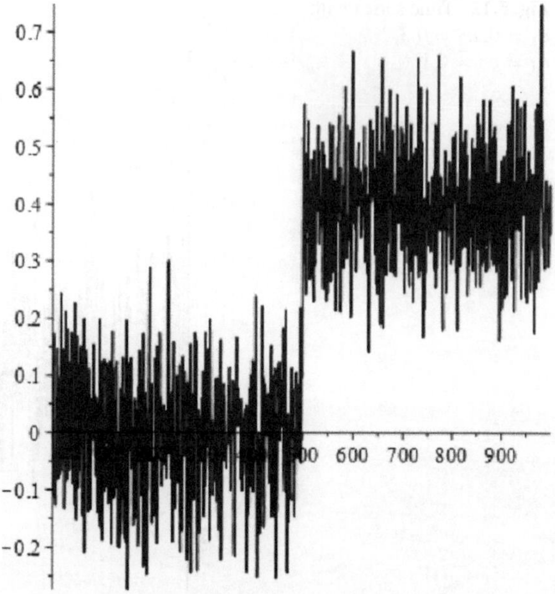

Fig. 5.13 Time series with $a_1 = 0, a_2 = 0.4,$ $\sigma_1 = \sigma_2 = 0.1$

Table 5.12 The results of the experiment with $a_1 = 0, a_2 = 0.4, \sigma_1 = \sigma_2 = 0.1$

Window width, points	Relative mean error for number of change-points			Absolute mean error of coordinates of change-point		
	P	KS	W	P	KS	W
10	0.00	0.01	0.00	3.72	3.43	3.17
40	0.06	0.02	0.00	10.45	8.58	6.82
70	0.09	0.00	0.00	14.31	10.31	9.11
100	0.07	0.00	0.00	21.25	13.79	10.57
Average	0.05	0.00	0.00	12.43	9.04	7.42

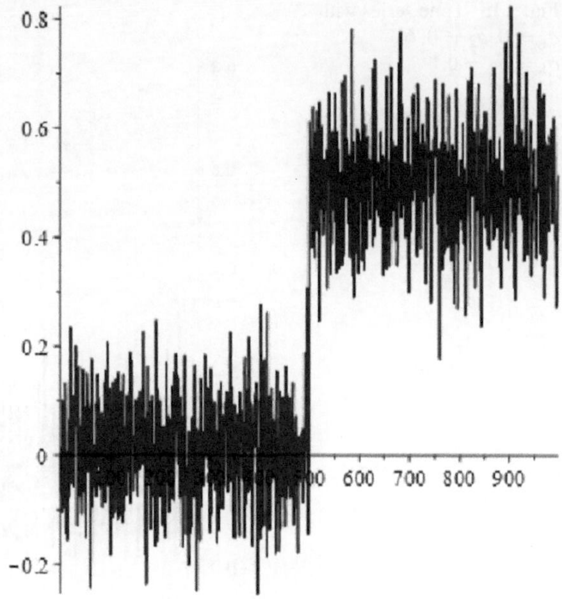

Fig. 5.14 Time series with $a_1 = 0$, $a_2 = 0.5$, $\sigma_1 = \sigma_2 = 0.1$

Table 5.13 The results of the experiment with $a_1 = 0$, $a_2 = 0.5$, $\sigma_1 = \sigma_2 = 0.1$

Window width, points	Relative mean error for number of change-points			Absolute mean error of coordinates of change-point		
	P	KS	W	*P*	KS	W
10	0.00	0.02	0.00	3.88	3.57	3.28
40	0.05	0.00	0.00	13.19	7.81	6.89
70	0.01	0.01	0.01	13.07	12.22	10.33
100	0.07	0.01	0.00	22.37	15.61	11.30
Average	0.03	0.01	0.00	13.13	9.80	7.95

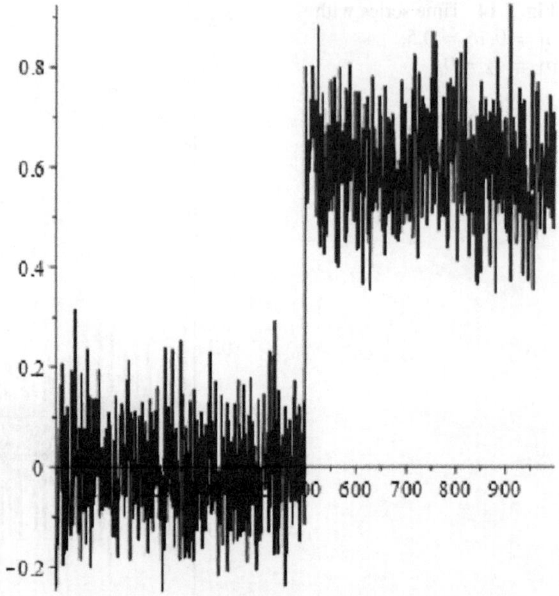

Fig. 5.15 Time series with $a_1 = 0, a_2 = 0,6$, $\sigma_1 = \sigma_2 = 0.1$

Table 5.14 The results of the experiment with $a_1 = 0, a_2 = 0, 6, \sigma_1 = \sigma_2 = 0.1$

Window width, points	Relative mean error for number of change-points			Absolute mean error of coordinates of change-point		
	P	KS	W	*P*	KS	W
10	0.01	0.01	0.00	3.93	3.34	3.06
40	0.07	0.01	0.00	12.98	8.07	6.63
70	0.07	0.01	0.01	16.77	11.33	9.23
100	0.08	0.00	0.00	23.64	12.93	10.62
Average	0.06	0.01	0.01	14.36	9.17	7.37

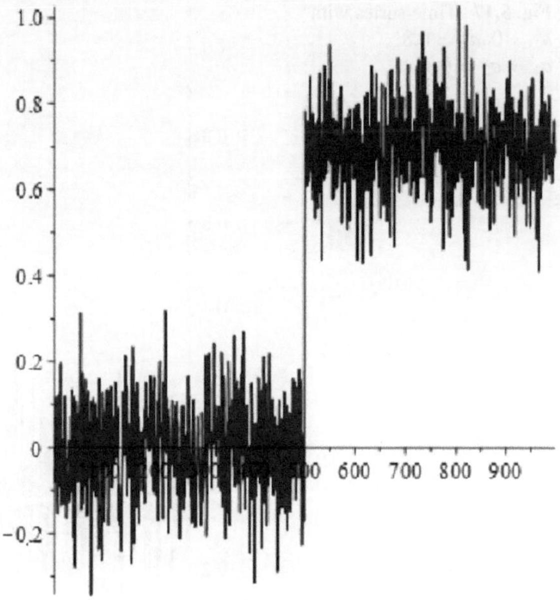

Fig. 5.16 Time series with $a_1 = 0, a_2 = 0.7, \sigma_1 = \sigma_2 = 0.1$

Table 5.15 The results of the experiment with $a_1 = 0, a_2 = 0.7, \sigma_1 = \sigma_2 = 0.1$

Window width, points	Relative mean error for number of change-points			Absolute mean error of coordinates of change-point		
	P	KS	W	*P*	KS	W
10	0.01	0.00	0.00	3.88	3.23	3.10
40	0.07	0.02	0.02	11.75	9.05	7.42
70	0.09	0.02	0.01	23.14	12.62	9.71
100	0.10	0.02	0.02	28.23	15.95	12.42
Average	0.06	0.02	0.01	16.75	10.21	8.16

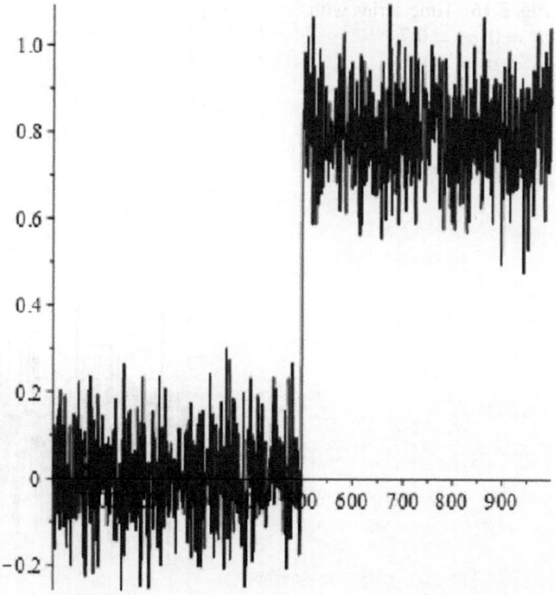

Fig. 5.17 Time series with $a_1 = 0, a_2 = 0.8$, $\sigma_1 = \sigma_2 = 0.1$

Table 5.16 The results of the experiment with $a_1 = 0, a_2 = 0.8, \sigma_1 = \sigma_2 = 0.1$

Window width, points	Relative mean error for number of change-points			Absolute mean error of coordinates of change-point		
	P	KS	W	P	KS	W
10	0.02	0.02	0.01	4.09	3.46	3.00
40	0.06	0.01	0.01	11.56	7.99	6.50
70	0.07	0.02	0.01	17.62	11.72	9.65
100	0.08	0.01	0.00	22.55	14.53	10.69
Average	0.06	0.01	0.01	13.95	9.94	7.46

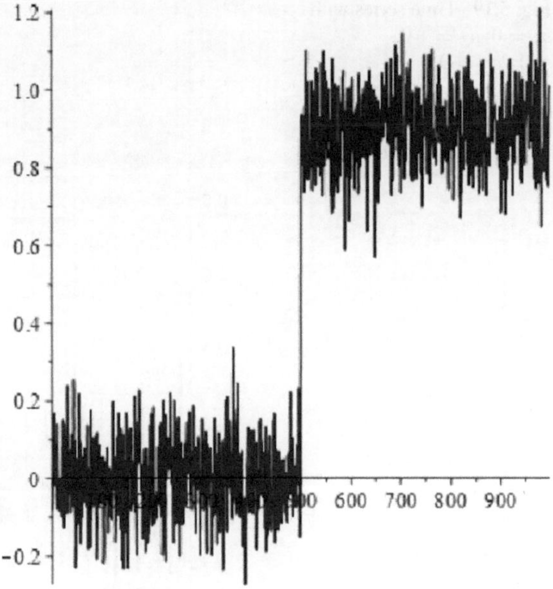

Fig. 5.18 Time series with $a_1 = 0, a_2 = 0.9,$ $\sigma_1 = \sigma_2 = 0.1$

Table 5.17 The results of the experiment with $a_1 = 0, a_2 = 0.9, \sigma_1 = \sigma_2 = 0.1$

Window width, points	Relative mean error for number of change-points			Absolute mean error of coordinates of change-point		
	P	KS	W	*P*	KS	W
10	0.01	0.02	0.01	3.94	3.35	3.09
40	0.08	0.02	0.01	13.11	8.38	7.11
70	0.06	0.01	0.00	14.75	11.82	9.32
100	0.05	0.00	0.01	19.26	14.19	9.89
Average	0.05	0.01	0.01	12.76	9.43	7.60

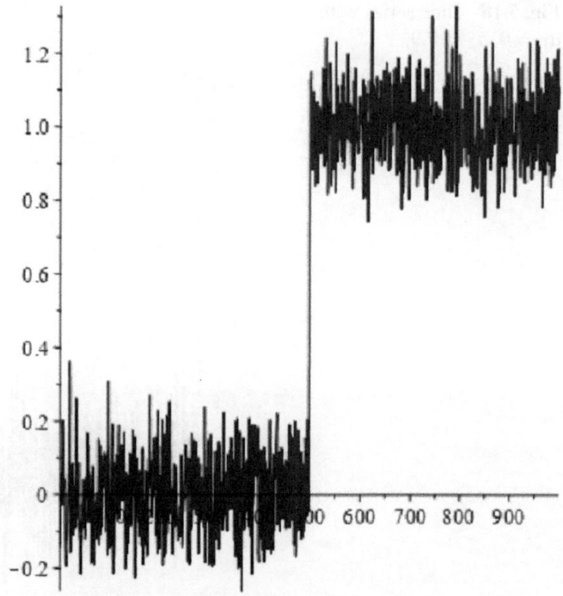

Fig. 5.19 Time series with $a_1 = 0, a_2 = 1.0$, $\sigma_1 = \sigma_2 = 0.1$

Table 5.18 The results of the experiment with $a_1 = 0, a_2 = 1.0, \sigma_1 = \sigma_2 = 0.1$

Window width, points	Relative mean error for number of change-points			Absolute mean error of coordinates of change-point		
	P	KS	W	*P*	KS	W
10	0.01	0.02	0.01	3.80	3.49	3.14
40	0.07	0.02	0.01	10.64	8.19	6.56
70	0.09	0.01	0.00	19.04	11.48	8.99
100	0.10	0.00	0.01	25.84	14.30	11.62
Average	0.06	0.01	0.01	14.83	9.36	7.58

References

1. Wang, D., Yu, Y., & Rinaldo, A. (2020). Univariate mean change-point detection: Penalization CUSUM and optimality. *Electronic Journal of Statistics, 14*(1), 1917–1961.
2. Suvorikova, A., & Spokoiny, V. (2017). Multiscale change-point detection. *Theory of Probability & Its Applications., 61*, 665–691.
3. Aminikhanghahi, S., & Cook, D. J. (2017). A survey of methods for time series change-point detection. *Knowledge and Information Systems, 51*, 339–367.
4. Smirnov, N. V. (1939). Estimate of difference between empirical distribution curves in two independent samples. *Moscow University Mathematics Bulletin,* 2(2), 3–14.
5. Smirnov, N. V. (1939). On the deviations of an empirical distribution curve. *Matematicheskii Sbornik,* 6(1), 3–26.
6. Dixon, W. G. (1940). A criterion for testing the hypothesis that two samples are from the same population. *The Annals of Mathematical Statistics, 11*, 199–204.

7. Wald, A., & Wolfowitz, J. (1940). On a test whether two samples ate from the same population. *The Annals of Mathematical Statistics, 11*, 147–162.
8. Mathisen, H. C. (1943). A method of testing the hypothesis that two samples are from the same population. *The Annals of Mathematical Statistics, 14*, 188–194.
9. Wilcoxon, F. (1945). Individual comparisons by ranking methods. *Biometrika, 1*, 80–83.
10. Mann, H. B., & Whitney, D. R. (1947). On a test of whether one of the random variables is stochastically larger than other. *The Annals of Mathematical Statistics, 18*, 50–60.
11. Wilks, S. S. (1961). A combinatorial test for the problem of two samples from continuous distributions. In: *Proceedings of Fourth Berkeley Symposium on Mathematical Statistics and Probability* (Vol. 1, pp 707–717).
12. Pitman, E. J. G. (1937). Significance tests which may be applied to samples from any populations. *Supplement to the Journal of the Royal Statistical Society, 4*, 119–130.
13. Lehmann, E. L. (1947). Consistency and unbiasedness of certain nonparametric tests. *The Annals of Mathematical Statistics, 22*, 165–179.
14. Rosenblatt, M. (1952). Limit theorems associated with variants of the von Mises statistic. *The Annals of Mathematical Statistics, 23*, 617–623.
15. Dwass, M. (1957). Modified randomization tests for nonparametric hypotheses. *The Annals of Mathematical Statistics, 28*, 181–187.
16. Fisz, M. (1960). On a result be M. Rosenblatt concerning the Mises–Smirnov test. *The Annals of Mathematical Statistics, 31*, 427–429.
17. Barnard, G. A. (1963). Comment on "The spectral analysis of point processes" by M.S. Bartlett. *Journal of the Royal Statistical Society B* 25:294.
18. Birnbaum, Z. W. (1974). Computers and unconventional test-statistics. In F. Prochan & R. J. Serfling (Eds.), *Reliability and biometry* (pp. 441–458). SIAM.
19. Jockel, K.-H. (1986). Finite sample properties and asymptotic efficiency of Monte Carlo tests. *Annals of Statistics, 14*, 336–347.
20. Allen, D. L. (1997). Hypothesis testing using L1-distance bootstrap. *American Statistician, 51*, 145–150.
21. Efron, B., & Tibshirani, R.J. (1993). An Introduction to the Bootstrap. *Monographs on Statistics and Applied Probability* (Vol. 57). Chapman-Hall.
22. Dufour, J.-M., & Farhat, A. (2001). *Exact nonparametric two-sample homogeneity tests for possibly discrete distributions* (pp. 707–717). Center for Interuniversity research in Quantitative Economics (CIREQ). Preprint 2001-23. California Press.
23. Klyushin, D. A., & Petunin, Y, I. (2003). A nonparametric test for the equivalence of populations based on a measure of proximity of samples. *Ukrainian Mathematical Journal, 55*(2):181–198.
24. Hill, B. M. (1968). Posterior distribution of percentiles: Bayes' theorem for sampling from a population. *Journal of American Statistical Association, 63*, 677–691.
25. Madreimov, I., & Petunin, Yu. I. (1982). Characterization of a uniform distribution using order statistics. *Teor Ver Mat Statist, 27*, 96–102.
26. Andrushkiw, R.I., et al. (2003). Construction of the bulk of general population in the case of exchangeable sample values. In: *Proceedings of the International Conference of Mathematics and Engineering Techniques in Medicine and Biological Science (METMBS'03)* (pp. 486–489), Las Vegas, Nevada, USA (2003)
27. Gravina, R., & Fortino, G. (2016). Automatic methods for the detection of accelerative cardiac defense response. *IEEE Transactions on Affective Computing, 7*(3), 286–298.
28. Andreoli, A., Gravina, R., Giannantonio, R., & Pierleoni, P. (2010). SPINE-HRV: A BSN-based toolkit for heart rate variability analysis in the time-domain. In: Lay-Ekuakille, A. (Ed.), Wearable and autonomous biomedical devices and systems for smart environment. Springer, Berlin, Heidelberg, pp 269–289.
29. Zhang, Y., Gravina, R., Lu, H., Villari, M., & Fortino, G. (2018). PEA: Parallel electrocardiogram-based authentication for smart healthcare systems. *Journal of Network and Computer Applications, 117*, 10–16.
30. Shao, X., & Zhang, X. (2010). Testing for change-points in time series. *Journal of the American Statistical Association, 105*(491), 1228–1240.

31. Mohr, M., & Neumeyer, N. (2020). Consistent nonparametric change-point detection combining CUSUM and marked empirical processes. *Electronic Journal of Statistics, 14*, 2238–2271.
32. Betken, A. (2017). Change-point estimation based on Wilcoxon tests in the presence of long-range dependence. *Electron Journal of Statistics, 11*, 3633–3672.
33. Wenger, K., & Less, V. (2020). A modified Wilcoxon test for change-points in long-range dependent time series. *Economics Letters, 192(C)*. Article 109237 (Elsevier)

Chapter 6
Internet of Things (IoT) and Vocabulary Learning in the English Language

B. R. Aravind and V. Rajasekaran

Abstract The present study has explored the importance of enhancing vocabulary through the Internet of Things (IoT) by learning English literature to 40 ESL learners (20 males and 20 females). This research also said that the vocabulary level increases when learners are confident in their learning activities. To improve the confidence level of the ESL learners, their understanding of vocabulary knowledge should be improved. Understanding will improve when the learners have enough vocabulary to convey their ideas, so it becomes the teachers' duty to improve their vocabulary. In order to ascertain the authenticities about ESL learners' vocabulary knowledge through the Internet of Things (IoT) by learning literature, a semi-structured interview was also conducted with 10 participants who completed the questionnaire. The results exposed the subsequent array of responses like 'able to learn vocabulary better when it was presented through the Internet of Things (IoT) activities', 'how significant is the Internet of Things (IoT) Technology in learning vocabulary', and 'Internet of Things (IoT) activities were useful in acquiring various kinds of knowledge and information about a particular word' scored maximum number of participants. The syllabus also should have task-based exercises and activities to evaluate the students so that they may be able to enhance their vocabulary through the Internet of Things (IoT) in their studies.

Keywords Internet of Things (IoT) · Vocabulary · ESL learners · Task-based activities · Confidence

6.1 Introduction

This article focuses on the troubles faced by undergraduates from rural backgrounds regarding their vocabulary enhancement. Vocabulary plays a vital role in improving the English language. This article identifies the problems in improving the students' vocabulary and gives suggestions and methods to improve the students' vocabulary

B. R. Aravind (✉) · V. Rajasekaran
School of Social Sciences and Languages (SSL), Vellore Institute of Technology, Chennai, Tamil Nadu, India

P. Kumar et al. (eds.), *A Fusion of Artificial Intelligence and Internet of Things for Emerging Cyber Systems*, Intelligent Systems Reference Library 210,
https://doi.org/10.1007/978-3-030-76653-5_6

level. The English language plays a prominent role in grooming one's personality. If he or she has mastered the language, they can face the corporate world. When they are good enough in the Internet of Things (IoT), they will have the self-assurance to do all the works assigned to them in their job. Reading and writing cannot be ignored, and it will get developed from listening and speaking. But the students from rural backgrounds are not fluent and accurate in terms of their vocabulary so they cannot master these LSRW skills in the undergraduate courses. When people talk about LSRW skills, they will not give importance to the listening skill but have to understand that only when a student listens properly can they speak well. It is a simple theory about how people started to speak a language as an imitation. In a nutshell, all the language skills are inter-related. These skills can be enhanced with the assistance of the Internet of Things (IoT) activities for better and long-term understanding.

6.2 Internet of Things (IoT) and Education

Ashton [1] used the term Internet of Things (IoT) for the first time in 1999. Since then the concept of connected things gave a new growth of the internet. The advent of the Internet of Things (IoT) has opened many avenues across industries through networking and artificial intelligence technologies. Especially in education, information sharing, resource management, global interconnectedness, smart classes, and smart labs were benefited. This Internet of Things (IoT) technology can increase educational quality and easy access. The course plan activities were designed to access quickly. The learning process will provide a multidimensional aspect and give a positive impact on education. The interactive smart classes were equipped with enhanced and updated learning materials that were readily available on the Internet of Things (IoT) platform. This allows an accurate picture of everything.

6.3 Survey of Literature

Many experts and researchers have spent valuable time retaining the learners' learned vocabulary in English Language Teaching. As a result, many strategies, methods, and techniques were introduced to uplift the ESL learners' vocabulary. TED wordlist improves noun hyponyms through WordNet, enhancing ESL learners' vocabulary level [2]. Sanaoui [3] remarked, "those students who had a more structured vocabulary learning approach performed better at recalling vocabulary". Ahmad Azman Mokhtar et al., [4] conducted a study, "Vocabulary Learning Strategies of Adult ESL Learners" with 360 students. Gu and Johnson's [5] questionnaire was used to gather the data. Results revealed that guessing and dictionary strategies were least used for most metacognitive regulation, note-taking, rehearsal, encoding, and activation strategies. ESL learners' vocabulary knowledge can be diagnosed through Schmitt's

[6] memory strategies [7]. Technological modality enhances learners' communication and conveys the intended message effectively through speaking skill. TED talks are useful in building confidence in the learners' academic performance, particularly in communication [8].

Coanca [9] connected the teaching and learning English to the ecosystem of the Internet of Things (IoT) in her study. She explored the Internet of Things (IoT) on the light of technological modality for language educators. The findings revealed that digital technologies should act as a tool for curriculum designers to design sustainable strategies. Veeramanickam [10] Internet of Things (IoT) created a change in education institutions by implementing the infrastructure changes. The impact of technology has transformed the traditional teaching practices. Internet of Things (IoT) has intertwined education and technology to enhance the academic production of the learners in the educational field. Generally speaking, teachers are in pressure in completing the syllabus. They are not ready to give the Internet of Things (IoT) activities to the students to evaluate their understanding level. They should provide opportunities for the students to participate in such activities. This will help the students to get success in their life. It becomes the teachers' duty to motivate them to recognize the vocabulary and responsiveness of mastering the other language skills. It becomes complicated for a teacher to monitor all the students' language skills in a class. So, they have to check their understanding of the Internet of Things (IoT) activities. When they have the Internet of Things (IoT) teaching methodology, they can develop the students' vocabulary level and language skills.

6.4 Significance of the Study

In the present situation, the undergraduate course syllabus gives less importance to the Internet of Things (IoT) in English, so the students do not get enough chance to improve English language learning. Students also do not show much interest, in the beginning, to improve their language skills in English. Writing the students' talent is only evaluated through the framed syllabus, the remaining three skills are not given much importance. But a student's comprehension level can be improved through the Internet of Things (IoT) activities. So the Educational institutions should try to accept the fact and start to reframe their outcome of the courses by having equal importance to all the four skills with the Internet of Things (IoT) assistance. So this article will concentrate on improving students' vocabulary through the Internet of Things (IoT). The findings of this investigation can be an information base for further research in the related fields. Implications on the design and teaching of pedagogical curricula can be formulated.

6.5 Research Objectives

In a country like India, many students study English as a second language they are not aware of mastering the language to get success in their life. Teachers always think about completing the syllabus on time and believe that is their success. They forget that when students lack their vocabulary level, they may not have good results, and it will give the students a severe problem in building their confidence. Hence the following research objectives were formulated:

- To highlight the importance of improving vocabulary through the Internet of Things (IoT).
- To identify the obstacles of ESL learners' in improving vocabulary

6.6 Research Questions

Literature is a reflection of life, and it came from the unique cultures which are more than thousands of years old, and it is an application of language. So, it is not surprising that it will help people to have the opportunity to expand and enrich their language skills. Whether it is learned by the child or by an adult, it will give interest and maintain enthusiasm towards the language. The current study answers the following research questions based on the understanding as mentioned above.

- How the Internet of Things (IoT) explores ESL learners' vocabulary knowledge by learning literature?
- Is there a relationship between the Internet of Things (IoT) and ESL learners' vocabulary knowledge?
- Whether the Internet of Things (IoT) will improve ESL learners' vocabulary knowledge?

6.7 Research Methodology

The descriptive and analytical analysis is used to study the problem. A questionnaire was circulated to know the feedback of the Internet of Things (IoT) activities posed by the researcher. The questionnaire data were collected through Google Forms 2020. The rating scale for the frequency of use of the Internet of Things (IoT) activities as: Helpful, Not Helpful, Not used but Helpful, and Not Aware respectively.

Table 6.1 Description of samples

Category	Number
Male	20
Female	20
Total	40

6.7.1 Sample and Sample Size of the Research Study

To study this research's objectives, the undergraduate syllabus of arts stream is taken and sample size, students who were doing undergraduate in English. To select the samplings, the Radom sampling method was used. The current study conducted with the help of 40 ESL learners who were randomly selected. The ESL learners were selected from Chennai as the sample of the study. The samples comprised 20 (50%) male respondents and 20 (50%) female respondents ranging from 18 to 22 years old. The proficiency level of the ESL learners ranges from intermediate to advanced level. In the sample, 20 (50%) were first-generation graduates, and the remaining 20 (50%) were second-generation graduates are shown in Table 6.1.

6.8 Discussion of the Study

Vocabulary teaching for the English language needs to be taught with utmost care. Vocabulary cannot be taught individually, and it can be taught to the students in an integrated way with literature and the Internet of Things (IoT). When students start to ask a question immediately during the lecture, they have developed their confidence level, which can be developed through effective teaching. Literature as poetry, novels, history, biography, and essay gives us pleasure and wisdom. All human emotions like anger, happiness, sadness, passion, love, and hatred can be expressed through literature. People communicate through spoken or written form of language with words and signs in a structured and conventional way. The great masters of literature always have to cudgel in their mind in choosing accurate vocabulary because they want to give their ideas impressive and everlasting to their readers. If they are not brilliant in their vocabulary, they cannot express their ideas or views in an enthralling way. The language and narrative techniques reveal the authors' personalities. If one wants to master their vocabulary, they have to read the genius of the past like Shakespeare, Milton, Shelley, Bacon, Addison or A.G.Gardiner, and so on. The writings of these writers will give so much to the young minds. Here, the Internet of Things (IoT) activities and literature were coupled for the effective teaching–learning process concerning vocabulary. For example, the poem The Solitary Reaper by William Wordsworth [11] experimented for the study. Usually, when a teacher teaches this poem to the students, they will give the poet's background, and when they explain they will give meaning for each line, the teacher will say he is a nature poet. Here, the role of a teacher becomes essential [12]. They should have the Internet of Things (IoT) activities for this poem to improve their vocabulary [13].

6.8.1 Interactive Whiteboard

The interactive whiteboard was used in the language classroom for a lively learning environment. Mayer [14] mentioned the impact of multimedia and deep learning was more from words and pictures than from words alone [15]. The interactive whiteboard was loaded with instantly accessible lesson materials of the class by the teacher. The digitized content can be displayed to the learners [16]. The teacher asks the learners to read William Wordsworth's [11], The Solitary Reaper poem aloud, and the audio played for the students. The teacher concentrating on the students' vocabulary asked questions relatively to know the word knowledge's comprehension [17].

6.8.2 E-books

The learners were asked to download E-books and collect biographical information of the poet, William Wordsworth [11]. Additionally, they were instructed to prepare four lines about the poet from the background information and collected [18]. The assigned write-up had to be typed and stored for classroom activity evaluation [19].

6.8.3 Tablets and Mobile Devices

Tablets and mobile devices were ubiquitous for students nowadays. This activity makes the students work actively with their gadgets. The teacher asks the students how many –ing words are there, and what are they? This type of simple question will give the learners interest to come with the answer quickly. When the students say that there are seven words, they are reaping, singing, overflowing, thrilling, breaking, ending, and bending. The next question they may ask for the meaning of the words. The next question they may ask to identify the root words, and after that, they may ask for the three forms of verbs. The entire activity would be noted in students' gadgets and E-mail their responses to the teacher for evaluation [20].

6.8.4 Web-Recording

To develop the learners' vocabulary, a web-recording activity was performed in the class. The teacher asked the students to express the literary analysis of the poem. This would help them to improve their vocabulary while conveying their thoughts on their own. Apart from the performer, the rest of the students may comment on the performance lively online. This web-recording activity learning will improve learners' vocabulary through performing and commenting. When we talk about speaking and listening, its outcome becomes communication, for the communication we need words.

The prime motive of this article is to inculcate and instil the Internet of Things (IoT) technology to the students. Because the mere talking on theories will not improve, students should have the confidence to face the known and unknown technological modality for effective learning. The confidence only will give fluency and accuracy in their English language learning. Questioning in the classrooms plays a vital role in the development of the students' learning. When they come out with questions without fear, they will improve their language skills. The results proved that most of the students said that the Internet of Things (IoT) activity-based teaching methodology helps them learn English vocabulary. They have confidence while learning vocabulary through the Internet of Things (IoT) activities. They have improved their vocabulary through the Internet of Things (IoT) activities in the class. They have started to explore the different learning technologies to foster their knowledge in language classrooms too. It is proved that the Internet of Things (IoT) technologies have become one of the mediums of learning English vocabulary.

The teacher controls language so he or she can have controlled activities. The best-controlled activity is drilling even though it is an ancient method; the results will be good. The study employs a drilling method with the Internet of Things (IoT) activities. Interactive whiteboard, E-books, Tablets and Mobile devices, and Web-recording were used to design the Internet of Things (IoT) activities for learning and improving ESL learners' vocabulary knowledge by learning literature. But these guided activities and learning situations should be creatively created by the teacher with the Internet of Things (IoT). The following results reveal the ESL learners' Internet of Things (IoT) activities after the teaching–learning process.

From the above Table 6.2, it is proved that activity-based learning helps improve the vocabulary of ESL learners through the Internet of Things (IoT) by learning literature. The results are shown in Figs. 6.1 and 6.2.

To certify this study's results about vocabulary knowledge through the Internet of Things (IoT) activities by learning literature, participants were asked open-ended questions in the form of a semi-structured interview. The results were mentioned above Figs. 6.3, 6.4 and 6.5. The questions posed to the ESL learners' at the semi-structured interview as follows. The first question, 'Do the rapid development of IoT platform in English language learning has empowered vocabulary knowledge?' Eight out of ten interviewees stated that they empowered their vocabulary knowledge with IoT activities during the learning process. The second question, 'Do you agree IoT could be integrated to support language and literature for effective comprehension?' Six out of ten interviewees strongly believed that incorporating IoT technologies with language classrooms would bring a greater understanding among learners. The third question, 'Do you think IoT can engage learners more actively in the field of English language learning?' Seven out of ten interviewees felt that IoT activities were more

Table 6.2 Responses from ESL learners' Internet of Things (IoT) activities

Items in the questionnaire	Helpful	Not helpful	Not used but Helpful	Not aware
1. How significant is the Internet of Things (IoT) Technology in learning vocabulary	35	2	3	0
2. Internet of Things (IoT) activities helped to clear up things that were confusing	28	5	2	4
3. Able to have a wider range of vocabulary activities through the Internet of Things (IoT) platform	27	8	4	1
4. Using the Internet of Things (IoT) activities to check my vocabulary learning was more reliable	25	10	5	0
5. Internet of Things (IoT) activities was useful in acquiring various kinds of knowledge and information about a particular word	34	6	0	0
6. Internet of Things (IoT) interactive activities helped to understand the corresponding topics	22	3	7	0
7. Able to learn vocabulary better when it was presented through the Internet of Things (IoT) activities	38	1	1	0
8. Internet of Things (IoT) supports peer and collaborative learning among learners	26	4	9	1
9. Keep vocabulary smart notebook in the Internet of Things (IoT) activities	23	9	8	0
10. Use English Language Media	24	7	6	3
Mean	28.2	5.5	4.5	0.9
Median	26.5	5.5	4.5	0
Standard deviation	5.24	2.87	2.87	1.37

engaging for learning. They said that the IoT activities could engage the learners actively in language classrooms too. The interviewees' collective responses were more opportunities, and the learning atmosphere was created in the IoT platform. The interviewees also realized that confidence level was stable while initially reacting to the activities inside the classroom. Once they gain their confidence by regularly

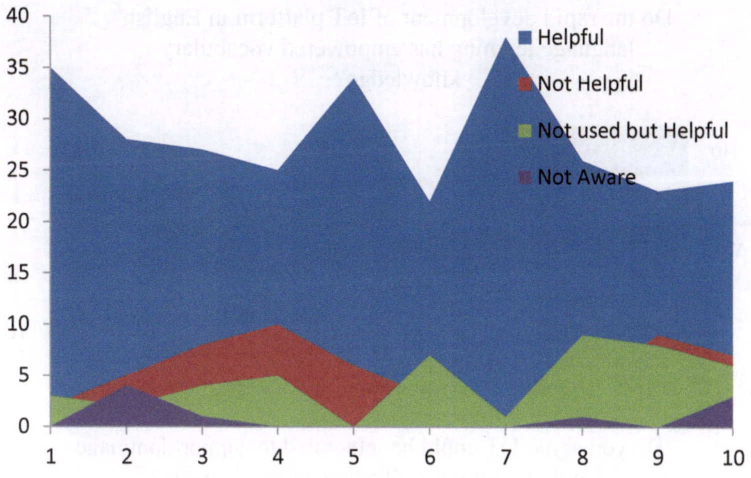

Fig. 6.1 Graphical representation of ESL learners' responses to the Internet of Things (IoT) activities

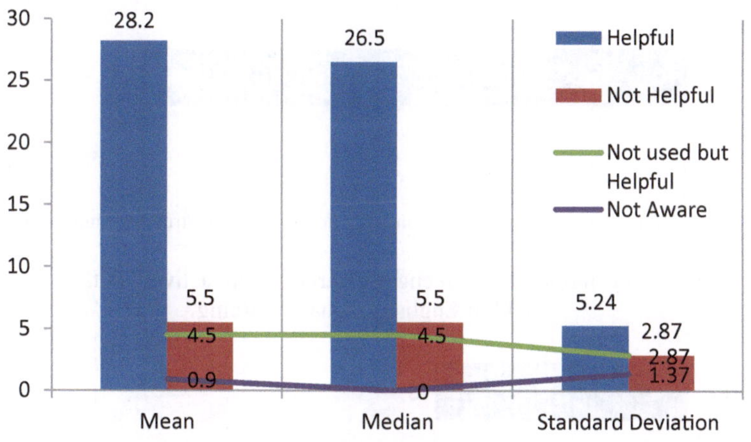

Fig. 6.2 Graphical representation of statistical data of ESL learners' responses to the Internet of Things (IoT) activities

practising the activities, they will utilize the learning activities even outside the classroom atmosphere.

The English syllabus for the undergraduate students must be included with travelogues, forms, pamphlets, advertisements, and newspaper articles with the Internet of Things (IoT) platform. All written, recorded format in literature can be included in IoT, and these current update in the syllabus arrests the students' attention. Because in the same classroom students are differentiated under knowledge, background, culture, and so on. So accepting a concept indeed will get varied according to their

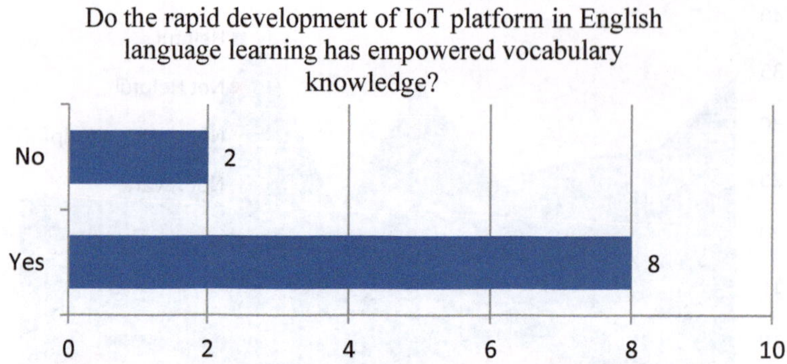

Fig. 6.3 ESL learners' responses to the first question in the semi-structured interview

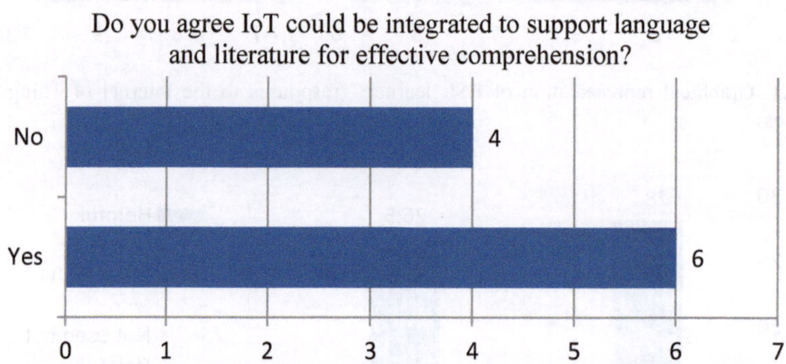

Fig. 6.4 ESL learners' responses to the second question in the semi-structured interview

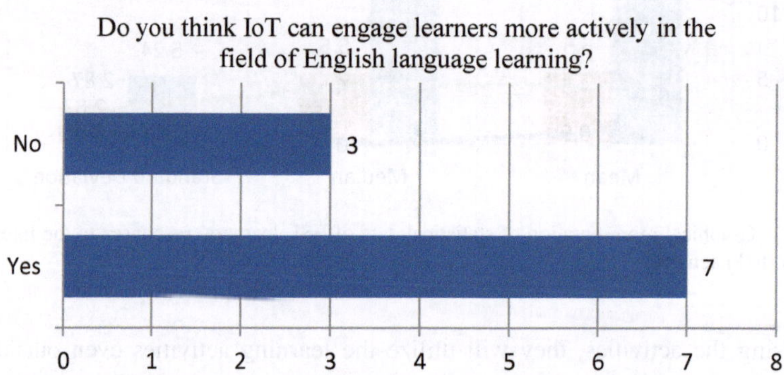

Fig. 6.5 ESL learners' responses to the third question in the semi-structured interview

understanding capacity. Two people getting the same idea for the given text is very rare. So teaching language in this aspect gives so much of an advantage in improving vocabulary. The second advantage is that there is so much space for discussion. So it is imperative that while selecting a literary text for the undergraduate students, the board members should consider the needs, interests, cultural background, necessity, and language level of the students. When students read a literary text, the impact on them will be high. It will be long-term, so when the learner improves their vocabulary by the Internet of Things (IoT), activities through literature will be unimaginable. Most of the students fear speaking because they do not have many words to translate their ideas into English. This becomes a significant obstacle to face the audience. To get rid of this obstacle, students should be given continuous support, motivation, and encouragement.

6.9 Conclusion

One of the main objectives of the literature is that it will develop one's vocabulary level unknowingly. When a reader hears or reads a literary work, they will come across early by the new words. To know the story or the content, they will search for the meaning of the new words they encounter. So, from a language point of view, their vocabulary level will get developed. This article will become an eye-opener to the language educators, experts in the related field, policy-makers, and curriculum designers to appreciate and acknowledge technology use. Enormously, the Internet of Things (IoT) has widened the learning opportunities for effective thoughts to convey. Internet of Things (IoT) activities gave comfortable circumstances and advancement for the teaching–learning process. The present study revealed the usefulness and effectiveness of IoT activities in vocabulary learning through literature. The ESL learners' responses also showed that IoT activities in English classrooms played a pivotal role in a steady improvement in vocabulary acquisition. The comprehension learning activities were achieved through IoT based modalities like Interactive Whiteboard, E-Books, Tablets and Mobile Devices, and Web-recording. To enhance students' vocabulary level through the IoT platform, when the students are free enough to share their thoughts and ideas through activities. The students' development will happen when they participate in activities such as speeches, write-ups, and debates. When students participate in all the activities will inculcate sportive skill among the students. This reinforcement is needed for the present young generation.

References

1. Ashton, K. (2009). That 'internet of things' thing. *RFID Journal, 22*(7), 97–114.
2. Aravind, B. R., & Rajasekaran, V. (2019). Diagnosing ESL learners' vocabulary knowledge through memory strategies. *Humanities & Social Sciences Reviews, 7*(5), 41–47 (2019). https://

doi.org/10.18510/hssr.2019.755.

3. Sanaoui, R. (1995). Adult learners' approaches to learning vocabulary in second languages. *The Modern Language Journal, 79*(1), 15–28. https://doi.org/10.1111/j.1540-4781.1995.tb05410.x.
4. Mokhtar, A. A., Rawian, R. M., Yahaya, M. F., Abdullah, A., Mansor, M., Osman, M. I., Zakaria, Z.A., Murat, A., Nayan, S., & Mohamed, A. R. (2010). Vocabulary Knowledge of Adult ESL Learners. *English Language Teaching, 3*(1), 71–80. https://doi.org/10.5539/elt.v3n1p71
5. Gu, Y., & Johnson, R. K. (1996). Vocabulary learning strategies and language learning outcomes. *Language Learning, 46*, 643–679. https://doi.org/10.1111/j.1467-1770.1996.tb01355.x
6. Schmitt, N., & McCarthy, M. (Eds.). (1997). Vocabulary: Description, acquisition, and pedagogy. Cambridge University Press.
7. Aravind, B. R., & Rajasekaran, V. (2019). Improving noun hyponyms using Select TED wordlist through WordNet: An investigational study. *Universal Journal of Educational Research, 7*(11), 2495–2500. https://doi.org/10.13189/ujer.2019.071129.
8. Aravind, B. R., & Rajasekaran, V. (2020). Using technological modality to learn vocabulary incidentally and intentionally for effective communication. *International Journal of Intelligent Enterprise, 7*(1/2/3), 166–175. https://doi.org/10.1504/IJIE.2020.104653.
9. Coanca, M. (2018). Connecting the English Class to the Internet of Things. *European Scientific Journal, ESJ, 14*(8), 30. https://doi.org/10.19044/esj.2018.v14n8p30.
10. Veeramanickam, M. R. M., & Mohanapriya, M. (2016). IoT enabled futurus smart campus with effective e-learning: i-campus. *GSTF Journal of Engineering Technology (JET), 3*(4), 8–87.
11. Wordsworth, W. (2014, November). Lyrical Ballads with Pastoral and other Poems. I (4 ed.). London: Printed for Longman, Hurst, Rees, and Orme, by R. Taylor. (1805). Retrieved 13 2014. via archive.org.
12. Sharma, S., & Obaid , A. J. (2020). Contact-mechanics and dynamics analysis of three-different ellipsoidal raceway geometries for deep Groove ball bearing using Abaqus 6.13 version FEA simulation for high load-bearing as well as speed-rotating applications. *International Research Journal of Multidisciplinary Science and Technology, 3*(5), 36–43.
13. Obaid, A. J., & Sharma, S. (2020). Recent trends and development of heuristic artificial intelligence approach in mechanical system and engineering product design. *Saudi Journal of Engineering and Technology, 5*(2), 86–93.
14. Mayer, R. E. (2019). *The Cambridge handbook of multimedia learning*. Cambridge University Press.
15. Alkaam, N. O., Mohammed, D. M. Q.. & Obaid, A. J. (2018). A hybrid technique for object detection and recognition using local features algorithms. *Journal of Advance Research in Dynamical and Control Systems, 10*(2), 2330–2343.
16. Al-Yasiri, D. S. Q., & Obaid, A. J. (2018). A new approach for object detection, recognition and retrieving in painting images. *Journal of Advance Research in Dynamic and Control System, 10*(2), 2345–2359.
17. Shelby, M. Improving Vocabulary Skills through Reading in English (2019). Retrieved from https://www.thoughtco.com/improving-vocabulary-skills-1211728.
18. Al-asadi, T. A., & Obaid, A. J. (2016). Object detection and recognition by using enhanced speeded up robust feature. *International Journal of Computer Science and Network Security (IJCSNS), 16*(4), 66–71.
19. Suman Rajest, S., & Suresh, D. P. (2018). Impact of 21st century's different heads of learning skills for students and teachers. *International Journal of Multidisciplinary Research and Development, 4*, 170–178.
20. Suresh, P., & Rajest, S. S. (2019). An analysis of psychological aspects in student-centred learning activities and different methods. *Journal of International Pharmaceutical Research, 46*(1), 165–172.

Chapter 7
Clustering Based Energy Load Analysis Model (CBELAM) in Wireless Sensor Networks

S. David, V. P. Ajay, P. T. Vanathi, Nesasudha, and S. Murugan

Abstract Sensor nodes of wireless sensor networks (WSNs) find it hard to prolong the network lifetime, which is also very expensive. In WSNs, reliable data transmission and prolonged network lifetime have to be achieved. Energy-efficient routing protocol has to be developed, which is a challenging issue in a network. The objective focused in this paper is to balance and minimize the energy consumed by a node during data transmission in WSN, thereby attempts in prolonging the network lifetime. Clustering-based energy load analysis model is presented in this paper to analyze the energy model. The experimental results of the proposed method show better results over the existing methods.

Keywords Wireless sensor network · Clustering · Routing protocol · Energy efficiency · Network lifetime

S. David · V. P. Ajay
Department of ECE, Kumaraguru College of Technology, Coimbatore, Tamil Nadu, India
e-mail: David.s.ece@kct.ac.in

V. P. Ajay
e-mail: ajay.vp.ece@kct.ac.in

P. T. Vanathi
Department of ECE, PSG College of Technology, Coimbatore, Tamil Nadu, India

Nesasudha
Department of ECE, Karunya Institute of Technology and Sciences, Coimbatore, Tamil Nadu, India
e-mail: nesasudha@karunya.edu

S. Murugan (✉)
Department of Computer Science and Engineering, Sri Aravindar Engineering College, Villupuram, Tamil Nadu, India

P. Kumar et al. (eds.), *A Fusion of Artificial Intelligence and Internet of Things for Emerging Cyber Systems*, Intelligent Systems Reference Library 210,
https://doi.org/10.1007/978-3-030-76653-5_7

7.1 Introduction

Wireless Sensor Network comprises many small and cost-effective sensors broadly applied in several applications with different characteristics and requirements. However, usually, devices are operated with only batteries but are expected to withstand for a longer duration. Further, it is not feasible, and the cost is too high to replace after deploying sensor nodes (SNs). Hence, in WSNs, the primary concern is to focus on energy efficiency [1]. Investigation proves that the existing models consume more energy while transmitting data. The performance of data transmission is largely based on the routing scheme utilized [2]. All these conclude that there is a necessity to develop and implement routing protocol with an energy-efficient for saving energy and prolonging the lifetime of the network.

Generally, WSN is further classified as small networks known as clusters. A protocol based on the clustering technique employs cluster head (CH) rotation in a random manner to appropriate the energy equally among the sensor nodes of the organization. The more stable and relatively reliable node than the rest of the cluster nodes is treated as Cluster Head node (CH) [3]. Each cluster presents within the region that is monitored as one CH. The node selected as CH is the most efficient in the cluster. Data is collected byCH from the nodes present within the same cluster [4]. After then, the data collected by CH is forwarded to the base station (BS). During collecting and sending data through CH, node requires energy. Hence, the major role in WSN is the selecting CH at the end of every round.

SNs transmitting data to BS makes use of a probabilistic function so that the communication is reliable. Moreover, the clustering approach is an effective technique to drag out the network lifetime by adjusting the nodes' energy [5]. Using the clustering method, route traffic delay is reduced, and data transfer control is enhanced when it depends on the routing protocol type [6]. These schemes convert the nodes as CH and generate clusters with unequal size depending on the energy conditions of nodes; there by reducing the number of dead nodes and increasing the network lifetime. Due to the energy limitations of the sensors, each sensor monitors the environment over its lifetime and sends the obtained information through intermediate nodes to the target node. After the setup, these networks' main goal primarily gathered information and increased the network lifetime [7]. The protocols of WSN should carefully use energy resources. Generally, SNs do not directly communicate with the destination since they require more energy for transmission to be reliable [8]. But, SNs forwards messages to the sink node (destination) through the multihop network. By the by, routing efficiently in these multihop networks has to achieve energy efficiency as well.

Moreover, reducing the energy requirements to communicate through multihop, requires a routing protocol which efficiently decreases the energy consumption when transmitting data from SNs to sink node. Recently, several works related to WSNs focus on minimizing energy, which considers a single sensor node's performance. Only a few works focus on balancing traffic throughout multihop WSNs and its impact on the network's lifetime. Thus, energy efficiency is the most important feature for WSN routing protocols, enhancing communication and network lifetime [9].

The rest of this paper is arranged as: literature survey is discussed in Sect. 7.2, Sect. 7.3 ellovorates the contributed methodology in detail. Section 7.4 presents the performance analysis and conclusion in Sect. 7.5.

7.2 Literature Survey

Several works exist that concentrate on reducing the energy consumed by nodes while transmitting data in WSN.

Montoya et al. [10] formulated a mathematical model and developed a heuristic approach to support constraints of flow conservation, connectivity, information limit per interface, limitations for communication and the reach of sensing for SNs, and splitting constraints to balance the energy consumed while transmitting data and increasing the network lifetime. This method improved the transmission rate of the sensor and increased the lifetime of the network. Moreover, they assured the path of communication between SNs and sinks.

Ahmed et al. [11] integrated the newly developed energy-efficient Clustering algorithm with hierarchical routing algorithm and named the new model as Energy-Efficient Scalable Routing Algorithm (EESRA). This algorithm's major objective is to improve the network's lifetime, even with the increase in network size. The three-layer hierarchy was employed selecting CHs and for reducing the load of CHs. Further, EESRA utilized multihop transmissions to communicate within the clusters' clusters and implemented a hybrid MAC protocol. EESRA was compared against other routing protocols with network performance by varying the network size. It was proved that EESRA outperformed the other protocols with measures like energy efficiency and load balancing for large scale WSNs.

Chu et al. [12] improved ant colony approach and compared the performance with other methods. Nodes in WSN which used this approach consumed low energy, and thus SNs had higher residual energy. Besides the energy model, the model that balanced the data transmission was established and verified with the transmission target method of the WSN. By applying an improved ant colony algorithm, the location information of the public node was easily determined. An effective routing protocol was developed with this information that had the discriminating ability to find the target node's location effectively. Consequently, this technique improved the existence cycle and limited the energy utilization of WSNs, which also effectively increased the rate of data packet transmission.

Li et al. [13] designed a novel Energy-Efficient Load Balancing Ant-based Routing Algorithm (EBAR) which integrated pseudo-random route discovery approach with improved pheromone trail update method for balancing the consumed energy. Efficient heuristic update approach was employed, dependent on the greedy expected energy cost for optimizing the discovered route. In the final sage, for reducing the energy consumed due to the control overhead, the EBAR model utilized energy-based opportunistic broadcast approach. WSNs were deployed in several applications for evaluating the performance of EBAR with various metric like energy consumption along with efficiency as well as network lifetime. It was observed from the results that EBAR produced significant improvement than existing EEABR, Sensor Ant, and IACO approaches.

Hermano Pereira et al. [14] introduced a new network interface average power metric (NIAP)based on the network's average power consumption interface. This approach contributed towards providing the reliable path, balancing the load and energy and so increased the network lifetime. The results demonstrated that NIAP promisingly replaced ETX approach as it is simply implemented with no modifications of RPL standard.

Chuan Xu et al. [15] designed a novel energy-efficient region source routing protocol (ER-SR) which dynamically selects a node with higher residual energy as source routing node in the network. At that point, that source directing node found the ideal source routing path for each basic node, in this manner empowering the incomplete nodes to partake in routing and consequently energy devoured by SNs were adjusted. Further, energy devoured by the hub during sending information must be diminished. In this way, distance-based insect province streamlining was actualized to discover the worldwide ideal transmission course for each hub. The trial results delineated that ER-SR created higher energy productivity, better organization lifetime, conveyance delayand bundle conveyance proportion, than other directing conventions in WSNs.

7.3 Research Methodology

The proposed architecture model is represented in Fig. 7.1. The node can be sent in the network, at that point nodes are gathered, which are called clusters. A cluster with node is picked as Cluster Head (CH).The data are transmitted over the network, and the clusters are maintained properly. By using Cluster-based energy load analysis model (CBELAM), energy load can be analyzed.

7.3.1 Cluster-Based Energy Load Analysis Model

Once the cluster size is known, clusters are formed, and then data is transmitted over the network. Generally, Clustered routing protocols have three phases: cluster formation, the transmission of data, and cluster maintenance. During the first phase, the network is divided for numerous clusters where every cluster has a CH. While transmitting data, its mode of transfer is evaluated. In the final stage, every CH finds out whether the rotation process of CH should be continued.

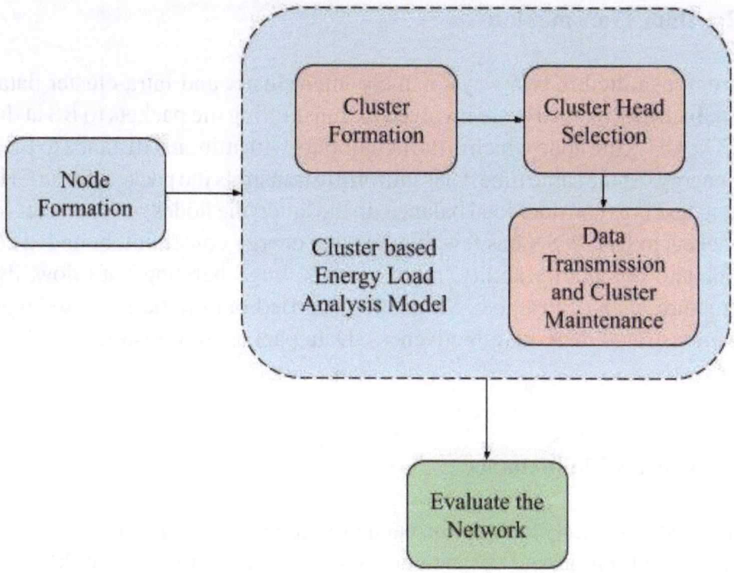

Fig. 7.1 The architecture of the Proposed Model

7.3.1.1 Cluster Formation

Sensor nodes available in the network are segmented as clusters where every cluster have one CH and several SNs. The data gathered by CH from other nodes of the cluster and then transmits them to BS. This can be done until the network is alive. Energy is required to form clusters and select CHs. CHs exhausts energy while data is received from SNs, aggregating and transmitting data to sink node. Thus, CHs must have more energy due to the responsibility of transmitting and receiving data. When CH dies, that particular cluster is disconnected from the network, and there is a possibility of losing important events.

An optimal node is selected as CH in every cluster. An investigation on optimizing the cluster's sizeby considering the reduced communication overhead was discussed in [16]. The parameters considered were the cluster's size, total BS, CH locations, and node density. It was determined that the position of optimal CH was the center of the cluster. It is observed that the CH at the center of the cluster consumes an approximately 15% low energy that at the random locations or closer to BS. In this study, CH is the SN which is to the center of the cluster. Initially, every node has equal energy. Hence the cluster's node having minimal distance from the center of the cluster is selected as CH. CH sends a notification to the nodes in the clusters and the CHs of other clusters. If the notification is not received within the stipulated time, SN requests to discovery a node with minimum distance as CH. Likewise, the cluster formation process is carried out.

7.3.1.2 Data Transmission

Data are transmitted in two ways, namely inter-cluster and intra-cluster data transmission. In the former, CHs are involved in transmitting the packets to BS at different levels. The CH is the node which is the neighbours with minimal distance to BS having higher energy. At the same time, CHs uniformly transmits the packets to the CHs at the adjoining levels to consider load balance. In the latter, the nodes of the cluster forward their packets to CH. WSN has few features like energy constraint, bounded communications and computing ability, more sensors, huge bursting data flow, dynamic infrastructure, data-centric, etc. Simulations carried out on the proposed algorithm have demonstrated network effectiveness. Data packet are transmitted to BS in this manner.

7.3.1.3 Cluster Maintenance

CH has to be essentially rotated for balancing the energy consumed by nodes. The frequency of a rotation and the method used to perform rotation are the key points which are usually very complex in a few routing protocols. Generally, Snare compromised while selecting new CH. A new CH is a node which has higher residual energy. When control packets are exchanged among nodes, additional energy is consumed by the nodes. CH manages the cluster and maintains the energy information of all its nodes. Hence, the rotation process followed in this study commences when the residual energy of CH falls below thethreshold T whose value is computed using,

$$T = F \times E_c^{avg} \tag{7.1}$$

where, E_c^{avg} gives the average residual energy of nodes in cluster c, F is the rotation frequency.

The two ways to select a CH are the node having higher residual energy or the node closer to its CH. But both of these are not optimum. The firstcostshigh for intra transmission while the nextdo not balance energy reduction in the cluster. Hence this study considers energy as well as location information in selecting a new CH. Present CH defines thenew CH to which a notification packet is sent. After receiving the notification packet, the new CH forwardsit to other nodes in the cluster and the other CHs of the other levels.

7.3.2 Energy Model

This model estimates the total energy consumed by SNs, as given in [17]. Energy consumption of a cluster is assessed by calculating the energy consumed bynon-CH while transmitting datato CH and the energy consumption of the CH for forwarding the aggregated data to the BS [18]. Energy consumption of the single non-CH in transferring K bits to the CH in a cluster is given as [19]:

$$E_{non-CH} = k.E_{elec} + E_{amp}(k, d) \quad (7.2)$$

$E_{non\text{-}CH}$ is the Non-CH energy consumption, $E_{elec}$is electronic energy, $E_{amp}$is Energy consumed by the amplifier [20].

Energy consumption of CH involves energy consumption while receiving data from every non-CH [21, 22]. Energy consumed during data aggregation and data transmission to BS is given by;

$$E_{CH} = \left(\frac{n}{k} - 1\right).k.E_{elec} + \frac{n}{k}.k.E_{DA} + E_{TX_{toBS}}(k, d) \quad (7.3)$$

where E_{DA}denotesenergy consumed during data aggregation, E_{CH} is the energy consumed by CH, $E_{TXto\ BS}$ is the energy consumed by the transmitter to base station.

Thus, in a cluster, the total energy consumed is expressed by;

$$E_{cluster} = E_{CH} + \left(\frac{n}{k} - 1\right).E_{non-CH} \quad (7.4)$$

Therefore, the energy model of the WSN is demonstrated with the energy consumed by the individual component in the network,which is as follows:

1. Energy consumption of CH (E_{CT}).

This is the energy consumed while forming a cluster as well as all thetasks performed by CH. The clusters' tasks are selecting CH, advertising CH, and transmitting and receiving several control packets /data packets.

2. Consumed Energy by the clusternodes (E_{MT}).

The nodes in the cluster are sensing data, transmitting as well as receiving several control packets /data packets. Hence, the nodes' energy dissipationis estimated with the expenses towards node setup and transmission procedures.

7.3.3 Load Analysis Model

In this research, the network is partitioned as numerous grid clusters whose length is l_i. With the parameters described previously, the load is calculated by

$$N_i = l_i \rho \quad (7.5)$$

where N_i and l_i represent severalSNsand length of the cluster at the level $q_i$and ρ is the density of the node in the network, and P is the packet. EverySN of the network generates a single packet within a period, say 5 s,with k bits. Hence, the number of bits generated by the cluster at level $q_i$is given by

$$P = N_i \times k \tag{7.6}$$

At level q_i, number of CHs is calculated by

$$S_i = \frac{W}{l_i} \tag{7.7}$$

where, l_i is the cluster length at the level q_i.

The packets handled by the cluster at level q_{i+1} are sent to the cluster at level q_i. The packets generated at dealt only the CH at level q_n. Thus, at level q_n CH deals with the total number of bits Which is given by

$$P_n = l_n^2 \rho k \tag{7.8}$$

At level, q_{n-1} CH deals with the total bits P_{n-1} is expressed by the equations below which involves the packets generated by SNs of the cluster at level q_{n-1} and the packets received from the cluster q_n. Packets at level q_n are evenly distributed to CHs at the level q_{n1}.

$$P_{n-1} \approx \frac{S_n \times P_n}{S_{n-1}} + l_{n-1}^2 \rho k \tag{7.9}$$

$$P_{n-1} = \frac{\frac{W}{l_n} \times + l_n^2 \rho k}{\frac{W}{l_{n-1}}} + l_{n-1}^2 \rho k \tag{7.10}$$

$$P_{n-1} = l_n l_{n-1} \rho k + l_{n-1}^2 \rho k \tag{7.11}$$

$$P_{n-1} = (l_n + l_{n-1}) l_{n-1} \rho k \tag{7.12}$$

At level q_{n-2}, every CH deal with the total number of bits P_{n-2} given by

$$P_{n-2} = \frac{S_{n-1} \times P_{n-1}}{S_{n-2}} + l_{n-2}^2 \rho k \tag{7.13}$$

$$P_{n-2} = \frac{\frac{W}{l_{n-1}} \times (l_n + l_{n-1}) + l_{n-1} \rho k}{\frac{W}{l_{n-2}}} + l_{n-2}^2 \rho k \tag{7.14}$$

$$P_{n-1} = (l_n + l_{n-1}) l_{n-2} \rho k + l_{n-2}^2 \rho k \tag{7.15}$$

$$P_{n-1} = (l_n + l_{n-1} + l_{n-2}) l_{n-2} \rho k \tag{7.16}$$

Hence, at level q_i, every CH handles the total number of bits which is given by

$$P_i = \frac{S_{i+1} + P_{i+1}}{S_i} + l_i^2 \rho k \tag{7.17}$$

$$P_i = (l_n + l_{n-1} + l_i) l_i \rho k \tag{7.18}$$

$$P_i = l_i \rho k \sum_{j=i}^{n} l_j \tag{7.19}$$

When the node's density and the packet-generation speed are fixed in the network, clusters deal with the number of packets at each level based on the cluster size.

7.4 Performance Analysis

The simulation is performed in NS2 platform, and the parameters considered for simulation are tabulated in Table 7.1 below where m and j represent meters and joules, respectively.

The factors considered for analysis are energy consumption, energy efficiency, network lifetime, packet delay and packet delivery ratio.

Table 7.2 analyzes the energy consumption of existing EBAR and ER-SR methods with proposed CBELAM method.

Figure 7.2 compares the energy consumption of the proposed and existing approaches. Number of nodes represents X-axis and the Y-axis and energy consumed

Table 7.1 Simulation parameters

Parameters	Values
Coverage area	100 m x 100 m
Threshold distance	70 m
Node count	100 nodes
Sizeof the packet	512 bytes
Transmission of node in range	20 m
Initial energy	2 J

Table 7.2 Analysis of energy consumption

Number of Nodes	EBAR	ER-SR	CBELAM
20	11	8	3
40	16	11	5
60	22	15	9
80	29	23	13
100	37	31	17

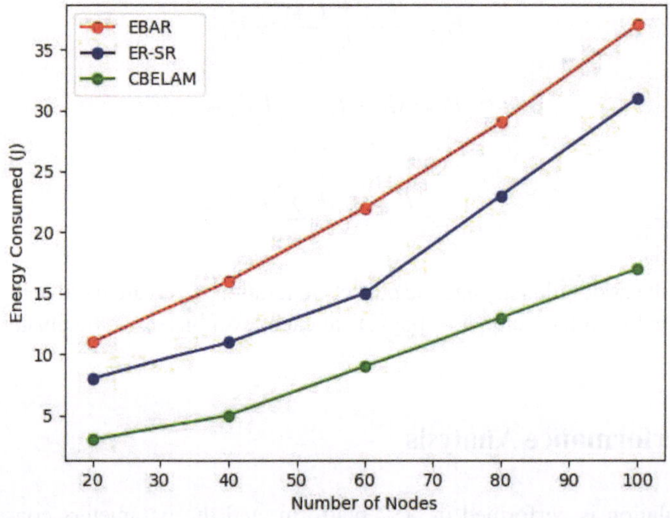

Fig. 7.2 Comparative analysis for energy consumption of the proposed approach against existing approaches

in joules accordingly. The red, blue, green lines indicate the existing EBAR and ER-SR and proposed CBELAMapproaches. The proposed approach shows better results than the existing approaches used for comparison.

Table 7.3 analyzes the energy efficiency of the existing EBAR and ER-SR methods with the proposed CBELAM method.

Figure 7.3 compares the energy efficiency of theproposed and existing approaches. Number of nodes represents X-axis and Y-axis and energy efficiency in joule,accordingly. The red, blue, green lines indicate existing EBAR and ER-SR and proposed CBELAMapproaches respectively. The proposed method shows better results than the existing approaches used for comparison.

Table 7.4 analyzes the network lifetime of the existing EBAR and ER-SR methods with proposed CBELAM method.

Figure 7.4 compares the network lifetime ofthe proposed and existing approaches. The number of nodes show X-axis and Y-axis and energy in joules accordingly. The red, blue, green lines indicate the respective existing EBAR and ER-SR and proposed

Table 7.3 Analysis of energy efficiency

Number of nodes	EBAR	ER-SR	CBELAM
20	7	11	16
40	14	19	24
60	28	31	39
80	36	42	52
100	43	54	64

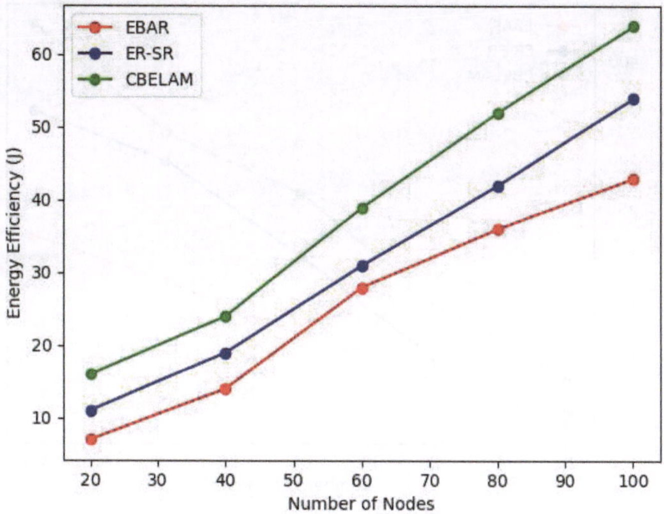

Fig. 7.3 Comparative analysis of energy efficiency with existing methods against the proposed method

Table 7.4 Analysis of network lifetime

Number of nodes	EBAR	ER-SR	CBELAM
20	32.4	43.7	76.8
40	76.4	98.1	154.2
60	121.5	201.4	265.2
80	176.5	298.4	326.4
100	221.4	348.5	437.1

CBELAM approaches. The proposed method shows better results than the existing approaches used for comparison.

Table 7.5 analyzes the packet delivery ratio of existing EBAR and ER-SR methods with proposed CBELAM method.

Figure 7.5 compares the packet delivery ratio of the proposed and existing approaches. X-axis and Y-axis show the number of nodes and packets, respectively. The red, blue, green lines indicate the existing EBAR and ER-SR and proposed CBELAM methods. The proposed method shows better resultsthan the existing ones used for comparison.

Table 7.6 shows the Packet Delay's analysis of existing EBAR and ER-SR methods with proposed CBELAM method.

Figure 7.6 compares the packet delay with existing and proposed approaches. The number of nodes show X-axis and Y-axis and packets, respectively. The red, blue, green lines indicate the existing EBAR and ER-SR and proposed CBELAM methods. The proposed method shows better resultsthan the existing ones used for comparison.

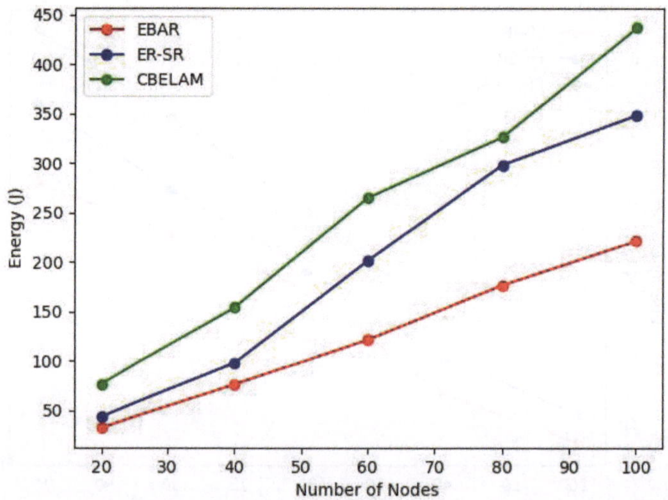

Fig. 7.4 Comparative analysis of network lifetimewith existing methods against the proposed method

Table 7.5 Analysis of packet delivery ratio

Number of nodes	EBAR	ER-SR	CBELAM
20	13	17	21
40	18	23	36
60	29	34	42
80	43	59	68
100	61	73	83

7.5 Conclusion

The major concern with clustering algorithms is forming balanced clusters and selecting cluster heads (CHs). After cluster formation, a node in the cluster having thehighest residual energy is chosen as CHs. This approach concentrates on not only conserving the energy of sensor nodes but on balancing the load also. Further, the energy consumed while transmitting data is also reduced and balanced in WSN and thus increased the network's lifetime. All the objectives were attained by introducing a Clustering-based energy load analysis model (CBELAM), and the results obtained were analyzed with the existing EBAR and ER-SR methods. It was proved that the proposed method achievedgoodoutcomes for all the parameters considered to analyze the proposed CBELAM approach.

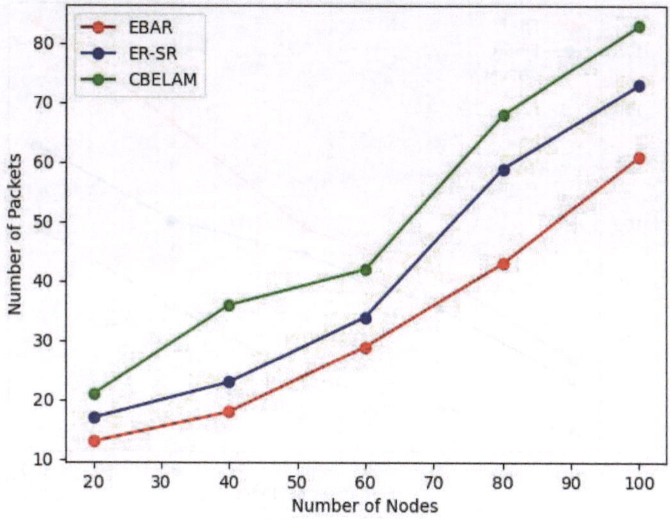

Fig. 7.5 Comparative analysis for packet delivery ratio with proposed method against existing methods

Table 7.6 Analysis of packet delay

Number of nodes	EBAR	ER-SR	CBELAM
20	41	32	11
40	78	64	21
60	98	87	34
80	144	102	65
100	187	134	98

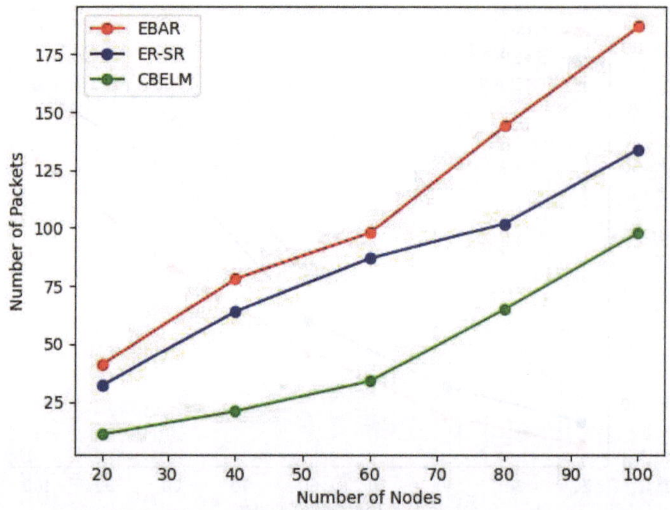

Fig. 7.6 Comparative analysis for packet delay with existing methods against proposed method

References

1. Hayes, T., & Ali, F. H. (2016). Location-aware sensor routing protocol for mobile wireless sensor networks. *IET Wireless Sensor Systems, 6*(2), 49–57.
2. Yan, J., Zhou, M., & Ding, Z. (2016). Recent advances in energy-efficient routing protocols for wireless sensor networks: A review. *IEEE Access, 4*, 5673–5686.
3. Rahim, R., Murugan, S., Mostafa, R. R., Dubey, A. K., Regin, R., Kulkarni, V., & Dhanalakshmi, K. S. (2020). Detecting the Phishing Attack Using Collaborative Approach and Secure Login through Dynamic Virtual Passwords. *Webology, 17*(2).
4. Rahim R., Murugan, S., Priya, S., Magesh, S., & Manikandan, R. (2020). Taylor Based Grey Wolf Optimization Algorithm (TGWOA) For Energy Aware Secure Routing Protocol.*International Journal of Computer Networks and Applications, 7*(93). https://doi.org/10.22247/ijcna/2020/196041.
5. Mittal, N., Singh, U., & Sohi, B. S. (2017). A stable energy efficient clustering protocol for wireless sensor networks. *Wireless Networks, 23*(6), 1809–1821.
6. Murugan, S., Jeyalaksshmi, S., Mahalakshmi, B., Suseendran, G., Jabeen, T. N., & Manikandan, R. (2020). Comparison of ACO and PSO algorithm using energy consumption and load balancing in emerging MANET and VANET infrastructure. *Journal of Critical Reviews, 7*(9)
7. Khalaf, O. I., Abdulsahib, G. M., & Sabbar, B. M. (2020). Optimization of wireless sensor network coverage using the bee algorithm. *Journal of Information Science and Engineering, 36*(2), 377–386.
8. Bouabdallah, F., Bouabdallah, N., & Boutaba, R. (2008). On balancing energy consumption in wireless sensor networks. *IEEE Transactions on Vehicular Technology, 58*(6), 2909–2924.
9. Sampathkumar, A., Murugan, S., Rastogi, R., Mishra, M. K., Malathy, S., & Manikandan, R. (2020). Energy Efficient ACPI and JEHDO Mechanism for IoT Device Energy Management in Healthcare. In Internet of Things in Smart Technologies for Sustainable Urban Development (pp. 131–140).
10. Montoya, G. A., & Donoso, Y. (2013). Energy load balancing strategy to extend lifetime in wireless sensor networks. *Procedia Computer Science, 17*, 395–402.
11. Elsmany, E. F. A., Omar, M. A., Wan, T. C., & Altahir, A. A. (2019). EESRA: Energy efficient, scalable routing algorithm for wireless sensor networks. *IEEE Access, 7*, 96974–96983.

12. Chu, K. C., Horng, D. J., Chang, K. C. (2019). Numerical optimization of the energy consumption for wireless sensor networks based on an improved ant colony algorithm. *IEEE Access*
13. Datta, D., Mishra, S., & Rajest, S. S. (2020). Quantification of tolerance limits of engineering system using uncertainty modeling for sustainable energy. *International Journal of Intelligent Networks, 1*, 1–8. https://doi.org/10.1016/j.ijin.2020.05.006
14. Pereira, H. (2020). Métrica de roteamento para prolongar o tempo de operação de redes de baixa potência e com perdas
15. Xu, C., Xiong, Z., Zhao, G., & Yu, S. (2019). An energy-efficient region source routing protocol for lifetime maximization in WSN. *IEEE Access, 7*, 135277–135289.
16. Förster, A., Förster, A., & Murphy, A. L. (2009). Optimal cluster sizes for wireless sensor networks: An experimental analysis. *International Conference on Ad Hoc Networks*, 49–63
17. Mahajan, S., Malhotra, J., & Sharma, S. (2013). Improved enhanced chain based energy efficient wireless sensor network
18. Obaid, A. J., Alghurabi, K. A., Albermany, S. A. K. & Sharma, S. (2021). "Improving extreme learning machine accuracy utilizing genetic algorithm for intrusion detection purposes." In Advances in intelligent systems and computing (pp. 171–177), Springer.
19. Sampathkumar, A., Maheswar, R., Harshavardhanan, P., Murugan, S., Jayarajan, P., & Sivasankaran, V. (2020, July). Majority Voting based Hybrid Ensemble Classification Approach for Predicting Parking Availability in Smart City based on IoT. In 2020 11th International Conference on Computing, Communication and Networking Technologies (ICCCNT) (pp. 1–8). IEEE.
20. Sharma, S., & Obaid, A. J. (2020). Mathematical modelling, analysis and design of fuzzy logic controller for the control of ventilation systems using MATLAB fuzzy logic toolbox. *Journal of Interdisciplinary Mathematics, 23*(4), 843–849.
21. Sharma, S., & Obaid, A. J. (2020). Contact-mechanics and dynamics analysis of three-different ellipsoidal raceway geometries for deep Groove ball bearing using Abaqus 6.13 version FEA simulation for high load-bearing as well as speed-rotating applications. *International Research Journal of Multidisciplinary Science and Technology, 3*(5), 36–43.
22. Obaid, A. J., & Sharma, S. (2020). Recent trends and development of heuristic artificial intelligence approach in mechanical system and engineering product design. *Saudi Journal of Engineering and Technology, 5*(2), 86–93.
23. Subramaniyan, M., Sampathkumar, A., Jain, D., Manikandan, R., Patan, R., & Kumar, A. (2021). Deep Learning Approach Using 3D-ImpCNN Classification for Coronavirus Disease. https://doi.org/10.1007/978-3-030-60188-1_7.

Chapter 8
A Novel Implementation of Linux Based Android Platform for Client and Server

M. Kiran Kumar, S. Kranthi Kumar, Ella Kalpana, Donapati Srikanth, and K. Saikumar

Abstract Android is an excellent operating system and offers many applications. Nowadays, mobile phones and tabs depend on the Linux platform. The Android working framework-programming stack base contains Linux Kernel, so it developed over the Linux Kernel. In a versatile market, it is a mainstream working framework. Because of the exponential increment in mobile use by the client's step by step, the Android cell phone has been communicated by the most extreme clients on the planet. There is a need to concentrate on the Android framework to give security to Android applications, information, gadgets, and the system. Linux® is an operating system that is open source (O.S.). The programme that directly controls the hardware and infrastructure, such as CPU, memory, and storage, is an operating system. The O.S. sits among hardware and applications links all of your apps with the physical resources that do the job. In the manuscript, please add one more keyword, and it should represent the complete technical sense. This paper focuses on researching the different genuine security issues associated with the Android devices; anticipating these issues; strategies have proposed alleviating the security hazards on the Android gadgets since the clients are putting away crucial data in their cell phones.

Keywords Linux platform · Android · SMS permissions · Web View · And cell phones

M. Kiran Kumar
Guru Nanak Institutions Technical Campus, Hyderabad, India

S. Kranthi Kumar
Department of CSE, Gokaraju Rangaraju Institute of Engineering & Technology, Nizampet Road, Bachupally, Kukatpally, Hyderabad, Telangana, India

E. Kalpana
Department of CSE, T.T.W.R.D.C., Hyderabad, Telangana, India

D. Srikanth
Department of CSE, Srikanth Tech Academy And Research Innovatives Pvt. Ltd., Hyderabad, India

K. Saikumar (✉)
Department of ECE, Koneru Lakshmaiah Education Foundation, Guntur 522502, India

P. Kumar et al. (eds.), *A Fusion of Artificial Intelligence and Internet of Things for Emerging Cyber Systems*, Intelligent Systems Reference Library 210,
https://doi.org/10.1007/978-3-030-76653-5_8

8.1 Introduction

The Linux Kernel associates the equipment of the Android cell phones. The Linux Kernel is responsible for a portion of assets, overseeing memory, memory store, distribution, de-allotment of remembrance for the record framework, booking of the procedure, and systems administration. The Android application conveys an incentive to the cell phone clients by utilising a tremendous measure of equipment and programming assets for innovation [1]. The University of Cambridge and Zimperium labs' surveyors researched that around 90% of keen gadgets of Android were influenced by unsafe vulnerabilities and hacked through SMS messages. There is a prospect of mounting malware and Android Apps created by the outsider through social engineering assaults by the assailant in Android telephones. Android displays an open-source-stage and sharpness condition for portable gadgets, and all Android applications keep running on the Application Sandbox. Regularly the restricted scope of the framework assets is accessed by the Android applications [2]. For the most part, the central running device depends on the Linux kernel [3] l. Usually, Android applications get written in Java programming language and continue running in the Dalvik virtual machine. In any case, applications are competent to be written in the usual code. Applications have mounted from the alone record with the—Apk record expansion. The application assets are perceived and isolated on the Android stage by using the Linux client based assurance. The Android working framework's upside is that it permits the Android application to keep running with a particular procedure with a unique client id, which is not the same as other operating frameworks [4]. The Android advancement group examined different security issues and vulnerabilities on other cell phones, work areas, and server stages and saw a need to give factual security shows in the Android location to empower a reliable utilisation ecosystem.

Its source code is the Android Open Source Project (A.O.S.P.), mainly licensed under the Apache Licenses free and open-source software. However, most Android devices ship with pre-installed additional proprietary applications, most notably Google Mobile Services (G.M.S.), including critical apps such as Google Chrome, the Google Play digital delivery platform and the related development platform for Google Play Services. Around 70 per cent of Android smartphones run the Google ecosystem; Fire O.S. (developed by Amazon) or Lineage O.S. are the competing Android ecosystems and forks. However, the "Android" name and logo are Google trademarks that enforce requirements to limit Android branding to "uncertified" devices outside their ecosystem.

8.2 Proposed Work

This paper centres around recognising the Android working framework's vulnerabilities by using the infusing the noxious app on the Android cell phone through SMS, email, and so forth and to test the effect of the helplessness on the gadget [5].

The Android working framework is in charge of recognising the gadget's vulnerabilities and should not permit introducing its vindictive app. This paper centres around the accompanying errands (i) Attacking and anticipation stage on the Android cell phone and (ii) Attacking and anticipation stage on the Android cell phone utilising Metasploit.

8.2.1 Assaulting and Anticipation Stage on the Android Cell Phone

8.2.1.1 Attacking Phase

The malignant Apps can be created by the assailant which can perform noxious exercises on the gadgets, for example, perusing the log record, propelling DoS assaults, perusing SMS messages, perusing exhibition data, dealing with the whole cell phone, taking or change of information, and so on. In the assaulting stage, the aggressor sends the pernicious app to the injured individual, which can peruse the person's log record in question and send it to the assailant's telephone through SMS/mail actuated. The activities that can perform in this stage are as per the following (i). Making a malignant App to peruse the log record and exhibition data from the cell phone [6]. The vindictive app shows the phoney promotion demonstrating the unique offers offered by Amazon with high limits. (ii). the noxious.apk record is infused into the unfortunate casualty gadget through SMS, WhatsApp message or email. (iii). the alluring message like limits offered on the items or message identified with welcome is s to be the client at whatever point the app is. (iv). When a typical client taps on the connection, the nasty application covered up inside the connection will be introduced on the injured individual gadget. (v). The app will run when it has actuated, or it can keep running out of sight. When the app has actuated, it peruses the log record from the compromised individual's gadget and advances it to the aggressor through an SMS message or email.

Steps Included During the Assaulting Stage:

(i). at the back end, the malignant app gets the call log data from the unfortunate casualty gadget utilising the Android Call Log Content Provider Managed inquiry [7]. This strategy restores the outcome set with lines and sections way, which contains various fields like telephone number, call type, call date, and contact span. (ii). Call code includes three kinds of calls, for example, approaching, active and missed calls. The calling code containing messages is assembled into three types utilising the strategies present in the cursor class [8]. (iii). the fields containing the call data have added to the String Buffer variable. Next, the information in the current String Buffer variable has changed over into string information type. The string message

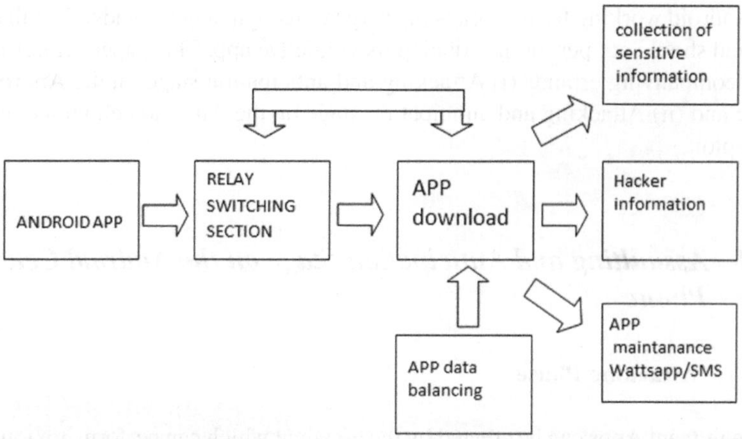

Fig. 8.1 Flow chart of assaulting stage

containing the injured individual call data has transmitted to the assailant through an SMS message or email utilising the strategies in the Sms Manager class.

The app has created to contain the malicious code to peruse the log record and store it in a document. Each column containing the following data, for example, telephone number, sort of call, and length of the market with information are changed over into string information type to advance it to the programmer's telephone through SMS or email as exemplified in the calculation underneath in Fig. 8.1 and Table 8.1.

8.2.1.2 Prevention of Malignant App

These days, every individual uses an advanced mobile phone, and there is a fast increment of utilising the applications in cell phones. Regularly, these P.D.A.s are in peril in light of some killer applications in a roundabout way introduced through malignant connections on our cell phones [9]. The Android gadget issue is that it enables us to download the Apps from outside other than Google play store, and there will be odds of downloading the malevolent Apps created by the evil programmers. Indeed, even these malignant applications take a portion of our private information like call data [10, 11]. Research is required to keep from taking the critical data on the cell phone [12]. This paper centres around executing applications to verify the Android gadget from taking the crucial data. This paper centres around following errands Scanning the device to inspect for malignant or tainted App, Examining the running application in the widget, Examine the Apps introduced on the gadget, Identifying and uninstalling the app that peruses the crucial data from the device and Identifying and uninstalling the app that utilises illicit authorisations.

Table 8.1 Algorithm to get critical data from a cell phone

```
Begin
Read the call log and store it in a file
                Read first row in a file
                while(!eof) do
                begin
                        Extract phonenumber, typeofcall,date,callduration from
the row
                        switch(typeofcall)
                        begin
                                case : OUTGOING
                                        calltype = "Outgoing Call"
                                        break
case : INCOMING
                                        calltype = "Incoming Call"
                                        break
                                case : MISSED
                calltype = "Missed Call"
                break
                        enddo
                Append phonenumber, calltype,date,callduration tostringbuffer
                enddo
        Forward the stringbuffer to the attacker's smartphone via sms or email.

End.
```

This effort centres around recognising the consents on the Android App. There are two sorts of authorisations typical and risky. Programmers utilise hazardous permissions on the app to take private data or perform pernicious exercises on the injured individual gadget [13]. This paper centres on distinguishing the unsafe authorisations on the app and uninstalling the app. When the app contains the perilous permissions, the Android working framework will alarm the client with warning messages. The malevolent app movements are seen by looking at the app's consent [14]. The accompanying activity will be framed by the aggressor utilising the unsafe authorisations. Stealing the secret phrase data and important log information, Reading the contact data and perusing/composing framework settings, Examining the active calls, Using the unfortunate casualty camera and following the area of the person in question, Sending SMS messages to aggressor telephone from injured individual gadget, Tracking the size of the regrettable casualty gadget examining or taking the display data of the incapacitated unique device [15].

The installation on a Mobile computer of a standard Linux release opened a whole new universe of possibilities. You can convert the Android computer into a full-blown Linux / Apache / MySQL / PHP database and run web-based frameworks thereon, uninstall and then use your favourite Linux software and even run a graphical desktop environment. In short, in so many cases getting a Linux version on an Android device will come handy [16].

Although K.B.O.X. provides a simple way of installing Linux on a Mobile computer, you can find it too restrictive for your needs. You should try downloading a standard Linux release in this situation, but you'll have to root your Android computer to do so. The exact rooting process depends on your Android system, and often it can be not easy. Nonetheless, several rooting tutorials are accessible online to support further along its route.

One choice is Linux Deploy when it comes to running Linux on a rooted Android computer. This open data application provides a simple way for a licensed Linux version to be activated and operated in the cheroots setting, a unique directory serving as a virtual directory. In other terms, a cheroot establishes an autonomous area that does not interact with the rest of the structure [17]. The unique thing is just behind a user-friendly G.U.I. Kernel Deploy masks all the morbid technological information, so you may not need to know all the nitty–gritty to install and operate the rooted Android computer software.

You need to load remaining two applications on your computer before you continue to instal Linux on Ios utilising Linux Deploy: console emulation and a V.N.C. (Virtual Network Computing) client. While many terminal simulators and V.N.C. client programmers are available on Google Play Store, VX ConnectBot and MultiVNC can't go wrong. Both of which are downloadable free, open-source software. Start the app and tap the Properties button to install one of the authorised Linux distributions utilising Linux Deploy. A collection of customisable choices found in the Properties pane. Begin by selecting the desired Linux distribution in the Deploy segment from the Package list. Many commons distress, like Debian, Ubuntu, Arch Linux, beret, and open S.U.S.E., are assisted by Linux Deploy. You may pick a special edition for some distributors as well. For example, select Debian, and you can choose the preferred version (stable, working, unstable, etc.) from the Deployment Suites list. Since the root directory is the top of the hierarchy of file systems, programmers operating in the cheroots setting cannot reach guides lower than the folder Fig. 8.2.

The app has created to recognise the noxious and uninstall the malicious app containing the hazardous authorisations as exemplified in the calculation underneath Table 8.2.

8.2.2 Attack on Android Telephone Utilising Metasploit

The Android cell phone's dangers are expanding step by step, as the telephones' utilisation is increasing [18]. Today the few assaults are performing on the Android working framework. The malware is made in-app by Metasploit device and once the app is introduced on the gadget types access on the widget to access to a gadget on a similar system which encourages to following activities on the individual device, for example, phishing, sending counterfeit SMS, picking up control on the unfortunate casualty gadget camera towards take pictures, and so forth. The means associated with playing out the assault utilising Metasploit instrument is.

Fig. 8.2 Linux based routing

Table 8.2 Algorithm to uninstall malevolent Apps

```
Begin
Check the Android version running on the device.
if( Android version is less than 6.0) then
send an SMS message to the user
endif
Check if permissions are granted to App
if(permission granted) then
        Check the App is requesting for dangerous permission
        if (the permissions are sendsms||readsms||receivesms||readexternalstorage||
                        writeexternalstorage||
writecalllog||getaccounts||processoutgoingcalls||receivemms)then
                if(above any of the permissions are granted) then
                        send notification message tovictim
                        uninstall the App
                endif
        endif
endif
end
```

Step 1: Retrieving the data of the Android gadget utilising Operating System Fingerprinting instruments as exemplified in the figure underneath:
Step 2: Build up endeavour apparatus in Kali Linux.
Step 3: Check the injured individual Android gadget is helpless against the endeavour apparatus.
Step 4: Build up payload utilising misuse apparatus.
Step 5: Implementing application to the departure the malevolent payload from Intrusion Prevention framework.
Step 6: Implementing the adventure on the unfortunate casualty gadget.

8.2.2.1 Prevention

One method for avoiding the noxious app on the gadget is by inspecting the applications containing the unsafe consents on the device physically and erase it physically if hazardous authorisations seen on the app [19]. The application created to shield from malevolent Apps executed from Metasploit instrument and work process of vindictive Apps of the Metasploit apparatus exemplified in Fig. 8.3 underneath [20].

Figures 8.3 and 8.4 explain Android structure by utilising the Linux window here the metadata is collected from the data server's framework. This server is sending the information through you might ask plot attack here target device Android application controlling the malaria preventions. Finally, the Android studio depending upon the versions these are and their analysis is controlled and giving the qualified outcome as an experimental result.

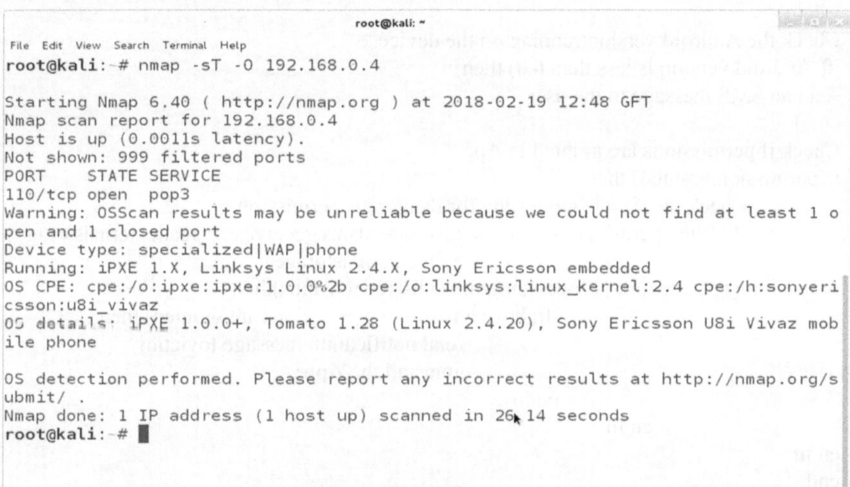

Fig. 8.3 Gathering data about smart phone-utilising port checking

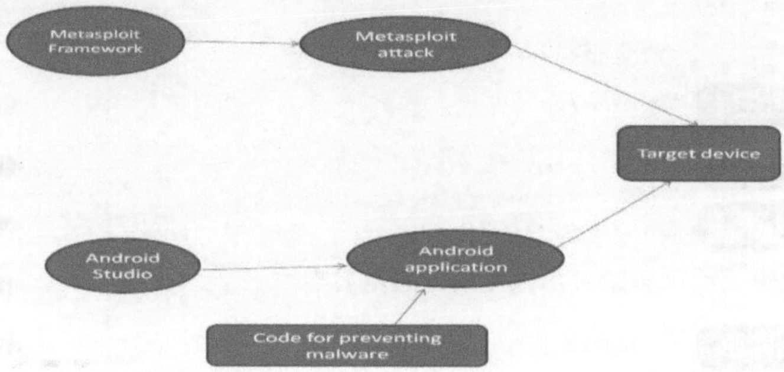

Fig. 8.4 Workflow handing Metasploit attacks

Figure 8.5 demonstrates a rundown of consents. When the client taps on SMS authorisations will show the rundown of Apps that contains SMS consents as exemplify [21] In the figure. Linux Would Cooke needs to retain the addition downstream similar to the Dash to Dock extension, i.e. upstream, with its potential inclusion in Ubuntu 18.10. This change will allow the function to acquire a more generic name (such as 'Computer Integration'), profit through Ubuntu-specific polishes, fixes, and integrations, and further facilitate Ubuntu project creation contributions [22].

Fig. 8.5 List of App Permission

Fig. 8.6 List apps containing SMS permissions

Linux: Linux. When you've been studying Linux mostly on the internet, it's incredibly possible that you've come through Ubuntu. Mix Cinnamon from Debian [23]. Linux Mint is already Linux release number one on Distrowatch for years.

1. Linux Zorin.
2. Linux Kindergarten.
3. Hazelnut-Mate Linux.
4. Kernel Manjaro.

Figure 8.6 demonstrates that the two applications in particular 'call log' and 'Amazon attractive offer' contain the SMS consents. On the off chance that any of these Apps are infused by the programmer to steal data from the gadget [24], the app can abuse the SMS authorisations by sending the unfortunate casualty gadget crucial data through SMS [25].

8.3 Results

The assailant will insert malicious app to the Victim Device through SMS/ WhatsApp / Gmail. In Fig. 8.6 below indicates that the attacker injects the hateful App through SMS messages.

Fig. 8.7 SMS message received by the victim

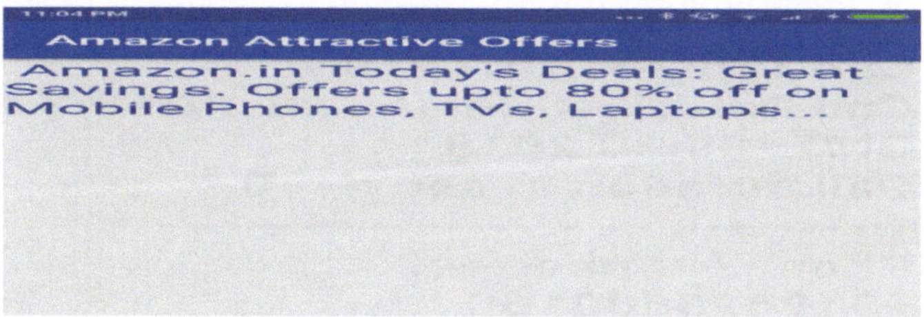

Fig. 8.8 Malevolent app is open

When the user clicks on the above link as shown in Fig. 8.7, the app will be activated and read the sensitive information from the victim such as call log file [26].

When the user clicks on the above link as the malicious app is inserted on the prey gadget. The hatefulApp shows a link, as shown in Fig. 8.8.

The app recovers the total subtleties of active, approaching and missed call data from the log record. The data will be sent to the aggressor gadgets. This movement will happen when the victim chooses the app or the app kept running out of sight and initiated through a specific timeframe. The symptoms for detecting this type of attack looking at the SMS messages sent on every day surpassing the limit esteem and Looking at Apps with unsafe authorisations running in the background shown in Fig. 8.9.

Three strategies to be produced for taking care of these sorts of assaults through distinguishing introduced Apps, Effected Apps and Running Apps. As appeared in Fig. 8.9 beneath the user can examine the List of Apps installed, running and effected Apps at any time. Figures 8.10 and 8.11 indicates the List of Apps installed on the device.

Figure 8.12 explains about different phases of running phases and their decision based on Linux commands.

When the client selects" List Affected apps" appear in Fig. 8.9, the application scrutinises the gadget and distinguishes suspicious app record in running applications. On selecting the influenced apps, a button which has utilised to uninstall the influenced app.

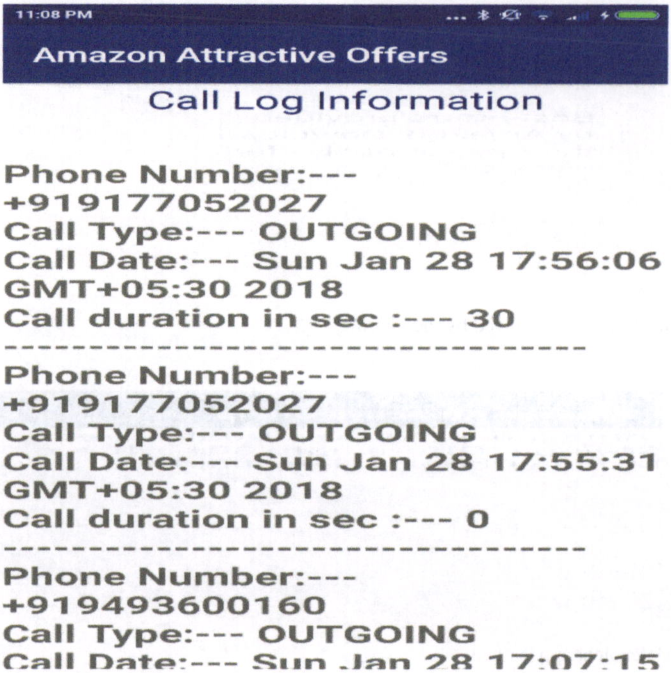

Fig. 8.9 Call log Information

Fig. 8.10 Preventing phase

When the client icon the button"Uninstall Affected Apps" appeared in Fig. 8.13, which sweeps to recognise the suspicious app records running on the gadget. It shows "click here to UNINSTALL" message as appeared in Fig. 8.13 to distinguish doubtful app record in running applications and uninstall the suspicious app documents when the client hits the contaminated app.

When there is an influenced application, it prompts an alarm box, i.e., Do you need to uninstall this application. The client can uninstall the influenced app by selecting on "OK" as appeared in Fig. 8.14 underneath. At that point, the controlled apps will be expelled from the injured individual's gadget.

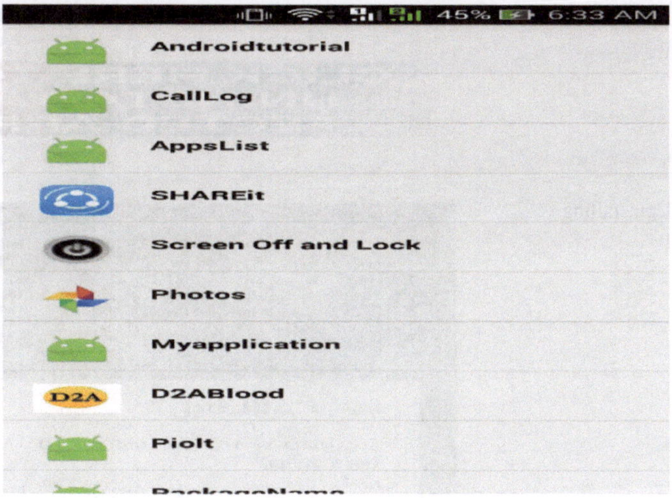

Fig. 8.11 Lists of apps installed on the device

Fig. 8.12 List of running app

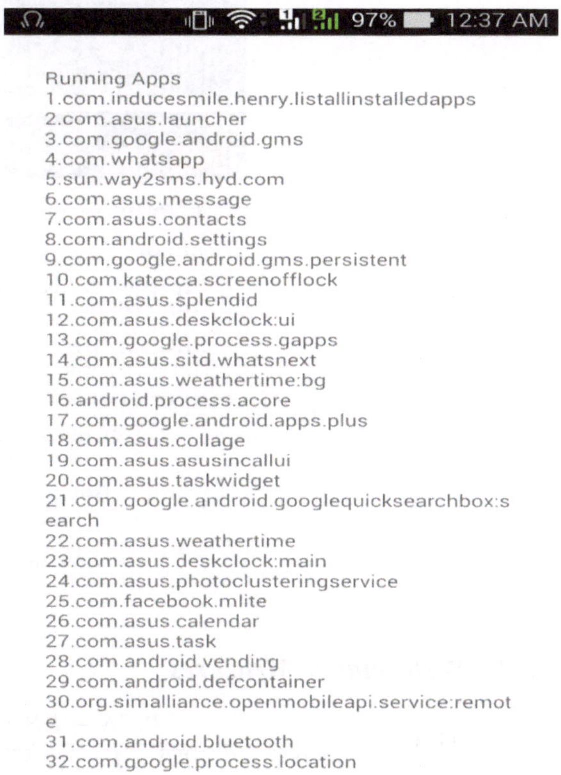

Fig. 8.13 App to uninstall affected app

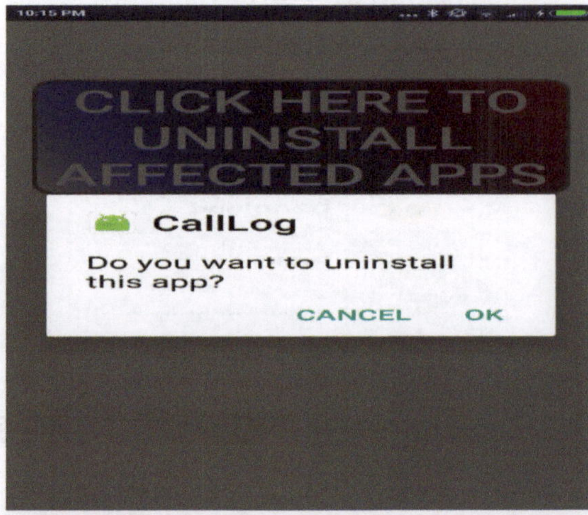

Fig. 8.14 Uninstalling effected app

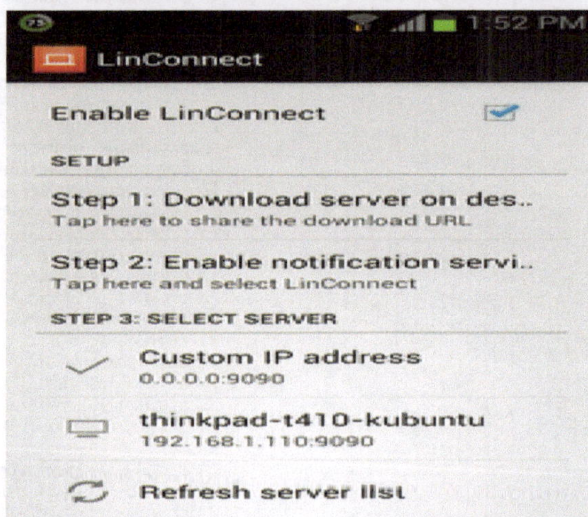

8.3.1 Performance Measures

$$\text{MCC} = \frac{TP.TN - FP.FN}{\sqrt{(TP+FP).(TP+FN).(TN+FP).(TN+FN)}}$$
$$(\text{MCC : worst value} = -1; \quad \text{best value} = +1) \tag{8.1}$$

$$F_1\,score = \frac{2.TP}{2.TP+FN+FP}$$
$$(F_1\,score: worst\ value = 0; best\ value = 1) \tag{8.2}$$

$$sensitivity = \frac{TP}{TP+FN}$$
$$(sensitivity: worst\ value = 0; best\ value = 1) \tag{8.3}$$

$$specificity = \frac{TN}{TN+FP}$$
$$(specificity: worst\ value = 0; best\ value = 1) \tag{8.4}$$

Above all equations are mainly used for calculating the performance estimation parameters; these are useful for comparing the designed application with existed methods.

Figure 8.15 explains performance measures of different models, in this although Android is providing consumers with a feasible alternative to other mobile operating systems, there are still some shortcomings. Coding complicated user interactions and interfaces on the developer side is an always daunting job requiring greater dependence on Java than impartial-C. The applications on the Android Platform appear just to have weaker expectations for consumers than similar app stores. In all other words, the applications have more insufficient protection profiles that render them more vulnerable to data breaches. Meanwhile, the absence of a voice-controlled associate from Interface and its firm advertisement reliance could overpower many consumers.

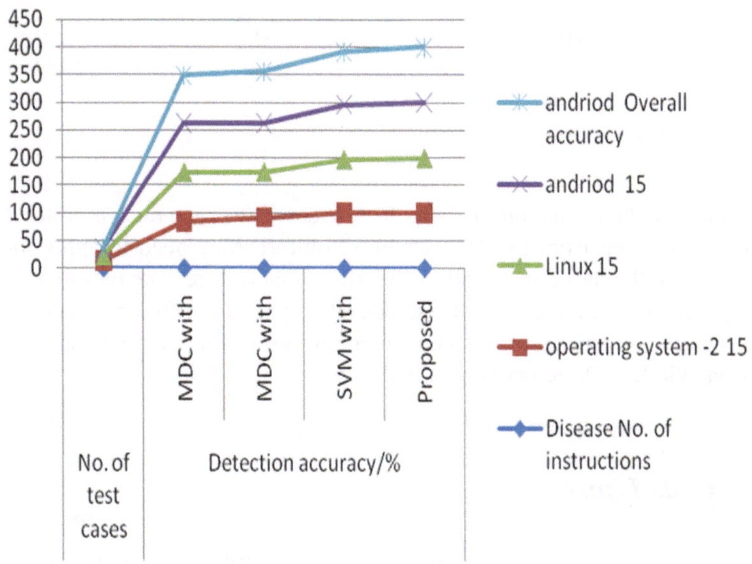

Fig. 8.15 Performance measures

Table 8.3 Accuracy analyses

Parameter	No. of instructions	No. of test cases	Detection accuracy/%			
			MDC with	MDC with	SVM with	Proposed
operating system -2	15	12	83.33	91.66	100	99.98
Linux	15	10	90.32	82.71	97.12	99.89
android	15	12	88.45	87.32	98.723	99.98
	Overall accuracy		86.54	93.63	95.71	99.98

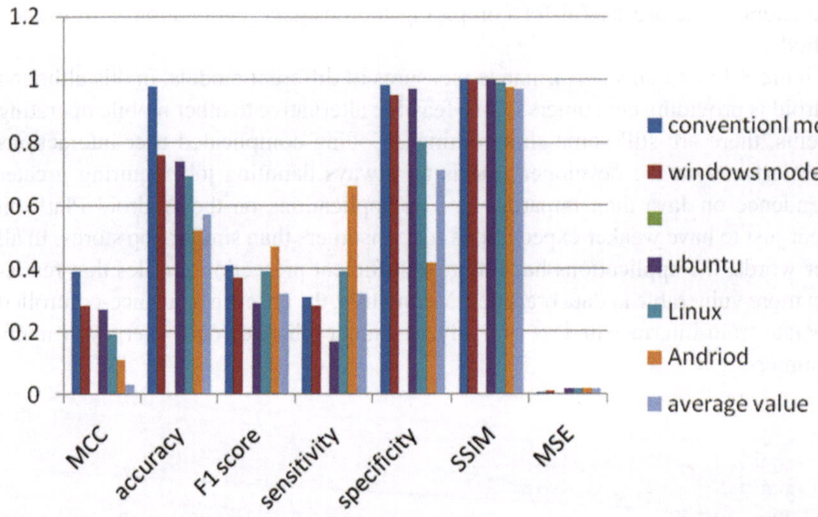

Fig. 8.16 Android analysis

For starters, the device observes its feature serves to keep an eye on the Mobile device's state again from the Java screen's comfort. May need to apply the KDE Link widget to the monitor and allow the use of this feature. The plug-in Clipboard makes the feature that connects the screen and Application folder contents. It may use the Dragon picture viewer to navigate the Android computer with activated web extension Wireless Directory shown in the Table 8.3 and Fig. 8.16.

8.3.2 Code Linux

Years at least —silent https:/raw.github.com/hauckwill/linconnect-server\
/LinConnectServer / install.sh.
+ x but seldom install. sh.
./.sh update.

The Transfer & Receiving system allows sharing of data here between Computer or Mobile computer. By right-clicking on it in Dolphin and choosing to Send to < DEVICE > via Java Link, it can transfer any file on a used computer it may even submit files the same way. The special control is introduced to google's sharing functionality by the Linux operating system Connection software so that you can push a file from almost any programme. The communication feature of KDE Attach is not restricted to files: You may transmit the hyperlinks from the mobile phone to the standard KDE search engine.

The android mobile system was invented by Linux, Inc., a Silicon Valley-based Tech Corporation till Google purchased it in 2005. The Android operating concept was developed by Apple, Inc., a tech corporation based in Silicon Valley until Google bought it in 2005. Investors and experts in the technology sector doubted facebook's real motives for joining the smartphone. After the takeover, stakeholders and observers in the technology industry have examined the real motives of Twitter to penetrate the smartphone business room. After that purchase, space. But since, designers of software applications have been able to create smartphone apps but are distributed via mobile apps, such as Google Play, utilising Smartphone technologies. And since Android users are made as a Google app, they are granted the ability to link their mobile devices to other Google items, such as data storage, internet and video sites shown in Fig. 8.17.

The Software for Linux comes out in an open-source framework to further promote minimum rules through portable apps. After being published as "accessible," whenever offered with smartphones Mobile, it is bundled within valuable place shown in Table 8.4.

New utilisation of innovation includes portable Apps with access to individual information this taken off by the specialist organisations and their utilisation employing advanced mobile phones in shoppers' hands. Thus, clears route for noxious Apps picking up guiltless portable clients' trust and subsequently imperilling essential individual resources or assets. Averting accidental information stream in cell phones draws in critical intrigue. An App that checks information robbery by malevolent Apps has displayed in the paper. The assault can be relieved at the source or the objective since the assault can be performed by advanced mobile phone on the devoted servers. In specific cases, it may not be possible to alleviate the ambush at the source. This paper also focused on mitigating the routing attacks at the target platform side performed through a smartphone.

8.4 Conclusion

In this research work, an advanced android Digital O.S. based platform created the android platform for use with all its digital computers, laptops, and smartphones. This operating system was initially developed by the Python-based tech firm Linux, Inc. until being purchased by a search engine in 2020. Although the original data for software is published in an open-source framework to accelerate standard practices

Fig. 8.17 Android screen

Table 8.4 Android analysis

Method	MCC	accuracy	F_1 score	sensitivity	specificity	SSIM	MSE
conventional model	**0.37**	0.923	0.58	0.41	0.89	0.9948	0.0084
windows model	**0.39**	0.78	0.47	0.38	0.94	0.989	0.0098
Ubuntu	**0.17**	0.73	0.48	0.27	0.92	0.989	0.0187
Linux	**0.29**	0.71	0.47	0.32	0.94	0.978	0.0196
Android	**0.21**	0.59	0.41	0.72	0.82	0.998	0.0198
average value	**0.14**	0.61	0.41	0.65	0.61	0.9785	0.023

worldwide, Windows ten also bundled whenever it's distributed on smartphones. This article focused on studying the numerous real protection challenges involved with android apps; predicting specific problems; solutions have suggested minimising the security threats on smartphones when consumers place sensitive details on their cell phones.

References

1. Mashhourand, A. S., Saleh, Z. (2015). "Community perception of the security and acceptance of mobile banking services in Bahrain: An empirical study". *International Journal of Advanced Computer Science and Applications*, *6*(9).
2. Pereira, A., Correia, M., Brandao, P. (2018). USB connection vulnerabilities on android smartphones: default and vendors' customisations". F.I.P. international conference on communications and multimedia security, communications and multimedia security, Lecture Notes in Computer Science, L.N.C.S. (pp. 19–3) vol. 8735.
3. Sareen, B., Sharma, S., Arora, M. (2014). "Mobile cloud computing security as a service using android". *International Journal of Computer Applications*, *99*(17).
4. Walnycky, D., Baggili, I., Marrington, A., Breitinger, F., & Moore, J. (2015). "Network and device forensic analysis of android social-messaging applications." *Digital Investigation 14*, http://dx.doi.org/https://doi.org/10.1016/j.diin.2015.05.009
5. Bucerza, D, Ratiu, C., & Manolescu, M.J. (2013). "SmartSteg: A new android-based steganography application", *International Journal of Computers Communications and Controls,*" *8*(5).
6. Cho, G., Cho, J., Song, Y., Choi, D., & Kim, H. "Combating online fraud attacks in mobile-based advertising", E.U.R.A.S.I.P. *Journal on Information Security*. https://jiseurasipjournals.springeropen.com/articles/https://doi.org/10.1186/s13635-015-0027-7.
7. Friedberg, I., McLaughlin, K., Smith, P., Laverty, D., & Sezer, S. (2018). STPASafeSec: Safety and security analysis for cyber-physical systems. *Journal of Information Security and Applications, 34*, 183–196.
8. Liu, J., & Yu, J. (1–3 Nov 2011). "Research on development of android applications", *4th International Conference on Intelligent Networks and Intelligent Systems* (*I.C.I.N.I.S.*), http://dx.doi.org/https://doi.org/10.1109/ICINIS.2011.40.
9. Datta, D., Mishra, S., & Rajest, S. S. (2020). Quantification of tolerance limits of engineering system using uncertainty modeling for sustainable energy. *International Journal of Intelligent Networks, 1*, 1–8. https://doi.org/10.1016/j.ijin.2020.05.006
10. Lim, J., & Yi, J. H. (2012). "Structural analysis of packing schemes for extracting hidden codes in mobile malware", E.U.R.A.S.I.P. *Journal on Wireless Communications and Networking*, pp. 1–12. http://dx.doi.org/https://doi.org/10.1186/s13638-016-0720-3.
11. Deypir, M., & Horri, A. (2018). Instance-based security risk value estimation for Android applications. *Journal of Information Security and Applications, 40*, 20–30. https://doi.org/10.1016/j.jisa.2018.02.002
12. Yildirim, N., & Varol, A. (Oct 2016). "Android based mobile application development for web login authentication using fingerprint recognition feature", *International Journal of Computer Science and Mobile Computing*, *5*(10), 61–68, ISSN 2320–088X.
13. Khan, S. R., & Dristy, F. S. (Mar 2015). "Android based security and home automation system", *International Journal of Ambient Systems and Applications*, *3*(1).
14. Zaidi, S. F. A., Shah, M. A., & Kamran, M. (2016). "A survey on security for smartphone device", *International Journal of Advanced Computer Science and Applications*, *7*(4).
15. Vijaykumar, G., Gantala, A., Gade, M. S. L., Anjaneyulu, P., & Ahammad, S. H. (2017). Micro controller base heartbeat monitoring and display on P.C. *Journal of Advanced Research in Dynamical and Control Systems, 9*(4), 250–260.

16. As'habi, K., Vafabakhsh, A., & Borji, S. "Data transmission security in clouds computing." *Indian Journal of Fundamental and Applied Life Sciences 6*, 2231–6345.
17. Santos, L. W., Singh, B., Suman Rajest, S., Regin, R., & Kadhim, K. H. (2021). A genetic programming approach to binary classification problem. *EAI Endorsed Transactions on Energy, 8*(31), 1–8. https://doi.org/10.4108/eai.13-7-2018.165523
18. Rao, S. V. R. K., Saritha Devi, M., Kishore, A. R., & Kumar, P. (Nov–Dec 2018). "Wireless sensor network-based industrial automation using internet of things (IoT)," *7*(6).
19 Manoj, B., Sasikanth, K. V. K., Subbarao, M. V., & Jyothi Prakash, V. (Nov–Dec 2018). "Analysis of data science with the use of big data," *7*(6).
20. Saritha Devi, N., Raju, K. S. R., Madhu, A., Raja Sekhar, R. (Nov–Dec 2018). "Safety and security for school children's vehicles using G.P.S IoT technology" *7*(6).
21. Saikumar, K., & Rajesh, V. Coronary blockage of the artery for Heart diagnosis with D.T. Artificial Intelligence Algorithm. International Journal of Research in Pharmaceutical Sciences, 11(1), 471-479, (2020).
22. Al-asadi, T. A., & Obaid, A. J. (2016). Object based image retrieval using enhanced SURF. *Asian Journal of Information Technology, 15*(16), 2756–2762.
23. Meshram, C., Ibrahim, R. W., Obaid, A. J., Meshram, S. G., Meshram, A., & Abd El-Latif, A. M. (2020). "Fractional chaotic maps based short signature scheme under human-centered IoT environments." *Journal of Advanced Research.*
24. Obaid, A. J., Alghurabi, K. A., Albermany, S. A. K., Sharma, S. (2021). "Improving extreme learning machine accuracy utilizing genetic algorithm for intrusion detection purposes." In *Advances in intelligent systems and computing* (pp. 171–177), Springer.
25. Obaid, A. J. (2021). "Critical research on the novel progressive, JOKER an opportunistic routing protocol technology for enhancing the network performance for multimedia communications." In *Research in intelligent and computing in engineering. Advances in intelligent systems and computing*, Vol. 1254 (pp. 369-378). Springer.
26. Sharma, S., & Obaid, A. J. (2020). Mathematical modelling, analysis and design of fuzzy logic controller for the control of ventilation systems using MATLAB fuzzy logic toolbox. *Journal of Interdisciplinary Mathematics, 23*(4), 843–849.

Chapter 9
Recognizing Unusual Activity with the Deep Learning Perspective in Crowd Segment

Bhoomeshwar Bala, Raja Shekar Kadurka, and Galeta Negasa

Abstract The data's unusual behavior suggests finding patterns in data that are not consistent with the expected behaviour. A developing prerequisite for more intelligent video vigilance of secure and open space utilizing shrewd vision frameworks can separate semantically significant behaviour towards the human spectator as typical behavior or anomalous behavior. Now presenting a novel-based energy approach for strange behavior recognition utilizing deep learning procedures. Utilize a versatile optical stream model to work on moving particles rather than articles and circuit highlights with data's shape and direction. We present and incorporate numerous behavioral models for precise irregular activity recognition in a complex crowd scene. We have to use the individual behavior model, yet also different social behavior models. The test results show that our proposed technique effectively recognizes the unusual behavior in a packed scene. This paper revolves around a bunch conducts assessment which can perceive regular conduct or abnormal conduct.

Keywords Group scene · Crowd behaviour analysis · Usual activity · Behaviour recognition

9.1 Introduction

The search for human exercises in the packed scene is one of the most testing PC vision undertakings. Since people's investigation of human activities is not a wholly tackled issue, swarm scene examination is a significant errand in PC vision. As of late, swarm conduct examination has gotten a ton of consideration since it is appropriate to new areas, such as the programmed discovery of mobs or disorderly acts utilized in the smart observation framework. These days, irregular conduct identification framework assumes a significant job in different regions, for example, jail, firefighting, open security, bank, and so forth. With the screen's promotion and a

B. Bala (✉) · R. S. Kadurka · G. Negasa
Department of Computer Science, Mettu University, Metu Zuria, Ethiopia

P. Kumar et al. (eds.), *A Fusion of Artificial Intelligence and Internet of Things for Emerging Cyber Systems*, Intelligent Systems Reference Library 210,
https://doi.org/10.1007/978-3-030-76653-5_9

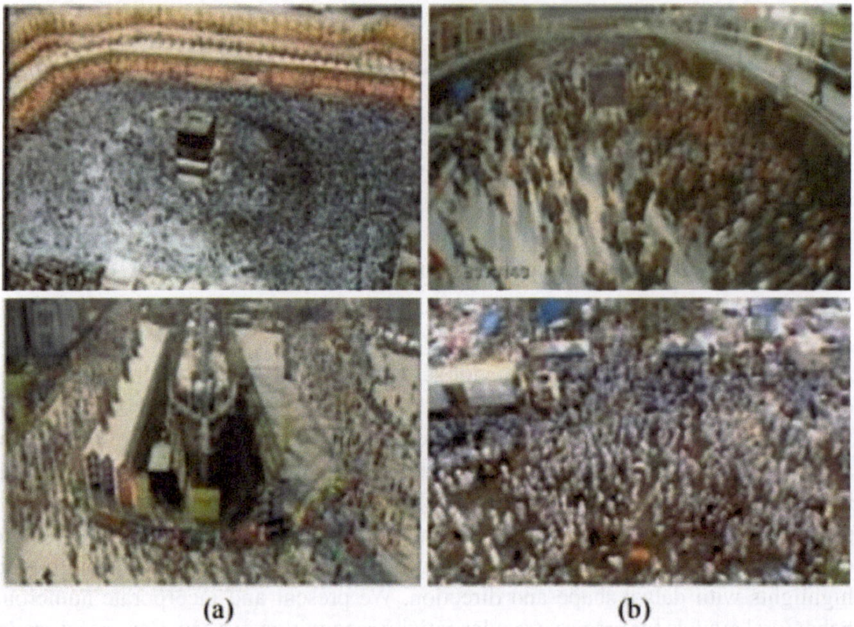

Fig. 9.1 Complex scene a Framed, b Unframed

wide range of unusual occasions happening, the inadequacy of customary observation framework is becoming increasingly self-evident. Truth is told, the standard observation framework can't early caution us when anomalous conduct is occurring. We have detected that some of the most labor-intensive work is useful when the video cuts. Furthermore, some of which may solve the problem of the division, as shown in Fig. 9.1.

The issue, for example, unpredictability and dynamic of recognizing and distinguishing strange conduct in a group scene, pull in numerous specialists. Taking care of a circumstance that relates to the anomalous in a group isn't simple. The most significant issue is the thickness of the packed scenes and the condition of being unusual or ordinary. Regardless, there are a few challenges in breaking down humans' conduct in the swarm scene, and one of the most widely recognized methodologies is led by utilizing video observation. With the Closed-Circuit Television (CCTV) innovation, video observation has gotten obvious right now to watch a few pieces of a procedure from control condition required in every single wise swarmed the scene. One of those themes in video observation is about group investigation. Group investigation is being utilized in video reconnaissance applications for programmed identification of abnormalities and alerts. For the most part, four segments are in the Crowd examination, and those incorporate group thickness estimation, swarm movement recognition, swarm conduct comprehension, and the group following. In a group examination, commonly, the application includes swarming the executive's methodologies, virtual situations, open space structure, visual reconnaissance, and astute condition. Be that as it may, here we are concentrating on the visual observation swarm investigation application. During a decade ago, investigate on distinguishing unusual conduct has effectively advanced exploiting ongoing improvements in some related fields, for example, Pattern Recognition (PR), Biomedical Information (BI), Soft Computing (PR), Computer Vision (CV), Mathematical Modeling (MM), Image Signal Processing (ISP), Data 9 Mining (DM), Computational Intelligent (CI), Artificial Intelligence (AI), Machine Learning (ML) and Deep Learning (DL). A survey of the ongoing advances in recognizing strange activity for identifying human in the packed scene is exhibited since 2000; except if that are some material realities that are important to express an exploration earlier than that.

9.2 Background and Key Issues

9.2.1 Macroscopic Modeling

Separating observation chronicles of a stuffed scene has been a working inquisitive about the field of PC vision over the latest couple of years. This strong interest is driven by the extended enthusiasm for open security at swarmed spaces, for instance, plane terminals, railroad stations, malls, fields, etc. In such scenes, conventional PC vision techniques for video observation aren't genuinely applicable to the stuffed location due to tremendous assortments of gathering densities, complex gathering components, and genuine hindrances in the scene. Right now, the most broadly perceived techniques for exhibiting swarm rehearses are analyzed. Dongping Zhang and also Min Sun suggested a game-plan of guide difficulty utilizing markers. Direct infers clinics happened in a comparable moment. At the moment, there is merely a guide, a time understood and neglected clinics. An instance of mixed lead is when fight direct occurs, by then tumble or furor direct moreover happened.

To pick up capability with the ordinary development plans in a jam-pressed scene, Jiang et al. [1] proposed that specific observation-based strategies utilize far-reaching properties of the stage, for instance, developments in close by spatio-short-lived cuboid. Using both the short and thick collection go directly to the following features and support system, which is to be separated, for example, thickness, speed, and flow. Horn, B.K.P. et al. [2] proposed a plan to select the optical stream. The development of space's optical properties in the stream is usually made between two successive edges of the field in the thick of the speed. Given the mass of video, data, video, and more video clips, Lucas B.D. [3] proposed change came as a series of frameworks using the pixel-wise between the edges of the stream of optical processing [4].

9.2.2 *Microscopic Modeling*

The minute show depends upon the assessment of video bearings of the moving components. This approach contains the going with propels: (1) Detection of the moving targets present in the scene (2) Continuous after the distinguished marks and (3) Analysis of the bearings to perceive winning streams and show ordinary development structures [5]. The multifaceted idea of following estimations depends upon the situation and condition where coming up next is performed. Concerning swarm video examination, following individuals inside a gathering leads into extra multifaceted nature in light of the correspondences and hindrances between people in the community if methods have been proposed to overcome the scene's challenges [6].

9.2.3 *Tracking of Pedestrians in Video Sequences*

We address the issue of location and following of people on foot in complex situations [7]. The incorporation of earlier information is increasingly more essential in scene examination to ensure adaptability and power, necessary to have unwavering quality in complicated settings. We expect to join picture preparing strategies with conduct models of the person on foot elements, aligned on truthful information. We present Discrete Choice Models (DCM) for the person on foot conduct, and we talk about their mix in discovery and the following setting. The acquired outcomes demonstrate how it is conceivable to join the two procedures to improve such frameworks' exhibitions in complex successions [8].

9.3 Proposed Methodology

9.3.1 *Control Flow Model*

Different stages for unusual activity revelation are preprocessing, following articles, and conduct acknowledgment in Fig. 9.2.

Strides for emphasize extraction. A Part of the analysts has been suggested Methods for preprocessing have been looked at in Fig. 9.3. Diagnosis is among those significant to research and gain competence with stranger's illustration [9]. Neural networks are very deep in the past few years (DNA) based object detectors have been proposed [10]. This research examines the performance of the SSD and the fastest RCNN's sophisticated DNN model, which is applied to the problem of classical detection, algorithms to determine where the animals were trained in a number of films; An example of the application of scientific research, rats were trained in tracking network to solve the problem YOLO.

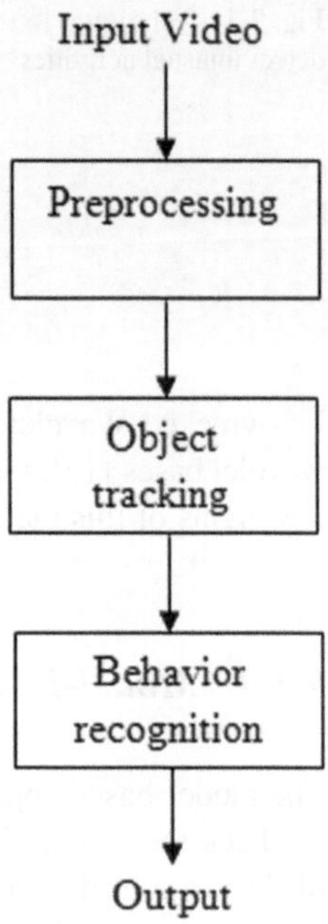

Fig. 9.2 Flow of crowd behavior process

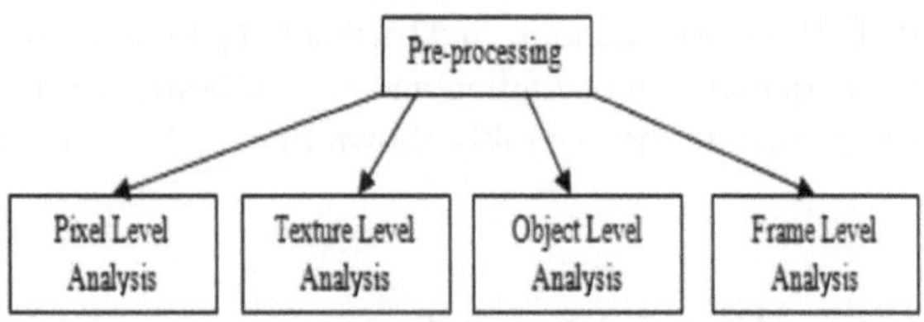

Fig. 9.3 Preprocessing techniques

9.3.2 Area-Wise Approach

It collects structural visualization PC scenarios; for example, thickness; Bearing, and the optical stream's speed are separated using a method [11]. It will be limited to changes in the shadows and the light of the establishment. Organizing the worst cardinality significant weight histogram is carried out using a lot of hiding, in contrast to the scene that cannot prove a substantial number of people [12].

9.3.3 Active-Counter-Based-Approach

Burdens of zone-based systems can comprehend by using the dynamic structure-based technique [13]. This technique is chiefly used when the scene has been defiled as a result of disturbance. For instance, systems used are chi-square, "Bhattacharyya,"

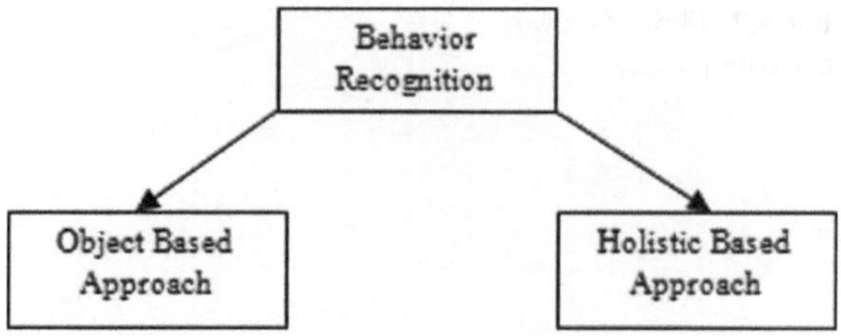

Fig. 9.4 Techniques to detect unusual activities

Un-wrecked Wavelet Packet Transform (UWVT), EM figuring, and bi-orthogonal wavelet bases [14]. Unfit to manage midway hindrance in the video course of action is weights of this methodology [15–17].

9.3.4 Model-Based-Approach

The model-based approach to the development of models of the scene and the scene to check the standard type [18–20]. Isolating the set and that the express rejection of the scale of the changes in the light zones. If anything, the problem is that after the changes are significant, then each pixel stimulates the current guide. GMM is used in the Bayesian course of action, EM estimation, correlations Topic Model (CTM), Sean Codebook, and tracker [21]. Finding the practices of the collection and its unique design or finding an array of bottles of the article is predicted. To reveal imbalances in the assembly shown in Figs. 9.4 and 9.5 there are two systems.

9.3.5 Direct-Approach

A group is dissected by gathering individuals to gauge the speeds, bearing, and irregular movement. The multifaceted nature happens when the impediment exists and that perhaps influenced the investigation procedure, for example, identifying the article, following directions, and perceiving exercises in a thick group [2].

9.3.6 Indirect-Approach

A tricky scene is investigated by getting a solitary element to gauge the speeds, bearing, and strange movement. The test is the view of the component of medium to high thickness. All the necessary analyzes can be used away from the set of symmetric housedress significantly, as shown in Fig. 9.6.

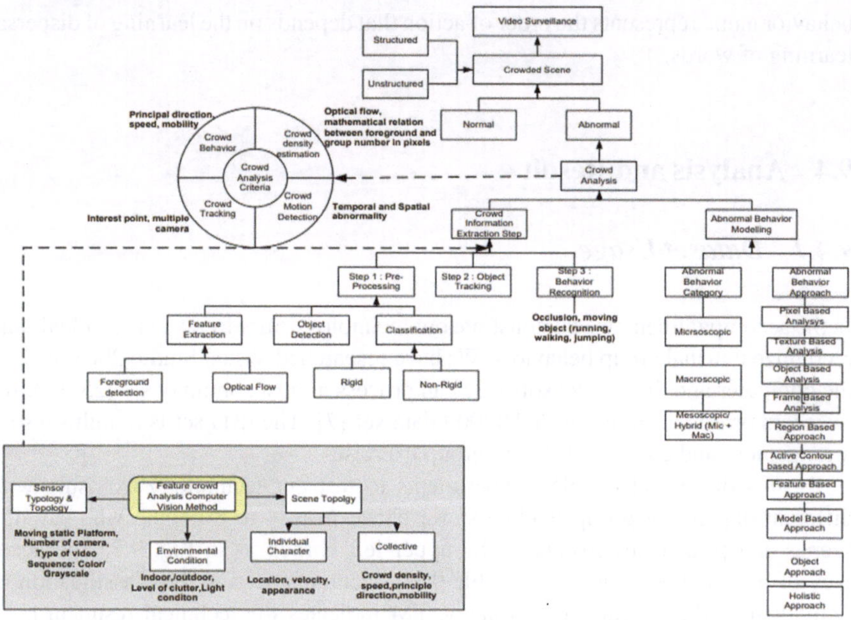

Fig. 9.5 Proposed model architecture

Fig. 9.6 Crowd specific behaviors

9.3.7 Behavior-Label-Distribution Algorithm

When irregular behavior is identified, the world generally faces mixing practices when some alien group practices are related to other actions. These informal practices co-occur. The models, for example, make the battle wobble or become frantic and worsen impulsivity. Such activity is called irregular mixing, in which only one behavior is distinguished, and the other is ignored. To dissect these practices better, name diffusion learning is used to distinguish between abnormal group behaviors. First, each action provides a sequential number. Each behavior group is related to some behavioral signs, and this behavior represents the level of activity after each behavior name represents the order of action that depends on the learning of dispersal learning of words.

9.4 Analysis and Result

9.4.1 Dataset Usage

In the accompaniment, we demonstrate our technology offer to select individuals and recognize unusual group behaviors. We have researched Visual Studio 2008 in C++, the computer and the project software can process almost continuously (25 fps). All tweezers originate from the PETS2009 data set [7]. The data set is a multisensory arrangement and contains different group exercises.

Results of several people's experiments to test our technology exposure with different degrees of group thickness; we chose from 4 to 8 people who saw the scenes and printed the results in the upper left corner, as in Fig. 9.7. The accuracy of our method appeared in Table 9.1—accurate speed in the classification of individuals. OM indicates our strategy. TM indicates the technical result in [22]. The M(N) characteristic of the baseline means that M individuals are in real conditions and remain in video images of N limits. In these tests, we use video from the PETS2009\S2\L1\Time12-34 [7] data set, and we set H = 170. Our strategy can obtain a higher resolution rate when the number of individuals is 4, 6, and 7 of the technique in [23]

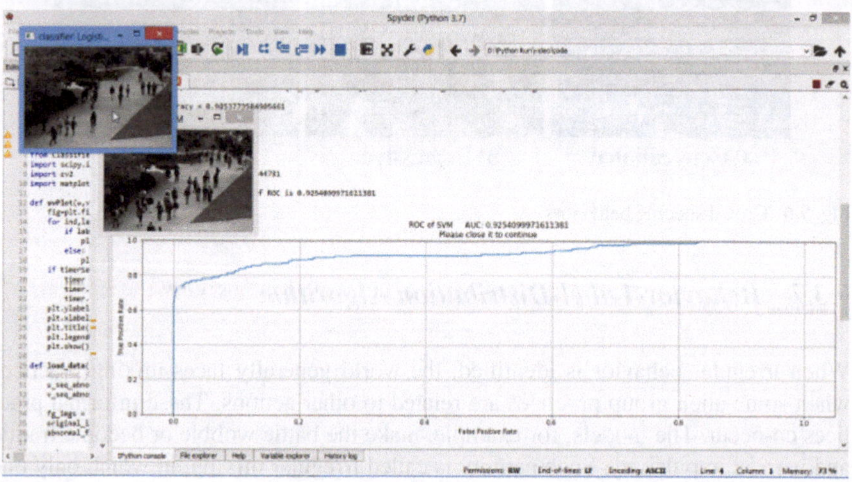

Fig. 9.7 Video clarifier using logistic

Table 9.1 The comparison of our method with method in [22] for people counting on the Pests2009 dataset

ER	4(32)		6(148)		7(60)		8(97)	
	OM (%)	TM (%)	OM (%)	TM (%)	OM (%)	TM (%)	OM (%)	TM (%)
4	72.3	12.3						
5	21.2	81.8	26.8	27.5				
6			58.4	27.5	16.4			
7			12.8	9.4	65.6	13.1	27.0	19.4
8					14.8	13.8	55.1	64.3
9							12.2	13.3

9.4.2 Results of Abnormal Activity Detection

In our research, we PETS2009 three sets of four tests conducted in a social event and a blast. Figure 9.8 shows the test results of the three groups. The views of the social contexts in different lighting conditions, isolated cases begin and end. Curve CD is well-suited to the assignment. At the point where people gradually entered the scene, the Curve CD rating becomes higher. At this point, individuals remain silent

Fig. 9.8 **a** Four people in the scene, **b** six people in the background, **c** seven people in the environment, **d** eight people in the ground

step by step in this scene, and in particular, the curved estimate becomes smaller at this point. In the next minute, as people begin to differentiate, the value of Curve CD increases rapidly and clearly.

9.5 Conclusion and Future Enhancement

This document presents a robust strategy to identify irregular supervised group behaviors. The medium-dark channel is used to reduce error detection speed and hint at a dual base model for a concrete frontal area. We can obtain precise effects for individual screening using the proposed potential vitality model. In the light of personal verification, a persuasive component swarm allocation curve is presented to visualize the group transfer and help identify anomalous behaviors. A reliable method to reap the results of our experiments shows little caution can prove false.

9.6 Future Work

For future work, different irregular group behavior is organized in battles, loading, etc. This increasingly specific characterization will allow professionals to manage conditions management gradually. Consequently, massive damage to human life can be expected using this framework.

References

1. Weidmann, U. (1993). Transporttechnik der Fussgänger. Technical Report 90, ETH Zürich.
2. Cem, D., Melike, S., & Noel, E. (2017). Abnormal crowd behavior detection using novel optical flow-based features.
3. Hiba, H., & Sreela, S. (2013). Detection of abnormal behavior in dynamic crowded gatherings.
4. Mehran, R., Oyama, A., & Shah, M.: *Abnormal crowd behavior detection using social force model.*
5. Helbing, D., & Molnar, P. (1995). Social force model for pedestrian dynamics. *Physical Review E, 51*, 4282.
6. Ali, S., & Shah, M. (2007). A lagrangian particle dynamics approach for crowd flow segmentation and stability analysis. In *IEEE conference on computer vision and pattern recognition. CVPR '07* (pp. 1–6).
7. David, Y. N., Blei, M., & Jordan, M. I. (2003). Latent dirichlet allocation. *Journal of Machine Learning Research, 3*, 993–1022.
8. Jiang, Y., Zhang, P., Wong, S., & Liu, R. (2010). A higher-order macroscopic model for pedestrian flows. *Physica A: Statistical Mechanics and its Applications, 389*(21), 4623–4635.
9. Horn, B.K.P., & Schunck, B.G. (2011). Determining optical flow. *Artificial Intelligence, 17*, 185–203.
10. Lucas, B. D., & Kanade, T. (2002). An iterative image registration technique with an application to stereo vision. In *Proceedings of image understanding workshop* (pp. 121–130).

11. Kratz, L., & Nishino, K. (2009). Anomaly detection in extremely crowded scenes using spatio-temporal motion pattern models. In *Proceedings of the conference on computer vision and pattern recognition* (pp. 1446–1453).
12. Mahadevan, V., Li, W., Bhalodia, V., & Vasconcelos, N. (2010). Anomaly detection in crowded scenes. In *Proceedings of IEEE international conference on computer vision and pattern recognition* (pp. 1975–1981).
13. Reddy, V., Sanderson, C., & Lovell, B.C. (2011). Improved anomaly detection in crowded scenes via cell-based analysis of foreground speed, size and texture. In *2011 IEEE computer society conference on computer vision and pattern recognition workshops (CVPRW)* (pp. 55–61). IEEE.
14. Zhang, T., Ghanem, B., Liu, S., & Ahuja, N. (2012). Robust visual tracking via multitask sparse learning. In *2012 IEEE conference on computer vision and pattern recognition (CVPR)* (pp. 2042–2049). IEEE.
15. Chang, M.-C., Krahnstoever, N., & Wingate. (2011). Probabilistic group-level motion analysis and scenario recognition. In *2011 IEEE International conference on computer vision (ICCV)* (pp. 747–754). IEEE.
16. Obaid, A.J. (2021). Critical research on the novel progressive, JOKER an opportunistic routing protocol technology for enhancing the network performance for multimedia communications. In *Research in intelligent and computing in engineering. Advances in intelligent systems and computing* (vol. 1254, pp. 369–378). Springer.
17. Sharma, S., & Obaid, A. J. (2020). Mathematical modelling, analysis and design of fuzzy logic controller for the control of ventilation systems using MATLAB fuzzy logic toolbox. *Journal of Interdisciplinary Mathematics, 23*(4), 843–849.
18. Sharma, S., & Obaid, A. J. (2020). Contact-mechanics and dynamics analysis of three-different ellipsoidal raceway geometries for deep Groove ball bearing using Abaqus 6.13 version FEA simulation for high load-bearing as well as speed-rotating applications. *International Research Journal of Multidisciplinary Science and Technology, 3*(5), 36–43.
19. Obaid, A. J., & Sharma, S. (2020). Recent trends and development of heuristic artificial intelligence approach in mechanical system and engineering product design. *Saudi Journal of Engineering and Technology, 5*(2), 86–93.
20. Alkaam, N. O., Mohammed, D. M. Q., & Obaid, A.J. (2018). A hybrid technique for object detection and recognition using local features algorithms. *Journal of Advance Research in Dynamical and Control Systems, 10*(2), 2330–2343.
21. Krausz, B., & Bauckhage, C. (2011) Analyzing pedestrian behavior in crowds for automatic detection of congestions. In: *2011 IEEE international conference on computer vision workshops.*
22. Lucas, B.D., & Kanade, T. (1981). An iterative image registration technique with an application to stereo vision. In *International joint conference on artificial intelligence* (pp. 674–679).
23. Ahad, M. A. R., Tan, J. K., Kim, H., & Ishikawa, S. (2012). Motion history image: Its variants and applications. *Machine Vision and Applications, 23*, 255–281.

Chapter 10
Implementation of Robust Privacy-Preserving Machine Learning with Intrusion Detection and Cybersecurity Protection Mechanism

R. Ravikanth and T. Prem Jacob

Abstract In this work, data mining, and cloud computing problems such as privacy protection, Intrusion detection, identification system have been realizing. Social-Cybersecurity is an evolving science-focused field for characterizing, interpreting, and predicting improvements in human behavior, social, cultural, political outcomes, and development. The cyber-infrastructure is required to sustain its critical existence cyber-mediated knowledge climate under shifting circumstances and cyber threats that are immediate or imminent. When the data moves from the local cloud to another cloud, some security issues automatically arise. At this stage, an advanced robust security system with advanced encryption or decryption algorithm is necessary. At the same time, intrusion may attack the cloud for hacking or modifying the existing information. There an advanced deep learning algorithm is essential to make the cloud efficient. The Logistic net regression Optimization is proposed for cybersecurity and protection. A final performance measures, estimating accuracy,0.952 sensitivity of 0.39, Recall, F1 score 0.54 and processing time.

Keywords Logistic net regression · Auto stack FCNN · AES · Cybersecurity · Machine learning

10.1 Introduction

The deep learning techniques offer many applications in different areas like cloud security, big data, cloud computing, image recognition pattern generation, and network security. Due to deep learning advancements, fast and complex problem solving and unstructured data analysis can be possible through deep learning interaction. Moreover, less economy and high performance of deep learning mechanism realize the cybersecurity operations. In deep learning, many limitations are crossover the outperformance and giving a less accurate decision. Therefore an advanced deep learning technology is required to make the cybersecurity decision accurate. The

R. Ravikanth (✉) · T. P. Jacob
CSE, Sathyabama Institute of Science and Technology, Chennai, India

P. Kumar et al. (eds.), *A Fusion of Artificial Intelligence and Internet of Things for Emerging Cyber Systems*, Intelligent Systems Reference Library 210,
https://doi.org/10.1007/978-3-030-76653-5_10

proposed model should decrease the misprediction rate at internal or external malicious effects. These limitations very sensitive and creates some complexity in real-time applications. So, the proposed deep learning model should robust Security and secrecy, and exposure at any types of risks. In this section, many conventional and modern privacy security and cybersecurity schemes are discussed briefly and challenges the problem statement.

Deep learning is often called hierarchical learning and deep-structured learning, and it consists of machine learning methods that are controlled or unsupervised. The Deep Learning definition stems from the human brain's configuration and functionality and the transmission of signals in the human mind by neurons. The advantages of artificial neural networks are now taken up by deep learning because it comprises an input, output, and several unknown layers. Based on the input layer's data, each layer of Deep Learning depends on the nonlinear response. The Deep Learning methodology has been mainly used in cloud security and cybersecurity for the last few years, visual recognition, exploration of the thing, and so many other fields, such as detecting attacks and intrusions [1]. Via a backpropagation method, Deep Learning built a mechanism towards interact by large data sets to illustrate how the system adjusts its core parameters to determine the representation in each layer of rendering in the previous layer [2].

Effective Deep Neural Networks may make a minimal difference amid training also test appearance, even though their immense complexity. The error of limited circularization is traced by conventional wisdom to the family's standard features or the management strategies used during training [3]. The encrypted knowledge that flows as of training then interface modules is DL's main challenge. Due to largely implemented DL models, protection and privacy concerns are very relevant in many applications, by way of described above. Also, Deep Learning, which prevails in all training component models, depends on many user big data, sensitive & confidential data, and training data. Holding this in mind, DL models do not reveal personal and confidential details. The comprehensive work reviewed in this paper was carried out on the challenges to deep learning stability, privacy risks to private data, then their corresponding established defensive techniques. Most protected strategies utilizing cryptographic primitives without the third party's indulgence and the description of possible problems and opportunities were also included in the e document. Many standard data analytics are done on the cloud; its data privacy issue has drawn many research interests. One of the strategies is keeping databases private and safe. A third party runs the cloud servers and cannot be fully trusted by users.

Consequently, it is a task to conduct privacy-preserving machine learning over cloud data from numerous service providers. The proposed privacy approach retains machine learning under multiple keys (PMLM) to solve this problem. Because stable multi-party computation (SMC) only supports computing on data protected under the same public key, it is essential to increase computing efficiency and accuracy. It requires time-consuming encryption and decryption operations on plain text to achieve privacy preservation in machine learning techniques. Results in the sluggish running speed of algorithms for machine learning and lowers precision as well. For several machine learning algorithms, data mining apps on arbitrarily partitioned data is a daunting job. We plan to say that both the implementation and principal domains in privacy-preserving machine learning can help our PMLM scheme. Privacy preservation in data mining is a relatively recent area. Privacy problems would also rise with the ongoing growth of e-data (digital data), and not enough work has been performed in this sector. As digital data grows, it can impose new security challenges that need to be resolved through ongoing PPDM (Privacy-Preserving Data Mining) study. There is an additional opportunity to incorporate efforts needed to minimize overheads for encryption operations and improve performance. And we can also recommend developing a different neural network to establish cloud protection by utilizing several keys in the Homo-morphic Encryption method, which is far more effective and has good efficiency even with large numbers of parties and can be extended to any Protected Multi-Party Computing assessment feature with adequate protection against internal and external competitors of dishonest majority [49].

10.2 Literature Survey

For high-dimensional operators kept by premium consumers, a stable selection protocol was implemented. In a centralized education through which users retain their libraries and types, these protocols can be used. Through safely gathering the User's learning notifications, the core system respects the intellectual knowledge model. The approach is based on the code's secret exchange then remains victorious against clients who relinquish the convention whenever. By applying ML to the subject of clinical, instructive, monetary, or different types of touchy information, CryptoNets demands detailed forecasts and special consideration to keep them safe and secure. By implementing leveled Homomorphic Encryption (LHC), CryptoNets is effectively created by the Microsoft Research community. The authors indicated that the activation functions be addressed using multiple-degree polynomials because of nonlinear activation functions that cannot be accomplished using LHE. Therefore, the neural network must be retrained to preserve reasonable prediction accuracy in plain text for the same activation feature. Another drawback of this method is that the number of serial multipliers placed by LHE is somewhat reduced, rendering the solution prohibitive. In comparison, to reach a higher degree of anonymity, Crypto Nets requires an anonymity/utility swap, and precision within the same computational resources must be diminished [4].

Table 10.1 clearly explains the advantages and disadvantages of earlier cyber-attack detection and detection techniques. This causative to assault based attack cannot solve the false-negative schemes' exploratory sector model. Overcoming the above method and giving the discriminator misclassification process. Coming to Thor attack, it is an honesty attack that is used by use capital modeling. The breach of privacy e and targeting are the two types of different attacks. These are giving quick training and testing database classification, but the time taking process is more [5, 6]. Therefore the above discussing literature methods facing different types of limitations and disadvantages. An advanced machine learning and deep learning-based

Table 10.1 Earlier models and counter techniques

S no	Attacking	Advantage	Disadvantage	Counter technique
1	Causative Assault [2]	Influence on training data and misclassifications of vulnerabilities	Consuming hours	[3–6]
2	Exploratory Sector [3]	Modification of discriminate outcomes Misclassifies favourable study	Not suitable for broad datasets	[6, 7]
3	An attack in Honesty [4]	The false-negative goes through the scheme	Use Capital	[5–8]
4	Disposability [5]	In the blocking of documents, false-positive findings	Detected quickly	[7–9]
5	Breach of Privacy [6]	Exploit the training dataset quickly	Period and usage of energy	[8–12]
6	Targeting [7]	For every arbitrary class misclassified	Its performance is not accurate since it is dependent on iterations	[9–14]
7	Assault indiscriminate [8]	Extremely effective decent trade-of	Excitation is elevated	[12–15]

cybersecurity e detection system are compulsory to design [7]. This DL represents the Deep learning model; it is an advanced mechanism in many data mining fields.

Table 10.2 clearly explains various cybersecurity attacks. Here testing or training database techniques and Taxonomy parameters are analyzed through Deep learning models.

10.2.1 Malware Deduction

Clients search for any connection strikingly. Not all organization traffic information produced by malignant applications relate to pernicious traffic [8]. Numerous malware appears as repackaged kind applications; accordingly, malware can likewise contain the essential elements of a favorable application [9, 10]. Therefore, the organization traffic they create can be described by blended favorable and vindictive organization traffic. We analyze the traffic stream header utilizing the N-gram strategy from the regular language preparing (NLP).

Table 10.2 Comparisons of attacks with DL

S no	Attack	Technique	Training/ testing	Taxonomy
1	Attack	Flips with Adversarial Mark	Train	Confidentiality, efficiency, and Honesty
2	Attack	Enchanting Beautiful	Train	Attack aimed
3	Attack	Obscurement	Train	Attack aimed
4	Attack	Toxicity	Train	Assault indiscriminate
5	Attack	Impersonating	Train	Boosts data analysis and prevention of intrusions
6	Defense	Education adversarial	Train	Boosts data analysis and prevention of intrusions
7	Defense	Distillation of Protection	Train	Boosts data analysis and prevention of intrusions
8	Defense	Form of Ensemble	Train	Protects the protection of knowledge
9	Defense	Privacy Difference	Training and testing	To guarantee secrecy, it ensures the safety of data base
10	Defense	Encryption of Homomorphs	Training & testing	To guarantee secrecy, it ensures the safety of data base

10.2.2 Junk Code Insertion Attacks:

Garbage code infusion assault is a malware hostile to legal procedure in contradiction of Attack type investigation. Name proposes, garbage code addition may incorporate the expansion of amiable Attack type groupings, which don't route in malware or consider directions (for example, NOP) that don't affect malware exercises. Garbage code inclusion procedure is commonly intended to jumble vindictive Attack type successions and decrease the 'extent' of malevolent Attack types in malware. EIGEN VECTOR (Matrix) will be created with all available malware features called a training model [11].

10.2.3 Feasibility Study

Practicality study is gone with once the troublesome is perceived. The possibility study is an extraordinary level tablet form of the entire framework examination and plan system [12]. The free is to characterize whether the arranged framework is conceivable or not. It benefits us to minimal cost of determining the issue and overseeing if the problem is riches settling [13].

10.2.4 Technical Feasibility

In the particular feasibility study, one needs to assess whether the completed system can be developed using existing advancement or not [14]. It is relied upon to execute the completed system in JSP [15, 16]. The endeavour enabled is speculatively feasible since going with reasons.

- All required advancement exists to improve the system.
- The existing system is malleable to the point that it might be advanced further.

10.2.5 Budgetary Feasibility

As an aspect of this, the expenses and advantages related to the realized structures are to be connected [17]. The undertaking is mindfully functional, just if undeniable and irrelevant helps balance the cost. I can say the completed structure is possible to set up on the going with grounds [18].

- The energize of building the filled system is sensible.
- The cost of hardware and programming for the application is less.

10.2.6 Operational Feasibility

This endeavour is operationally viable, for there is central assistance from the endeavour affiliation and the customers of the executed structure. Implemented system thoroughly doesn't mischief and confirmation not make the savage results and no troublesome will move after use of the system [19, 20].

10.2.7 Easy to Understand

Customers will use the structures for their various trades, for instance, for including new courses and seeing the course's nuances [21]. Moreover, the Customer needs the reports to see the various trades subject to the necessities. These structures and reports are delivered as simple to use to the Client [22, 23].

10.2.8 Unflinching Quality

The pack will get current trades online. Concerning old customs, the User will enter them into the system.

10.2.9 Security

The web labourer and database specialist should be protected from hacking, disease, etc., it is an opportunity from potential harm achieved by others [24]. Beneficiaries of security may be of individuals and parties, things and foundations, natural frameworks, or some other component or wonder feeble against unfortunate change by its condition [25, 26].

10.2.9.1 Accessibility

This product will be accessible consistently. High accessibility programming is programming used to guarantee that frameworks are running and accessible more often than not. High accessibility is a high level of time that the framework is working [27, 28]. It very well may be officially characterized as below.

Algorithms:

1. Logistic net regression
2. Auto stack FCNN
3. AES (adaptive advanced encryption)
   ****** Auto stack FCNN— deep learning**
   ******* Logistic regression — machine learning**.

10.3 Methodology

The Fig. 10.1 block diagram clearly explains the logistic net regression-based machine learning model and FCNN deep learning model used to attend the cybersecurity issue.

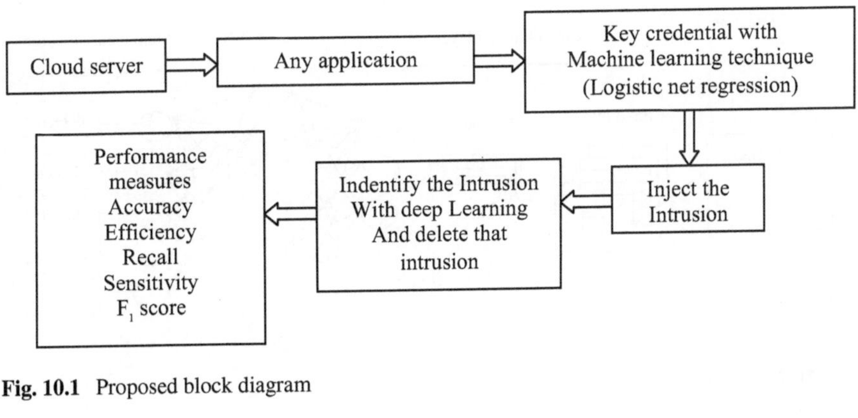

Fig. 10.1 Proposed block diagram

Algorithm: logistic net regression
Step 1: weight determination
Step 2: weight estimation
Step 3: Standard weight calculation
Step 4: Condition for weight
Step 5: Perform the training
Step 6: Stop the process

$$\gamma_{im} = -\left[\frac{\partial L(y_i, F(x_i))}{\partial F}\right]_{F(x)=F_{m-1}(x)} \quad \text{for i} = 1, \ldots, \text{n} \tag{10.1}$$

Algorithm: 03 FCNN
Step: 1 input dataset and real time images
Step: 2 decision tree adjustment
Step: 3 classifications
Step: 4 predictions
Step: 5 output measures

Also, customizing privacy and protection policies for DNNs is a fascinating and open field of study to create a feasible solution [29]. In the implementation of Deep Learning, authors are also intended to conduct their study, especially in the fields of cloud, cybersecurity Fig. 10.2.

Example of Eigenvector.

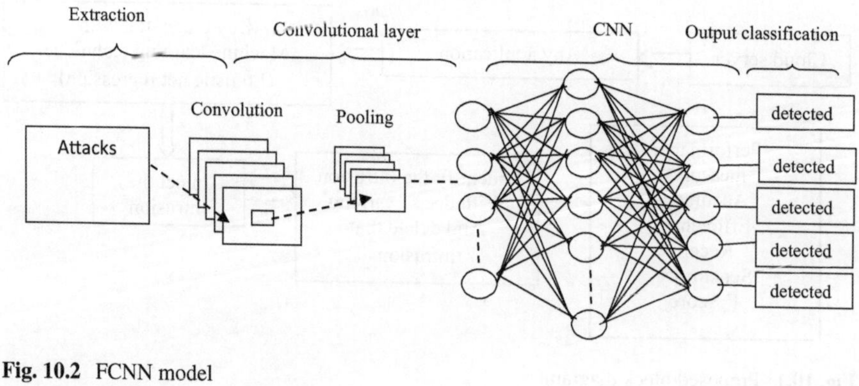

Fig. 10.2 FCNN model

	Attack type1	attack type2	attack type3	attack type4
Device1	e1a0c00d	e1a0c00e	e1a0c00f	e1a0c00a
Device2	e1a0c00d	e1a0c00e	e1a0c00a	e1a0c00b

Above is the ATTACK TYPES (Will understand by computers only) dataset.

To build matrix first all unique ATTACK TYPES will come in matrix header and their count will go matrix values.

Example of train dataset.

```
e1a0c00d  e1a0c00e  e1a0c00f  e1a0c00a  e1a0c00b  //unique ATTACK TYPES
1         1         1         1         0
1         1         0         1         1
```

For above devices matrix will form base on number of ATTACK TYPES they executed and both matrix will get multiply to get EIGEN values. Suppose if above matrix contains malware features and if test data also contains above features then that test data will be predicted as malware by checking closer distance between train and test data using EUCLIDEAN DISTANCE.

Above ATTACK TYPES took from dataset, below are some dataset data.

```
Disassembly of section.init:
000080b4 < _init > :
80b4:        e1a0c00d   mov     ip, sp
80b8:        e92ddff0   push    {r4, r5, r6, r7, r8, r9, sl, fp, ip, lr, pc}
80bc:        e24cb004   sub     fp, ip, #4
80c0:        e91baff0   ldmdb   fp, {r4, r5, r6, r7, r8, r9, sl, fp, sp, pc}
```

Data in bold is ATTACK TYPE and mov (move) is the meaning of that code And sub means subtraction and push mean push data.

It is quite challenging for the writers to precisely describe the techniques' efficiency due to various datasets and measurements after learning the above techniques. It is necessary to say that these techniques/methods differ in efficiency across protection areas. The computer protection domain has a wide variety of data gathered to implement Deep Learning experiments across multiple outlets. Since many databases are not freely accessible, e studies/studies may not be performed and produce reliable findings. The bulk of the references for the dataset are limited and obsolete. It is essential to validate the system on massive, modified, and consistent datasets to establish a protection approach by way of a meaningful system. It is necessary to equate the outcomes of the strategies with each other via real-time scenarios shown in Figs. 10.3 and 10.4

Click on the 'upload dataset' button and upload the dataset folder to get below screen.

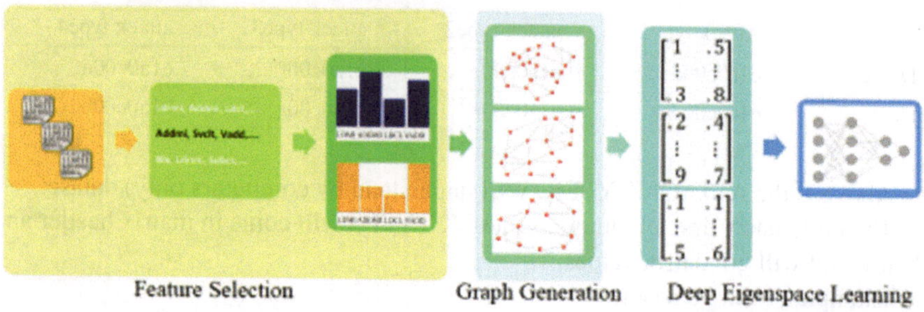

Fig. 10.3 LNR-FCNN models

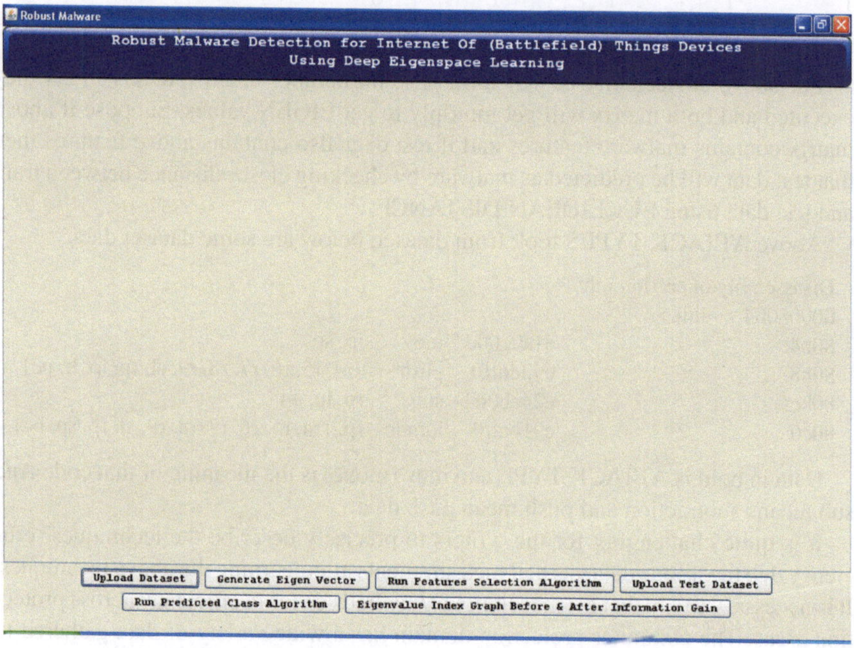

Fig. 10.4 Robust malware attack detection

These implementations have rudimentary cryptographic elements and various applications. It should be remembered that there is no full capacity for private intrusion systems to provide protection and privacy for DNNs Fig. 10.5.

We define the specifics of various kinds of security attacks on Deep Learning [30, 31]. Several forms of aggression invested in taking advantage of Deep Learning outcomes. Model details retrieved, or awareness of training data such as sample inversion, sample extraction, and membership inference can be acquired [32].

Now click on the 'Generate Eigen Vector' button to generate a matrix.

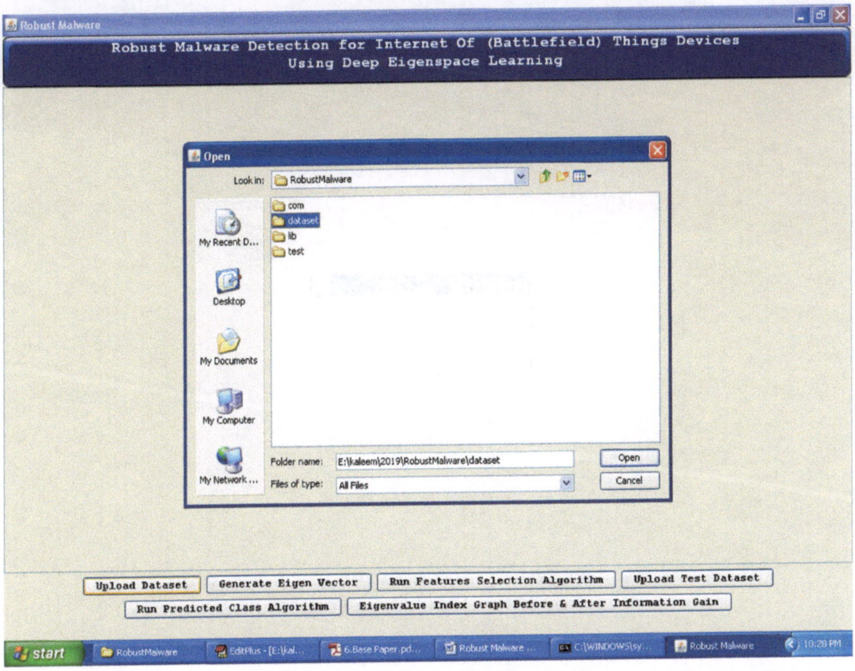

Fig. 10.5 Robust malware attack detection on IoT cloud

Which attacks steel training data and produces anticipated outcomes? Compared to the interface, Deep Learning's private training portion has more processing overhead. Therefore, in this direction, further focus and analysis are needed to build a more effective solution for data privacy protection while preserving models. Above is the Eigen matrix, and all ATTACK TYPES are in the column header, and their values are in column [33]. Now click on the 'Run Features Selection Algorithm' button to remove junk or useless features from the dataset using INFORMATION GAIN ALGORITHM. By applying this algorithm, we can reduce the attribute size of the dataset, which can increase system performance shown in Figs. 10.6 and 10.7.

We can see a total of 184 attributes found in the above screen, and after applying the feature selection algorithm, it drops down to 177. Now click on the 'Upload Test Dataset' button to upload the test files and predict its class as malware or benign (standard). Due to the different features of Deep Neural Networks, which depend on many input training data, privacy concerns often remain in Fig. 10.8 [34, 35].

Now click on the 'Run Predicted Class Algorithm' button to predict the uploaded test dataset class.

We have addressed potential privacy risks to sensitive and confidential data from the Deep Learning model in this segment [36, 37]. Using Deep Learning, numerous experiments have been performed on privacy retention assaults [38, 39]. It is important for researchers to extensively explore different rudimentary cryptographic solutions for DNNs for future studies [40, 41]. The computational overhead for encryption

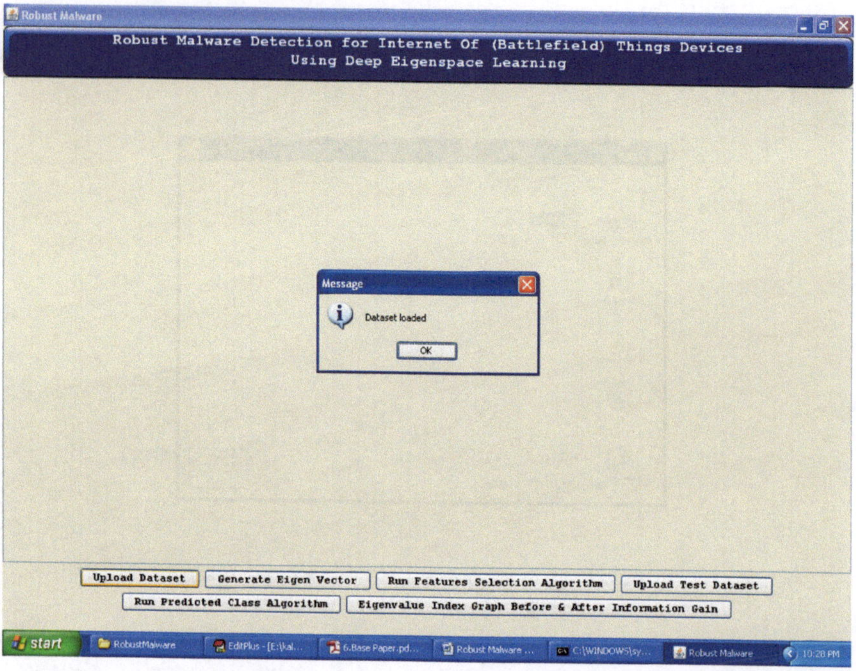

Fig. 10.6 Detections of malware

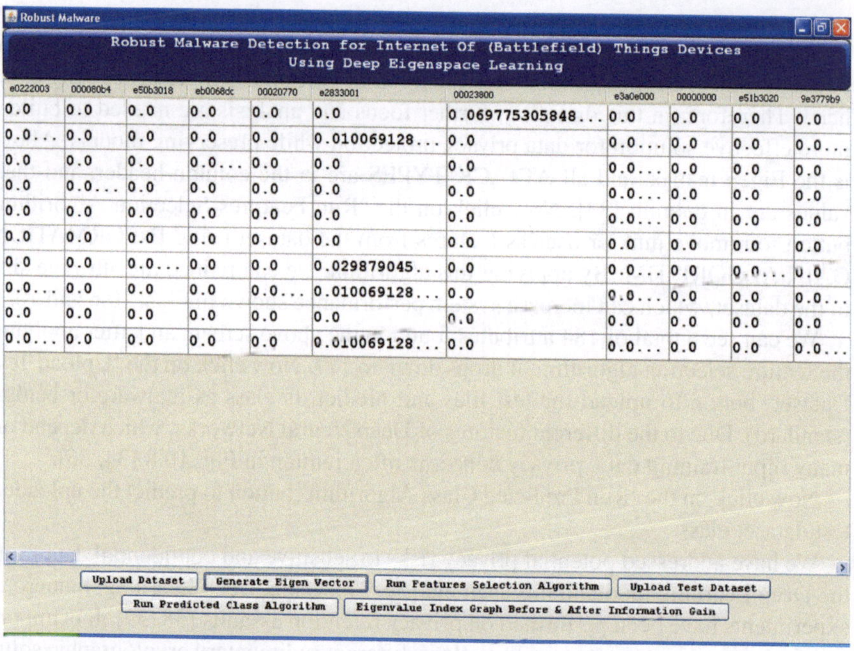

Fig. 10.7 Feature extractions

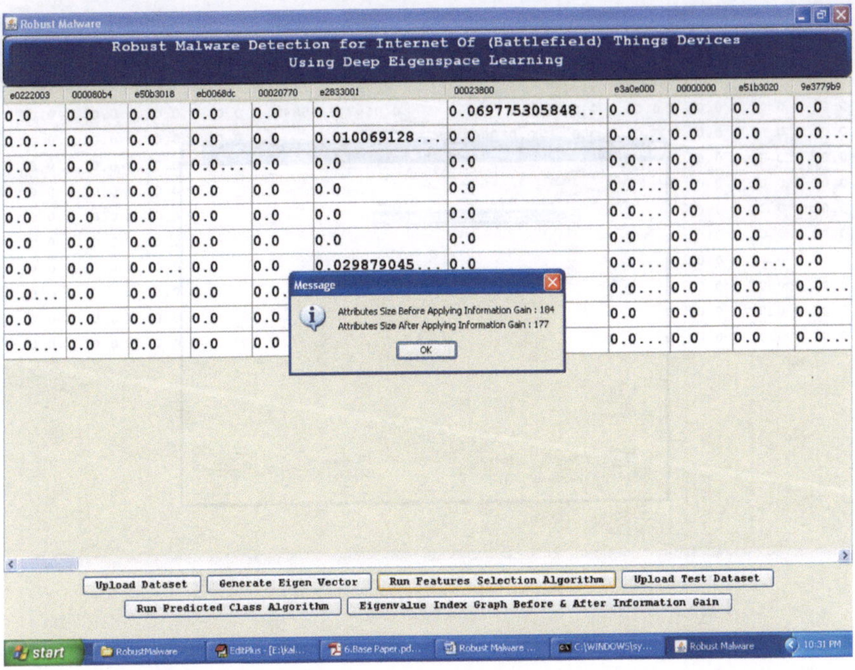

Fig. 10.8 Training analysis

and privacy protection solutions can be minimized by a mixed protocol technique [42, 43]. Also, customizing privacy and protection policies for DNNs is a fascinating and open field of study for creating a feasible solution shown in Figs. 10.9, 10.10 and 10.11.

In the above screen application saying the predicted class for the uploaded test file is malware. Now click on.

We are getting the distance of similarity between the uploaded test file and the training file [44–48]. Now click on 'Eigenvalue Index Graph' to see an index of Eigen before and after the feature selection algorithm is shown in Fig. 10.12.

In the above graph, the x-axis represents the technique name, and the y-axis represents the count of eigenvalues shown in Fig. 10.13.

Table 10.3 and Fig. 10.14 explains the comparison of results with the proposed method.

10.4 Conclusion

This research work is mainly focusing on cybersecurity issues and protection schemes. Here, various cybersecurity models are discussed, and their problems are shootout in the literature survey. With this identified that and accurate cybersecurity

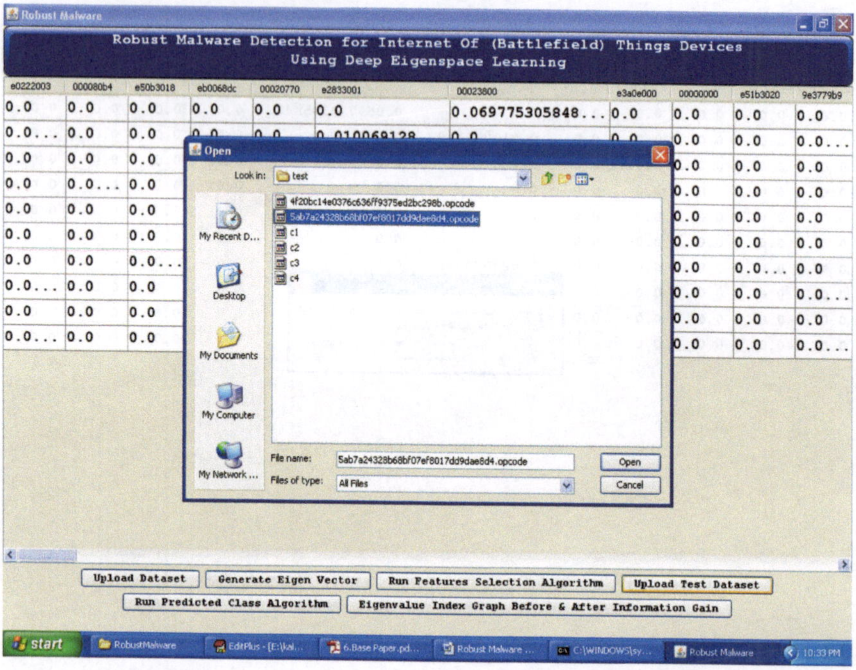

Fig. 10.9 Test analyses

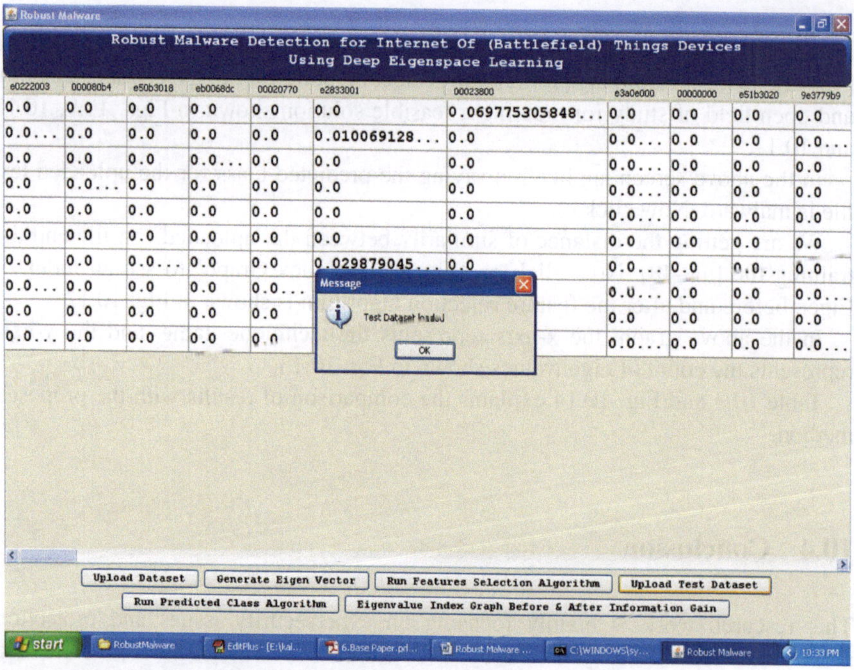

Fig. 10.10 Test dataset analyses

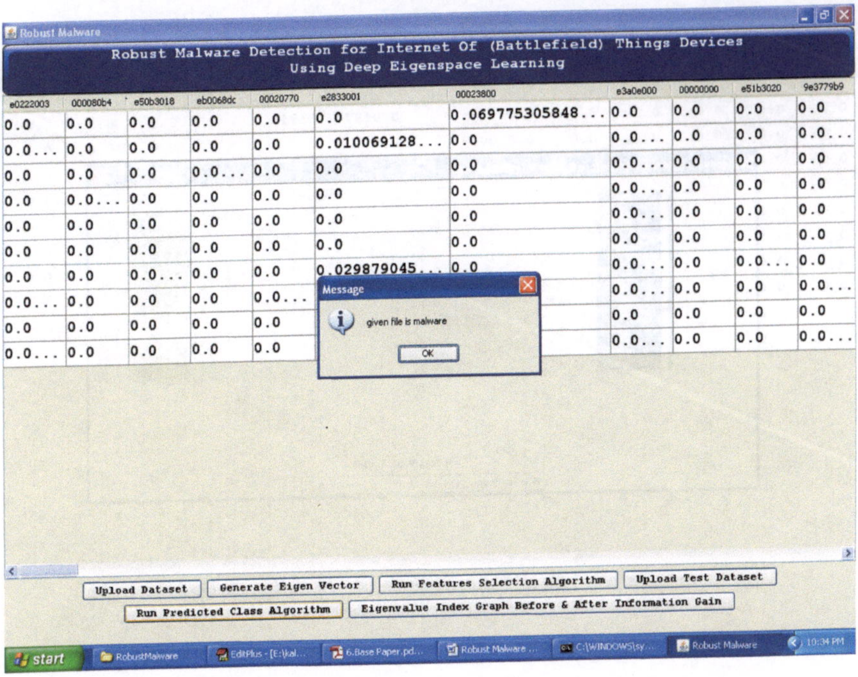

Fig. 10.11 Malware detection

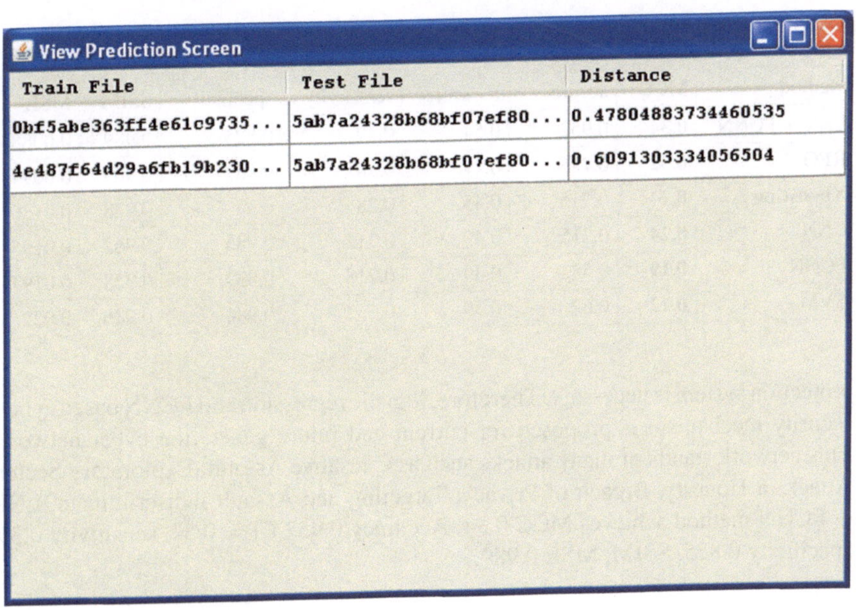

Fig. 10.12 Testing detection

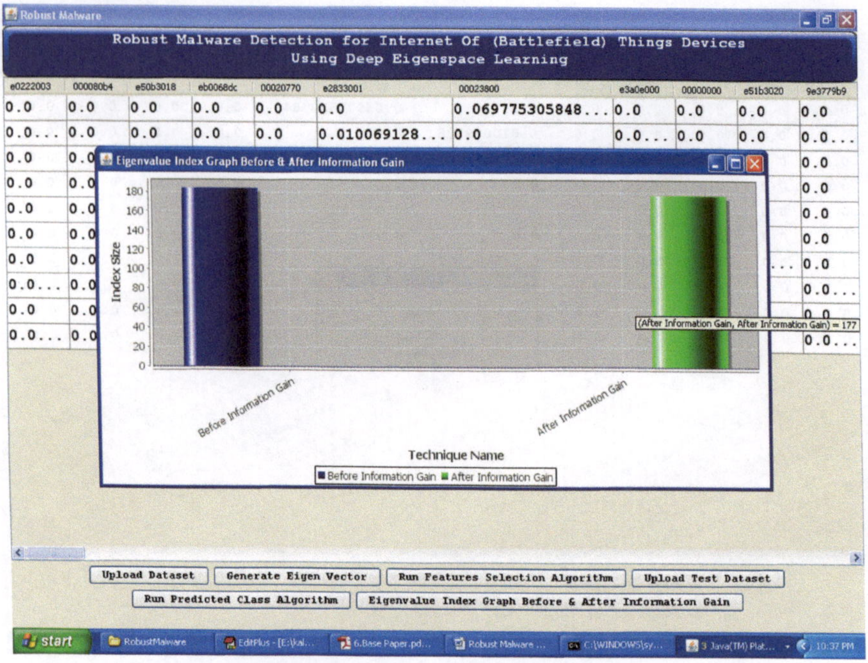

Fig. 10.13 Malware detection on database

Table 10.3 Performance measure

Method	MCC	accuracy	F_1 score	sensitivity	specificity	SSIM	MSE
LNR + FCNN	**0.34**	0.952	0.54	0.39	0.825	0.989	0.0088
RFO	**0.32**	0.745	0.49	0.38	0.932	0.989	0.0094
Xboosting	**0.14**	0.715	0.45	0.28	0.93	0.928	0.0186
CNN	**0.24**	0.715	0.49	0.312	0.985	0.962	0.0195
FCNN	**0.19**	0.58	0.39	0.718	0.839	0.928	0.0197
SVM	**0.12**	0.62	0.39	0.632	0.986	0.986	0.022

protection system is necessary. Therefore, logistic regression and FCNN based cyber-security mechanism is proposed for current and future generation cyber network. This network stands of many attacks such as Causative Assault, Exploratory Sector, Attack in Honesty, Breach of Privacy, Targeting and Assault indiscriminate. LNR + FCNN method achieves MCC 0.34, Accuracy 0.952 F1 = 0.54, sensitivity 0.39, specificity 0.825, SSIM, MSE 0.989.

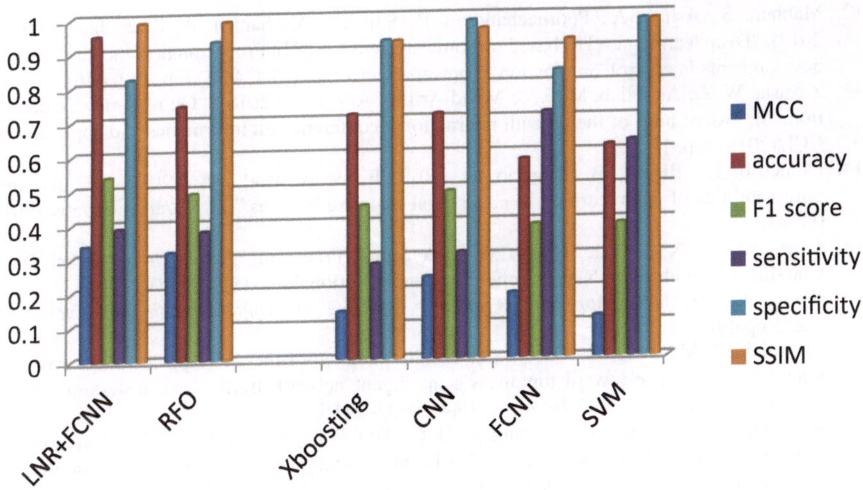

Fig. 10.14 Comparison of results

References

1. Tramer, F., Kurakin, A., Papernot, N., Goodfellow, I., Boneh, D., & McDaniel, P. (2018). "Ensemble adversarial training: attacks and defenses." https://arxiv.org/abs/1705.07204.
2. Qiu, X., Zhang, L., Ren, Y., Suganthan, P. N., & Amaratunga, G. (Dec 2014). "Ensemble deep learning for regression and time series forecasting." In Proceedings of the 2014 IEEE symposium on computational intelligence in ensemble learning (CIEL) (pp. 1–6).
3. Abadi, M., Chu, A., Goodfellow, I. et al. (2016). "Deep learning with differential privacy." In Proceedings of the 2016 ACM SIGSAC conference on computer and communications security—CCS'16 (pp. 308–318)
4. Phong, L. T., Aono, Y., Hayashi, T., Wang, L., & Moriai, S. (2018). Privacy-preserving deep learning via additively homomorphic encryption. *IEEE Transactions on Information Forensics and Security, 13*(5), 1333–1345.
5. Papernot, N., McDaniel, P., Wu, X., Jha, S., & Swami, A. (May 2016). "Distillation as a defense to adversarial perturbations against deep neural networks." In Proceedings of the 2016 IEEE symposium on security and privacy (SP) (pp. 582–597).
6. Hardy, W., Chen, L., Hou, S., Ye, Y., & Li, X. (July 2016). "DL4MD: A deep learning framework for intelligent malware detection." In Proceedings of the international conference on data mining (DMIN) (p. 61).
7. Rhode, M., Burnap, P., & Jones, K. (2018). Early-stage malware prediction using recurrent neural networks. *Computers and Security, 77*, 578–594.
8. Kalash, M., Rochan, M., Mohammed, N., Bruce, N. D., Wang, Y., & Iqbal, F. (Feb 2018). "Malware classification with deep convolutional neural networks," In Proceedings of the 2018 9th IFIP international conference on new technologies, mobility and security (NTMS) (pp. 1–5).
9. Wang, X., & Yiu, S. (2016). "A multi-task learning model for malware classification with useful file access pattern from API call sequence." https://arxiv.org/abs/1610.05945.
10. Chalapathy, R. & Chawla, S. (2019). "Deep learning for anomaly detection: A survey." https://arxiv.org/abs/1901.03407.
11. Chen, L., & Ye, Y. (Aug 2017). "SecMD: Make machine learning more secure against adversarial malware attacks." In Proceedings of the Australasian joint conference on artificial intelligence (pp. 76–89).

12. Maniath, S., Ashok, A., Poornachandran, P., Sujadevi, V., Sankar, A. P., & Jan, S. (Oct 2017). "Deep learning LSTM based ransomware detection." In Proceedings of the 2017 recent developments in control, automation & power engineering (RDCAPE) (pp. 442–446).
13. Zakaria, W. Z., Abdollah, M. F., & Mohd Ariffin, A. F. (Nov 2018). "On ransomware detection." In Proceedings of the seventh international conference on informatics and applications (ICIA2018) (pp. 12–17).
14. Demetrio, L., Biggio, B., Lagorio, G., Roli, F., & Armando, A. (2019). "Explaining vulnerabilities of deep learning to adversarial malware binaries." https://arxiv.org/abs/1901.03583.
15. James, C. D., & Aimone, J. B. (2015). A Signal Processing Approach for Cyber Data Classification with Deep Neural Networks, Sandia National Lab.(SNL-NM).
16. Wang, Z. (2015). *+e Applications of deep learning on traffic identification* (Vol. 24). TechRepublic.
17. Fadlullah, Z. M., Tang, F., Mao, B., et al. (2017). State-of-the-Art deep learning: evolving machine intelligence toward tomorrow's intelligent network traffic control systems. *IEEE Communications Surveys & Tutorials, 19*(4), 2432–2455.
18. Aminanto, M. E., Choi, R., Tanuwidjaja, H. C., Yoo, P. D., & Kim, K. (2017). Deep abstraction and weighted feature selection for Wi-Fi impersonation detection. *IEEE Transactions on Information Forensics and Security, 13*(3), 621–636.
19. Aceto, G., Ciuonzo, D., Montieri, A., & Pescap'e, A. (June 2018). "Mobile encrypted traffic classification using deep learning." In *Proceedings of the 2018 network traffic measurement and analysis conference (TMA)* (pp. 1–8).
20. Mi, G., Gao, Y., & Tan, Y. (June 2015). "Apply stacked auto-encoder to spam detection." In *Proceedings of the international conference in swarm intelligence* (pp. 3–15).
21. Shi, C., Liu, J., Liu, H., & Chen, Y. (July 2017). "Smart user authentication through actuation of daily activities leveraging WiFi-enabled IoT." In *Proceedings of the 18th ACM International Symposium on Mobile Ad Hoc Networking and Computing* (p. 5).
22. Catak, F. O., & Yazı, A. F. (2019). "A benchmark API call dataset for windows PE malware classification." https://arxiv.org/abs/1905.01999.
23. Gibert, D. (2016). *Convolutional neural networks for malware classification.* University Rovira i Virgili.
24. Cha, Y. J., Choi, W., & B¨uy¨ukozt¨urk, O. (2017). Deep learning—based crack damage detection using convolutional neural networks. *Computer-Aided Civil and Infrastructure Engineering, 32*(5), 361–378.
25. Murata, M., & Yamanishi, K. (2017). *Detecting drive-by download attacks from proxy log information using convolutional neural network.* Osaka University.
26. Vinayakumar, R., Soman, K. P., & Poornachandran, P. (Sept 2017). "Applying convolutional neural network for network intrusion detection." In *Proceedings of the 2017 International Conference on Advances in Computing, Communications and Informatics* (ICACCI) (pp. 1222–1228).
27. Wang, W., Zhu, M., Zeng, X., Ye, X., & Sheng, Y. (Jan 2017). "Malware traffic classification using convolutional neural network for representation learning." In *Proceedings of the 2017 International Conference on Information Networking* (ICOIN) (pp. 712–717).
28. Datta, D., Mishra, S., & Rajest, S. S. (2020). Quantification of tolerance limits of engineering system using uncertainty modeling for sustainable energy. *International Journal of Intelligent Networks, 1*, 1–8. https://doi.org/10.1016/j.ijin.2020.05.006
29. Maleh, Y. (2019). "Malware classification and analysis using convolutional and recurrent neural network." In *Handbook of Research on Deep Learning Innovations and Trends* (pp. 233–255), IGI Global, Harrisburg.
30. Kolosnjaji, B., Zarras, A., Webster, G., & Eckert, C. (Dec 2016). "Deep learning for classification of malware system call sequences." In *Proceedings of the AI 2016: Advances in artificial intelligence* (pp. 137–149).
31. Tobiyama, S., Yamaguchi, Y., Shimada, H., Ikuse, T., & Yagi, T. (June 2016). "Malware detection with deep neural network using process behavior." In *Proceedings of the 2016 IEEE 40th Annual Computer Software and Applications Conference (COMPSAC)*, vol. 2, (pp. 577–582).

32. Yu, Y., Long, J., & Cai, Z. (2017). "Network intrusion detection through stacking dilated convolutional autoencoders." *Security and Communication Networks, 2017*, Article ID 4184196, p. 10.
33. Hill, G. D., & Bellekens, X. J. A. (2017). "*Deep learning based cryptographic primitive classification.*" https://arxiv.org/abs/1709.08385.
34. Kang, M.-J., & Kang, J.-W. (2016). "Intrusion detection system using deep neural network for in-vehicle network security," *PLoS One, 11*(6), Article ID e0155781.
35. Potluri, S., & Diedrich, C. (Sept 2016). "Accelerated deep neural networks for enhanced Intrusion Detection System." In *Proceedings of the 2016 IEEE 21st international conference on mobile information systems 17 emerging technologies and factory automation* (*ETFA*) (pp. 1–8).
36 Sebasti'an, M., Rivera, R., Kotzias, P., & Caballero, J. (2016). "AVclass: A tool for massive malware labeling." In *Research in attacks, intrusions, and defenses* (pp. 230–253), Springer.
37. Santoso, L. W., Singh, B., Suman Rajest, S., Regin, R., & Kadhim, K. H. (2021). A genetic programming approach to binary classification problem. *EAI Endorsed Transactions on Energy, 8*(31), 1–8. https://doi.org/10.4108/eai.13-7-2018.165523
38 Mi, G., Gao, Y., & Tan, Y. (2016). "Term space partition based ensemble feature construction for spam detection." In *Data mining and big data* (pp. 205–216), Springer.
39 Anderson, H. S., Woodbridge, J., & Filar, B. (2016). "DeepDGA: Adversarially-tuned domain generation and detection." In *Proceedings of the 2016 ACM Workshop on Artificial Intelligence and Security —ALSec'16* (pp. 13–21).
40. Yu, B., Pan, J., Hu, J., Nascimento, A., & De Cock, M. (2018). "Character level based detection of DGA domain names." In *Proceedings of the 2018 International Joint Conference on Neural Networks* (*IJCNN*) (pp. 1–8).
41. Zhauniarovich, Y., Khalil, I., Yu, T., & Dacier, M. (2018). A survey on malicious domains detection through DNS data analysis. *ACM Computing Surveys, 51*(4), 1–36.
42 Alrawashdeh, K. & Purdy, C. (Dec 2016). "Toward an online anomaly intrusion detection system based on deep learning." In *Proceedings of the 2016 15th IEEE International Conference on Machine Learning and Applications* (*ICMLA*) (pp. 195–200).
43. Yuan, Z., Lu, Y., Wang, Z., & Xue, Y. (2014). "Droid-sec: Deep learning in android malware detection." In *Proceedings of the 2014 ACM conference on SIGCOMM—SIGCOMM'14* (pp. 371–372).
44. Wang, S., Liu, T., & Tan, L. (May 2016). "Automatically learning semantic features for defect prediction." In *Proceedings of the 2016 IEEE/ACM 38th International Conference on Software Engineering* (ICSE)—ICSE'16 (pp. 297–308).
45. Athiwaratkun, B., & Stokes, J. W. (Mar 2017). Malware classification with lstm and gru language models and a characterlevelcnn. In *2017 IEEE international conference on acoustics, speech and signal processing* (*ICASSP*) (pp. 2482–2486).
46. Al-Asdi, T. A., & Obaid, A. J. (2016). An efficient web usage mining algorithm based on log file data. *Journal of Theoretical and Applied Information Technology, 92*(2), 215–224.
47. Al-asadi, T. A., & Obaid, A. J. (2016). Object based image retrieval using enhanced SURF. *Asian Journal of Information Technology, 15*(16), 2756–2762.
48. Meshram, C., Ibrahim, R. W., Obaid, A. J., Meshram, S. G., Meshram, A., & Abd El-Latif, A. M. (2020). "Fractional chaotic maps based short signature scheme under human-centered IoT environments." *Journal of Advanced Research.*
49. Obaid, A. J., Alghurabi, K. A., Albermany, S. A. K., & Sharma, S. (2021) "Improving extreme learning machine accuracy utilizing genetic algorithm for intrusion detection purposes." In *Advances in intelligent systems and computing* (pp. 171–177), Springer.

Chapter 11
An Optimal Hybrid Solution to Local and Global Facial Recognition Through Machine Learning

K. Raju, B. Chinna Rao, K. Saikumar, and Nalajala Lakshman Pratap

Abstract Face recognition need is fine assured as enormous industrial relevance use them to implement one or another objective. As the programmes move closer to everyday usage to hold a database of actual events, an individual's identification primarily demanded as an instance of consistency. As facial recognition has beating advantages over other industrial applications and human eyes can quickly evaluate performance, improved algorithms and smaller computing costs are continuously improving this methodology. This research takes the conventional algorithms of recognition in the first stage and uses hybrid approaches to counter their limitations. The study starts with basic computation of global face features using Principal Component Analysis (PCA), Discrete Wavelet Transform (DWT) and Independent Component Analysis (ICA), with some standard classifiers like Neural Network (NN) and Support Vector Machine (SVM). As the learning rate is high in machine learning, then the system's accuracy goes high, but increases the area and cost overhead. Fusion-based methods have been proposed in further work to overcome that training limitation, based on Harris corner, Speed up Robust Features (SURF) and DWT + PCA system model where only 10% training sample has been taken on Essex database, and 99.45% accuracy is achieved. Creating the Fusion rule requires some hit and trial methods that may not be Universal in every database. To overcome this limitation further an efficient Hybrid method proposed which elaborates the local features Linear Binary Pattern (LBP), Histogram Oriented Gradients (HOG), Gabor wavelet and global features (DWT, PCA) of the face. Further, this feature

K. Raju (✉)
Department of Electronics and Communication Engineering, Narasaraopeta Engineering College, Narasaraopet, Guntur, Andhra Pradesh, India

B. Chinna Rao
Department of Electronics and Communication Engineering, NS Raju Institute of Technology, Visakhapatanam, India

K. Saikumar · N. Lakshman Pratap
Department of Electronics and Communication Engineering, Koneru Lakshmaiah Education Foundation, Vaddeswaram, Guntur, Andhra Pradesh, India
e-mail: lakshmanpratap@kluniversity.in

P. Kumar et al. (eds.), *A Fusion of Artificial Intelligence and Internet of Things for Emerging Cyber Systems*, Intelligent Systems Reference Library 210,
https://doi.org/10.1007/978-3-030-76653-5_11

trained with Neural Network classifier to obtained better accuracy nearly 99.40% with single image training from each class.

Keywords DWT · PCA · ICA · LBP · HOG · GLCM · Harris corner

11.1 Introduction

However, the face tends to be a simple target to identify and recognize through the eye, but artificial intelligence is not yet sophisticated enough to efficiently accomplish the role. As the source of a face is typically an entity that collects pictures, some variations and complexions remain with the image, such as noise, movement, etc. Several approaches use any or another algorithm to locate resemblance in the face model. The test picture and most of them effectively obtain more substantial similarities in the test [1]. However, a single algorithm cannot reach optimum performance everywhere given the varied scope of applications and method of image sourcing. After using the right algorithm for a specific mission, facial recognition problems must be countered by an application. The skill of recognition strategies fits the same trend as identification, and the increasing demands of businesses are finding changes in patterns of recognition. Four datasets-ORL, Essex, YaleB and Indian are tested in several experiments to achieve good results in multi-purpose domains. This research aims to conduct a comparative study of all the techniques with current precision methods and incorporate a real-time face recognition device on the Raspberry Pi 2 CPU (central processing unit) [2].

11.1.1 Problem Statement

Achieving high performance for a face recognition system in the real world has been a problem open to researchers for recent years. Inappropriately, uncontrolled conditions such as illumination variations, occultation, and facial expressions and poses variations significantly influence the performance of face recognition systems particularly those based on 2D information, as this kind of knowledge depends mainly on sources of light, in addition to the 2D image or the color image does not represent the shape of the face. It does not treat the beginning as an object, while these systems are sensitive to the real uncontrolled environment [3].

The variation of poses of the head is a significant problem for the recognition of a face. Correcting the posture and estimation of the head's rotation angle are necessary processes to solve these problems. However, because of the mathematical complexity and high costs in terms of memory and computing time, the development of a robust automatic face recognition system with the variation of poses is considered a big challenge for facial biometry researchers [4].

Another difficulty for automatic face recognition systems is the representation of facial images and the extraction of the most discriminating features to build people's biometric signature. In this research work, this signature is in the form of a character vector [5]. Several methods have been developed for extracting features that are global or local methods. Local descriptors are recently used as very effective image representation methods; they are proposed initially for texture classification. They are used for facial image analysis. This thesis involves the study and evaluations of five best descriptors, used recently in the literature.

More recently, the development of facial biometrics research has focused on using the 3D facial surface and the shape of the face to represent the most discriminating features with an increase in the dimensionality of the characteristic vectors [6]. To reduce the impact of such problems, dimension reduction is used in most face recognition systems. Dimensionality reduction plays an essential role in information processing, particularly in facial biometrics systems that require rapid execution and high accuracy. Several approaches have been proposed in the literature; note that the two algorithms, Principal Component Analysis (PCA) and Linear Discriminant Analysis (LDA) are basic algorithms for most of the proposed approaches [7].

The main problem of research work concerns uncontrolled conditions in the automatic face recognition system: the variations of illuminations, facial expressions and poses. As well as extracting the discriminative characteristics of facial images based on local descriptors using depth images. However, the proposed work examines the data reduction process's role in developing a robust, efficient and accurate automatic face recognition system.

11.2 Literature Review

The biometric recognition techniques also termed as biometrics, identify an individual with their physiological and behavioural patterns. Similarly, the process of identifying an individual by its face has called as face recognition. The motive of recognising a look is identifying an individual from a similar class of species. The individual has determined by several face features that are unique for an individual. Face recognition is a vast research area with numerous problems. These are due to its application in different fields such as Mugshot album, surveillance camera, identification card, access control card etc. The face classification when firstly introduced in [8]. In this proposed paper, the author started gathering the faces the database and classified it according to its deviation from the norm. This generated vectors which could compare with other database vectors. This is called multimodal classification. Due to the vast advancement, FR systems have used in real-time applications widely [9]. Such a vast improvement in face recognition is because of the following reasons:

- Development of various algorithms
- An extensive database for experiments (e.g. FERET, Yale, ORL etc.)
- Various performance evaluation methods.

As we know, each coin has two sides. Similarly, face recognition technology also has some drawbacks. If a system has a static image, it will be easily recognised, but it is quite difficult to recognise the shot if given a video frame. Similarly, suppose the system is commanded to identify one or more person from a single image. It is quite complex to compare all the faces from the database and recognise them accurately.

While comparing the queried image and database image, many aspects need to be considered. These are:

1. Firstly, a person who seems to be co-operative claims to be "someone". This means there is no need to verify the complete database images (model images) with the incoming/queried image (probe image). Only the database image whose identity is claimed by the probe image needs to be verified. For recognition, all database images have compared with the probe image, which is comparatively extensive.
2. Secondly, this automated FR system should consume less time to be accepted in real-time practical applications.
3. In the final step, the images of different persons from the training database have taken for comparison. Only when an imposter (most alike person) is present, needs to be recognised more carefully.

Face recognition technology belongs to biometrics, where the person's identity is matched with the database and thus verified. This verification depends upon the person's facial features. On general basis biometrics field uses fingerprint, iris-pattern also for identification [10]. The use of iris-pattern recognition is less than other biometric methods because there is a risk of protectiveness for one's eyes. Face recognition system is a passive system which does not interrupt the natural verification process. The face recognition process has carried out into three steps:

(1) Firstly an observation is done by the sensor. This observation gives some unique signature about a person's identity, also known as 'Biometric signature'.
(2) Secondly, the sensor and observed data depend upon the type of biometric device used. Now the biometric signature is normalised to get the database signature and biometric signature into the same format, i.e. image size, view, resolution etc.
(3) Finally, the matching process has carried out. Here the normalised signature is compared with the database signatures. And the scores are allotted for each match. The match having the highest score is declared to be the perfect match.

Each biometric system has a different way to deal with the similarity score. The face recognition process initiates with the detection of the face area in the image. Sometimes, the image's background is more crowded, leading to wrong face recognition if a face is not detected correctly [11, 12]. The face detection process includes detecting the face area, normalizing the image parameters (like illumination changes, views etc.), facial appearance etc. The results are then further processed using modelling schemes.

Face recognition technology has its applications mainly in two field's i.e.

- Law enforcement
- Commercial field.

FR technology is mainly used for mugshot albums in the law enforcement field and matches the surveillance images. Whereas, in commercial applications photograph of a person on ATM cards, credit cards, passport, photo identity, driving license etc. is matched with a real-time image for granting access to that person.

Though both fields have different face recognition uses but it faces certain constraints during its processing. The face recognition technology processes in two phases:

(1) Face detection and Normalisation
(2) Face identification/Matching.

An algorithm or a system consisting of both phases is known as fully automatic face recognition algorithm (For example, only the face image has provided). An algorithm or a system consisting of any one of the phases is known as partially automatic face recognition algorithm (For example, in this algorithm, the face image and the eye co-ordinates are also provided). The development of face recognition technology permits an organisation to perform frontal, view-tolerant and profile recognition. The type of award depends upon the image type and the type of algorithm used. The frontal recognition uses classical strategy, whereas the view-tolerant recognition uses a sophisticated manner to perform recognition based on some geometrical, physical or statistical features. Profile recognition is an individual recognition system which has a small significance in honour [13]. These recognition systems are efficient for fast and large databases to lessen the load and computability for a particular or hybrid algorithm. These hybrid algorithms have a special status among other face recognition algorithms because they combine various algorithms serially or parallel to avoid stand-alone components' drawbacks. Face recognition techniques can also categorise based on models or exemplars. Models have used differently in different researches. In [14], models are used to compute the quotient image, whereas in [15] models are used to check the active appearance. These models provide information about the class, and they also offer a way to deal with appearance variations. Exemplars have also been used for recognition such as in [16] both training and matching images have been used against probe image. The methods using models do not use exemplars because they cannot be employed together as they are not mutually exclusive. An approach was proposed in [17] to combine models and standards. These images were further used as exemplars in a learning phase in which models were used to synthesise training images.

Considering pose invariance again face recognition can be categorised as:

(1) Global approach
(2) Component-based approach.

In a global approach, the whole face image represented by a single vector has given as an input to a classifier. Various classifiers have used in previous research such as Fisher's discriminant analysis [18], Neural Networks [19], minimum distance classification etc. For front view images, global techniques have proved themselves well. But they are not immune to pose variations. Therefore, they get affected by the rotation of the face. To overcome this issue, the addition of an alignment before the classification stage is an option. To align the reference image and input image needs more computation. The correlation between the two ideas could found through critical points such as lip corners, nose, eye-balls etc. On this correlation, the input image can be deformed (warped) according to the reference image. Warping can be performed by computing affine transform. In [20], active-shape models with model faces are used to align images. In [21], a combination of SVM and partially automated alignment were proposed. An alternate for this approach can be the classification of local facial components. This component classification approach is to reimburse the pose variations by permitting flexible relations among the classification stage components. Face-recognition was complete through independent template matching of three critical points of the face, i.e. eyes, lips and nose. Because the geometrical characteristic was comprised in the system, configuring the classification stage components was abandoned. An extra alignment stage with a similar approach was proposed in [22]. The implementation of the geometrical model using two-dimensional elastic graph was shown in 42. This technology of face recognition still lacks in some aspects like pose and lighting variations. Though there are many methods and algorithms proposed until now and they give a promising performance, some problems remain. Hence, the performance of matching is degraded because of the reasons mentioned above and is low compared to other biometric matching performance (e.g. fingerprint, iris recognition). Error rates are in the range of 2–25%bwhich is very common. The system can perform better if it has used in combination with other biometric computations. According to ref. [23], existing recognition systems perform well if test images and training images have the same image capturing conditions. But if these conditions differ for test images and moving ideas, the system will not be capable enough to deal with these changes. These changes can be lighting variations, head-pose modifications, accessories, hairstyles, facial expressions etc. The approaches can be further categorised based on interpretations, i.e., canonical forms, invariant features, variation-modelling [24].

1. "Canonical form" approach "normalises" the changes by synthesising new test images in canonical/prototype form or by the creative transform. Then these prototypes are used for recognition. In the examples of this approach is depicted. In [25], the probe image captured under random lighting conditions has again processed from frontal illumination. Both states have then compared with one another.
2. An invariant feature approach, the characteristics that are invariant to variations are analysed. The example can see in reference [26], in which the quotient image is consistent with the lighting conditions and can recognise a person accurately under varying illumination.
3. In variant-modelling approach, the main motive is to study the extent of variation in a particular subspace region. These can cause to define and decide the parameters that are mandatory for subspace. Then a nearest (close to test image)

subspace is chosen for recognition after mapping. In spite, this recognition can estimate the pose and a person's identity. The example of this system can be seen in [27]. Even after developing so many techniques and research efforts, the face recognition field's problems remain unsolved.

All three approaches work well for a particular variation. For other variations, the performance of the system degrades gradually. It can explain well with the following example that a feature invariant technique is invariant to pose and face expressions until they are constant. Still, as soon as the pose or face expressions change, the method fails to remain consistent. In some cases, this is not an issue like granting access etc. because both the test image and training image have captured under the same circumstances. But it cannot recognise the idea in general conditions, and hence none of the three techniques is reliable.

Furthermore, it is also not sure that these techniques can be used in combination or not to avoid the drawbacks. Some of the methods have behaviour to remove others. Like in [28], the symmetric shape-from-shading way depends upon frontal face symmetry. But it is not clear how this technique will work in case of side-view of a face or a condition where balance is not present. From this research, two essential facts have observed, and they are:

(1) Firstly, no feature/set of features are correctly or entirely invariant for all the variations that can occur to an image.
(2) Secondly, if any system has more training images, then many techniques will perform better.

11.3 Face Recognition Techniques

National Institute of Standard and Technology (NIST) of the U.S. performed a survey to evaluate the FR system's performance. According to that survey, the best FR algorithm's ability experimented on 1.6 million images to recognise a person or object correctly (known as True Positive Identification Rate) is 92.3%. The experiment was conducted on various datasets started from the FERET database. Suppose face recognition is considered for criminal case investigations. In that case, the faces' evaluation can be affected by some factors like ageing, varying facial expressions (smiling, sad, tensed etc.), changing orientation angles and accessories such as (glasses, hat, moustache or beard etc.). These variations have depicted in Fig. 2.1 As per the growing development of face recognition technology, the FR system's accuracy is hardly affected by the above effects. NIST's evaluation results state that the current face recognition algorithm is much better than the earlier algorithm of time 2001. Thus the existing FR algorithms are reliable enough for being used in criminal case investigations [29]. As discussed above, face recognition technology required for criminal case investigations should be more advanced than general face recognition systems. This is because the images given for forensic analysis have infrequently shot under non-ideal conditions, leading to false rejection or false acceptation. In this paper

[30], the improvements/developments in the general face recognition system and the forensic face recognition system have discussed [31]. According to this article, to avoid false rejection or false acceptation, the following factors should be considered during forensic face recognition:

1. **Ageing**
   Human face changes according to age. Thus ageing is a factor that should be considered during face recognition [21].
2. **Facial Marks/Scars**
   Some marks or scars become an identity for a person. It may be from birth or due to some accidents. But in the FR system, it plays an essential role in identifying the correct person [10].
3. **Sketch**
   A sketch of a suspected person has generated if his/her photograph is not available. The illustration differs from an original picture of a suspected person because of inaccurate description from the witness or wrong understanding of an artist drawing the sketch. It can sometimes lead to false acceptance.
4. **Image Captured by Surveillance Cameras**
   The video frames or images captured by CCTV/surveillance cameras sometimes help to identify the person. But sometimes the image captured is burred not clear. It could mislead the face recognition system [32].
5. **Near-Infrared Face Recognition (NIR)**
   Near-Infrared Face Recognition (NIR) image method has proposed to overcome the illumination effects. According to Klare and Jain, the NIR method obtained 94% true accept rate and 1% false accept rate. Hence, the NIR technique should be considered for forensic face recognition. This paper studies the various face recognition algorithms such as trainable, non-trainable and commercial in 1, 2 and 3 algorithms in number. These algorithms have evaluated on a partitioned gallery, which has demographic cohorts in its every part. These parts are sub-categorised based on gender, colour complexion, age etc. The experiment conducted shows that non-trainable and commercial algorithms are less accurate in recognising similar cohorts. Whereas trainable algorithm is less affected by the demographic cohort. Hence the use of dynamic face matcher selection comes into the picture. In this method, various FR algorithms are available to select the relative demographic information extracted from the probe image. This method gives higher accuracy in many FR scenarios.

Mostly the frontal face images are considered for research of face recognition algorithms. The benefits and drawbacks of all methods such as Eigenfaces, Hidden Markov Model (HMM), Neural Networks (NN), Dynamic Link Architecture (DLA), Geometrical Feature Matching (GFM), Fisherfaces, Template Matching (TM), Elastic Bunch (EB) etc. are discussed. Each of these methods has discussed and analysed one-by-one as below.

11.4 Eigenfaces

One of the vital face recognition techniques is Eigenface (also called as Karhunen–Loeve expansion). The face pictures have also represented through principal component analysis in [33]. In this paper, according to the authors; it is possible to recreate the face image using a set of weights for each face and standard face image (also known as Eigen picture). The face image is projected on to the edge picture to achieve consequences for each face. Mathematically, eigenfaces can be defined as the principal components of the distribution of faces or the eigenvectors of the set of face images' covariance matrix. Eigenvectors ought to represent various changes for different looks. Eigenfaces are used in linear combination to describe each face. It can be estimated through the use of eigenvectors which have the highest eigenvalue. Such best' M' eigenvectors construct M dimension-space (known as face-space also).

An experiment conducted taking 2500 images of 16 different persons with different backgrounds shows a 96% accuracy for varying illumination, 85% accuracy for varying orientation, and 64% accuracy for varying sizes. Though this approach gives a powerful performance using the correlation between image and variable illumination, the correlation between image and running illustration does not satisfy the face recognition result. For Eigenface methodology, tilizedng illumination (lighting effect) is a necessary step. Another new approach has suggested three images having different illumination have considered for calculating the covariance matrix. Gu, Wenfei, et al., have worked further improved from Eigenface to eigenfeatures which incorporates face organs as features such as (lips, nose, eyes etc.). This modular eigenspace method is less affected by the change in appearance than other standard eigenface methods. Another experiment was conducted on the FERET database taking 7562 images of 3000 different persons, resulting in a 95% accuracy recognition. Hence it is proclaimed that an eigenface method is a practical approach with faster and simpler strategy.

Koc, Mehmet, and AtalayBarkana suggested that ears can also prove to be better than face for biometric recognition. Earlier similar work was done with ear and face using PCA (principal component analysis), but it resulted in low ear image recognition. Hence, in this proposed paper 58, the Eigenface and linear method is used, which results in 70.5% for face and 71.6% for ears when experimented. And when both were considered together for multimodal recognition, it gave 90.9% recognition result. The multimodal biometrics work has been extended. Another similar approach could be seen where face and fingerprint and face and voice have been tilized for multimodal face recognition.

11.5 Neural Networks

Neural Networks has grabbed so much attention because of its non-linear network recognize. Hence, it proved to be more capable of feature extraction as compared to other linear approaches. WIZARD is an adaptive technique among artificially generated neural networks with a single-layered architecture. It comprises of an individual network for storing every single information. The path of designing a neural network for efficient face recognition is quite tricky because it depends on the application for which it will use. In [34], MLP and convolutional NN are used respectively for face identification. Therefore, a neural network must be a biological neural network that consists of actual biological neurons and an artificial neural network that finds solutions with artificial intelligence (AI).

1. Face Detection: This is the step where a person's face has detected by skipping the background image.
2. Eye Localization: This is the step where both the eyes' position is determined using feature vectors.
3. Face Recognition: PDN network is not interconnected completely. It partitions the network into 'K' number of subnets. One subnet recognizes one person from the database. The face subnet has a total of the neuron outputs. It uses Gaussian function for neurons to estimate the likelihood density.

The PDNN in its learning stage can be considered as:

Phase I—In this phase, each sub-network has trained by its face image.

Phase II—This phase is known as decision–based learning phase. Here the subnetwork parameters are trained by some other face images using particular image samples. It does not use all images for training, but the wrongly classified patterns have only been used.

11.6 Geometrical Feature Matching

Here, the set of geometrical features has computed from the face image. This technique is thus known as a geometrical feature matching technique. This is the technique because the face identification is possible at the low resolution of 8×6 pixels. After all, the general geometrical feature information is sufficient for recognising the image. The face organs such as eyes, eyebrows, lips, nose etc. have represented by the vectors, which describes the overall configuration. In [35], similar work on automated face recognition system was executed using geometrical features. This FR system's peak recognition rate was 75% taking 40 images of 20 different people (2 images per person—1 training/model image, 1 testing image). According to the reference manual feature extraction in the FR system could give satisfactory results. In geometrical features such as nose length and width, lips position co-ordinates etc. were automatically extracted. 35-dimensional vectors extracted around 35 components and recognised using Bayes classifier. When an experiment was performed taking 47 persons, it provided a recognition rate of 90%. Another technique known as the mixture-distance method was applied to a database of 685 people which provided accurate recognition of 95%. Thirty distance measures are extracted manually and represent each face image.

Gu, Wenfei, et al. makes use Gabor wavelet transform to decompose image and identify the feature points for face image, which reduces the storage space requirement. Usually, 35 to 45 such feature points have created. The topological representation of these feature points has used to match the test image with the database image. The recognition rate was 86% for identifying the right person and 94% cases are such that the person's face matches with the top three matches. Concluding this, it can be said that matching of geometrical features has based on the distance measurement between two elements which is more desirable in matching queried image with the Mugshot database. Although it depends upon feature location accuracy. Existing feature location methods need more computation time and still does not have rightful accuracy.

11.6.1 Fisher Faces

Another face recognition technique is, which has used widely. It is an appearance-based method. R.A. Fisher developed Fisher analysis in 1930. A study demonstrates the better results successfully obtained in face recognition. Another FR approach is the LDA approach, which determines the base image, increasing the between-class scatter and within-class scatter ratio. This ratio calculation consumes a lot of processing time. There is one drawback of this technique, and that is the scattering matrix is always single. If the illumination or pose conditions changes, it will maximise the error detection rate because of the number of pixels that are more in number than the number of images. Thus to avoid this issue, many other approaches were proposed. Fisher face approach is one of Belhumeur et al., which can overcome this problem of varying illumination or pose because it removes the first three principal components responsible for light variations. Hence it is more robust to lighting changes. Fisher's face makes use of within-class information which reduces variations within the class. Thus it proved to be a successful method to resolve the given problem [36]. The PCA and LDA were used with the Fisherface process, which generated a subspace projection matrix (similar to dimension-space/face-space in the eigenface method). Fisher face method uses within-class information and reduces the variation but increases the separation among classes. Identical to the eigenface method, the initial step in fisher face is to reshape the image array of size (NXM) into an image vector of length ((NXM)X1). Fisher face is the enhanced version of the eigenface method but with better classifying ability. Here, the training image set could be trained to handle different persons' images and other expressions. Fisher's face gives more accurate recognition as compared to the eigenface method. Determining projection face space is more complicated in Fisherface than in the eigenface method. This is one of the drawbacks of fisher face method.

11.6.2 Template Matching

In this technique, an image under test is represented by a 2D-array whose intensity values have been compared using a suitable measure unit like Euclidean distance. Here, the entire face has defined by a single template. Other advanced versions of template matching algorithm are also available. An individual's face could be represented from different pose/viewpoints using more than one feature template. It could also be defined from an individual view utilising several smaller templates. The grey level of the image needs to be dealt with appropriately before the matching process. Similar work was done using four templates to represent the nose, eyes, lips and the whole face. The experiment was conducted on 188 images of 47 different persons for template matching and geometrical matching. Both methods' performance outcome was compared, and it was seen that template matching (100%) has a superior performance over geometrical matching (90%) approach. Template matching algorithm has a complex computation and its complex description. In general comparison, the template matching approach is far better than the feature matching approach. Though the face recognition system bears some discrepancies, this tolerance yields a unique face identity for an individual. There is no such algorithm which is perfect and has no drawback. Thus further improvements are needed to enhance the FR system performance, which could be used in real-time applications.

11.6.3 Morphable Model

This technique is the vector representation for human faces which is shape and texture vectors combination. To recognise the three-dimensional object correctly, the use of 3D models is necessary. 3D Models could be used on images in two ways. In the first case, the recognition is based on model coefficients and is free from image conditions. The model coefficients here represent the shape and texture of the object/face. In the second case, the 3D-face could be reconstructed by generating synthetic views from trained images. These views have then transferred to other position-dependent face recognition system.

Li, Shaoxin, is proposed that the three-dimensional shape, image parameters and texture can be estimated from a combination of deformed 3D-model and computer graphics. In this algorithm, rotation and lighting changes are easy to be handled by a single model. Though, the illumination changes can cast shadows which will affect the human skin appearance features. Hence the 3D-morphable model is introduced, which can learn face properties during 3D-scan. It represents the shape and texture of the face through high-dimensional face vectors. The algorithm proposed automatically detects the head orientation, face orientation, lighting direction, camera position etc. This is possible because of the initialisation procedure, which uses co-ordinates between 6–8 feature-points. This, in turn, increases the reliability and robustness of the FR system. When the experiment was conducted on CMU-PIE database, it showed 95% accurate recognition for side-view images. In contrast, when the investigation was born on the FERET database, it showed 95.9% proper recognition for front view images with correct head-pose recognition.

11.7 Support Vector Machine (SVM)

In this approach, the key-points for face representation have selected on a manual basis. The feature vectors have then brought to a computable level. The SVM classifier is then used to handle both small and high dimensional samples to classify such feature vectors. In this research, a face recognition Human–Robot-Interaction system has implemented Hidden Markov Models (HMM) combined with SVM. Only the full pose have taken under consideration for this face detection technique. The two significant benefits of this algorithm are:

(1) It does not require any manual selection for face or head pose estimation.
(2) To improve the normalisation performance, the image is classified based on whether it is in the front pose.

The experiment was conducted on various datasets, and as a result, the system was able to identify the face accurately with the recognition rate of 99.67%.

In [73], the two conventional Support-Vector-Machine methods have used. Those are:

(1) One-against-One
(2) One-against-Rest.

This research work has divided into three sections.

(1) The Gabor wavelet process eliminates the illumination and robust effects of images.
(2) The large size of the feature data has reduced by using the bilateral two-dimensional-LDA method.
(3) As a final step, the classification of face recognition features has carried out using an SVM classifier. The actual aim of SVM classification is to handle binary/multiple classifications.

The One-against-One and One-against-Rest method has applied in the matching process. The experiment was conducted on the ORL database, and it was found that the One-against-Rest way gives more accurate recognition than the One-against-One method. Kamal has proposed here to use PCA for feature extraction and SVM for pattern classification. SVM is a widely used classifier introduced to tackle recognition patterns. The experimental setup contains 400 ORL database images of 40 persons captured under distinct circumstances like light variation, varying facial expressions, different face orientations, with or without glasses etc. Various SVMs have used till now like Polynomial-SVM, Linear-SVM, Radial Base Function SVM

(RBF-SVM). These SVM classifiers have compared to one another. The SVM classification compared to MLP (Multilayer perception, a standard eigenface classification approach) category shows that SVM has a better performance over MLP. But when compared among SVMs, P-SVM and RBF-SVM have a better understanding of L-SVM.

11.7.1 Harris Corner

Shang, Li et al. have proposed Harris Corner method used for palmprint recognition. The corner points have lined using orientation features and distance. Matching is done based on fuzzy logic. This algorithm filters the resulting corner points. The algorithm identifies the palm print using orientation and distance features. This algorithm results in a higher recognition rate with more than 50% less memory requirement. A new pyramid-based-FAST-corner detection has proposed a superior version of FAST detector. This pyramid-based-FAST-corner detector detects the face organ corners, and its result has compared to other corner detection techniques such as (FAST, SUSAN, Harris, SIFT etc.) As an initial step, face corners and other unnecessary corner points have filtered. In the next step, based on organ corner points the organ such as eyes, eyebrow, lips etc. ten such points have highlighted.

Further in the final stage, similar ten key attributes have used to locate the organ's centre. In this manner, the face's model has constructed through triangulation and geometric features. The experiment has conducted over 200 sample images, and it is seen from the results that this algorithm can locate key points accurately for both grey and coloured illustrations. The recognition rate for JAFFE is 94.95% and for MMI is 90.95%.

Mathe, Stefan, and Cristian Sminchisescu [37] have worked on the Harris Corner detection to identify the human eyes from the given coloured images. In the first step, the detection has done through YCbCr conversion. The Cb component here is useful eye detection. The human face has extracted from the preprocessed image, and the eyes have detected by applying the Harris corner method in the second step. This study reveals that the Harris corner method uses only two corner points to see the eyes with the accurate detection rate of 88% and proves itself suitable enough to be used for further detection.

Eye detection plays a significant role in Human–Robot interaction (HRI) system. But in this paper, eye detection is performed using shape and colour information about skin (known as skin patches) in combination with Harris corner method. Here Harris corner method is used to figure out the region-of-interest in the skin patch. This work has influenced from the previous work on "skin patch detection". In this proposed scheme, an RGB image has been used to detect the eyes with the relative information of corner points, colour and edge (shape). The experiment was conducted on the AR database and the champion database, which shows that the proposed scheme is faster than the previous schemes with the average time of 2.53 and 0.21 s per image and the detection accuracy of 100% and 89.5% for the AR and champion database. Above we have seen the Harris corner related work for eye detection. But in Royce et al. has used Harris corner and FAST corner methods for recognising the smile automatically. In this proposed work, the lip corners are taken as reference points to detect the smile accurately. The experiment has conducted on Video, Image and Signal Processing (Visio) dataset, and the results of the two methods have compared to each other. The results show that the Harris corner detector gives a higher recognition rate of 77.5% than the FAST corner detector (recognition rate of 72.5%).

11.8 Speeded-Up Robust Features (SURF)

In the face recognition field, for Human–Robot-Interaction (HRI) system it is difficult to identify a personality accurately compared to humans due to vision problems encountered because of varying illumination, varying expressions, varying background etc. To resolve these issues, and efficient technique has proposed in. In this proposed algorithm, once the skin region is successfully detected, the face region is detected using the ellipse-fitting method. Features of the face region obtained from the ellipse-fitting process have extracted using SURF (Speeded-up-Robust-Features) classifiers. Further, the queried image has compared and matched to the database image. The proposed algorithm has experimented on Caltech's face database, and the recognition rate of 93.75% has achieved for different resolutions [38].

Li, Jianguo, Tao Wang, a SURF based novel boosting cascade framework has proposed. Three contributions are essential here:

- SURF features have used for faster face recognition
- AUC has used for cascade training as a single criterion, which also gives fast training convergence.
- A real-time example has shown to train fast and accurate cascade face detector from billion sample images within 1 h.

In [39], salient features have used for face recognition or object recognition. In the proposed algorithm, SURF descriptors have used to generate feature vectors, and SVM classifiers have used to classify feature vectors. One classifier verifies whether the feature vectors belong to face image or not [40]. This is done in the first layer. In the 2nd layer, each object in the face like eyes, lips, nose etc. has assigned labels. These are known as component labelling [41]. This technique gives high recognition rates. However, there are further chances of improvement in future [42].

Feature detection, Feature matching, transformation are the necessary steps required for image alignment [43]. These steps have based on image features and image reconstruction. Matching and feature extraction are two essential steps demanded in various image applications [44]. In this paper, the Scale Invariant Feature Transform (SIFT) and Speed Up Robust Features (SURF) methods have proposed. An experiment was conducted to compare SIFT and SURF, and it was seen that SIFT had detected more features than SURF. But SURF performs better than SIFT in terms of speed.

A feature detection scheme "SURF" has been researched in this article. SURF (Speeded-up-Robust-Features) was motivated and derived from SIFT (Scale-Invariant-Feature-Transform). Both SIFT and SURF have good performance over feature detection and are faster than previous detection schemes. Seeing its version, Xu et al. decided to study and implement SURF feature detector and SURF feature descriptor. Authors also suggest implementing SURF comparator to group similar feature pairs. Since this comparator is not featured dependent but based on code, hence we can use other feature types like SIFT etc. to evaluate its performance.

In this research paper, SURF has been introduced as a novel method. SURF proved itself as a robust technique which can perform better and faster than the previously submitted schemes. SURF can be implemented by.

- Convolving image through integral images
- Strengthening the current detectors and descriptors
- Simplication of the used methodology

11.9 Independent Component Analysis (ICA)

ICA (Independent Component Analysis) is a new method introduced in the field of face recognition. It is a superior version of PCA. The architectures of ICA were analysed using PCA. ICA has Architecture I and Architecture II. Study reveals that Architecture I has vertically centred PCA whereas Architecture II has whitened PCA with horizontal centring process. Using these two architectures as a baseline for ICA increases the robustness of the face recognition system. In this paper, the experiment has conducted on the FERET database, AR database and AT&T face database. Though there is no significant difference between ICA and PCA, a small number of changes could be seen in some cases. ICA entirely relies on PCA as the centring and whitening process has a great deal of impact on ICA performance.

A robust method known as locally salient ICA (LS-ICA) is proposed in this paper to improve the performance and eliminate distortion and occlusion. This algorithm divides the original image into parts and then extracts the image information from them. The salient features of the face part of the image have used, which benefits from using the idea "recognition by parts". In LS-ICA, the partitioned image is created by adding extra constraint information during kurtosis ecognizedn. When compared the LS-ICA technique's performance with LNMF (Localized Nonnegative Matrix Factorization), LFA (Local Feature Analysis), PCA, ICA Architecture I and ICA Architecture II; LS-ICA always shown a better result.

Thang, Nguyen Duc, et al. is concerned for dimensionality reduction for face recognition. ICA is one of the methods that decrease the dimension of the image. Also, the proposed method Locality Preserving Projection (LPP) is good at reducing image dimensions. These two methods have compared, and the results are analysed. An experiment was conducted on the Yale B database (with 64 varying illuminations) and AT&T database (with varying facial expressions). The performance evaluation states that ICA is better than LPP under the variable definition, whereas LPP is better than ICA under running facial expressions due to its better recognition rates. A standard face recognition method (ICA) that can opt according to the changing face expressions is proposed and implemented in this paper. This is due to the traditional method's ecogniz of treating the varying face expression as noise. Reconstruction of different facial expressions is done based on two KIM conditions and then trained to get a single sample. Then the whitening process is applied to these samples using eigenvalue decomposition. This method provides a better result and recognition rates. Can use this method for compressing image or for MPEG-4 animation etc.

Singha, Anu, et al., GaborJet-ICA technique has proposed. Here face image has convolved with 40 Gabor wavelets. Feature vectors have derived from GaborJets that are the facial key-points. These GaborJets have later on transformed as its ICA and PCA components have computed. The experiment was conducted on ORL database taking subspace and ICA as its parameters. Observations were taken for the varying range of subspace dimension and no. of independent components. GaborJet-PCA and GaborJet-ICA were compared, and it was found that GaborJet-ICA performs better with accurate recognition of 84.5% than GaborJet-PCA with special mention of 82.25%. In this paper, ICA aims to feature extraction of the original image from the raw image data. These components can't be entirely relied upon because ICA components do not appear in a particular order. Thus it can't be said that which feature is more important. To enhance the ICA performance, it is essential to select the components having a good impact, whereas the elements having a harmful effect or are noisy can be eliminated. Thus two methods are introduced in this paper [23] to get the better performance result of ICA by selecting and removing the respective ICA components. First is Component Subspace Optimization, and the other is the Sequential Forward Floating Selection (SFFS). In future, recognized of independent components can be expected in ICA face recognition system. In the FR system, there are numerous methods such as PCA, ICA, LDA, SURF etc. for dealing with front-view images even though there is a lack of information.

In some cases, the recognized face varies from user to user. Thus a modified version of ICA algorithm is proposed. In this modified ICA, the information about the face database has already known. But the test sample and the training samples are not similar. An experiment conducted on the FERET database reveals that the proposed algorithm saves the computation time and the system's memory.

11.10 Results

1. Acquisition of the image: Depending on the system used, it tends to be a still image, a video frame, a three-dimensional image, etc.
2. Face detection: Uses object recognition algorithms that detect whether there is a face in the image. Provides the location and size of the face. This uses Viola, Jones technique which is the standard one.
3. Preprocessing: When a face is resolved on the picture, the normalisation process is performed. Face components such as size, pose, and illumination is found. To

normalise face images, diverse guidelines can be taken after, such as the nose's position, the separation between the pupils of the eyes, and the lips' size. To improve the framework's execution, procedures, such as decreasing the picture size, changing the picture to grayscale, or a low-pass filter, are utilized if the image resolution is too high.

4. Extraction of the characteristics: After the picture is processed, the picture's characteristic vectors of coefficients will be calculated depending upon the method utilised.
5. Recognition: The technique mentioned above has used to detect the facial image and face features extracted using Local binary pattern (LBP) and PCA methods, which are an appearance-based approach depending on pixel values of facial images. At last, the extracted component vector has compared with the feature vectors extracted from the faces database. If it finds one with a high percentage of similarity, it returns the face's identity with that person's name; if not, it indicates an unknown face, as shown in Fig. 11.1.

Implementation testing of the system In this work, an application for real-time face recognition is implemented in Python over the RPi2. To perform this, the library OpenCV that supports image processing has been used. The algorithm implemented for the real-time face recognition problem is divided into two different, independent modules:

- The training stage, performed at distinct distances and various illuminations and orientations, as shown in Fig. 11.2a.
- The recognition (testing) stage, performed to identify the working range of the system, and its display's that person with his name, as shown in Fig. 11.2b.

11.11 Experimental Results

The outcomes of the categorisation for five different persons for frontal images are taken. To determine the correctness of the developed system in real-time, we examined some of the person images by capturing their facial images, through webcam connected to Raspberry Pi2 and got the expected outcome with the person's name below (Fig. 11.3).

11.12 Conclusion

Automatic face recognition has been an active area of research over the last four decades. Face recognition has many applications of considerable importance, such as biometric identification, surveillance, human–computer interaction and multimedia data management. Facial biometrics has played an essential role in improving our security by limiting criminals' mobility, preventing fraudulent activities, and

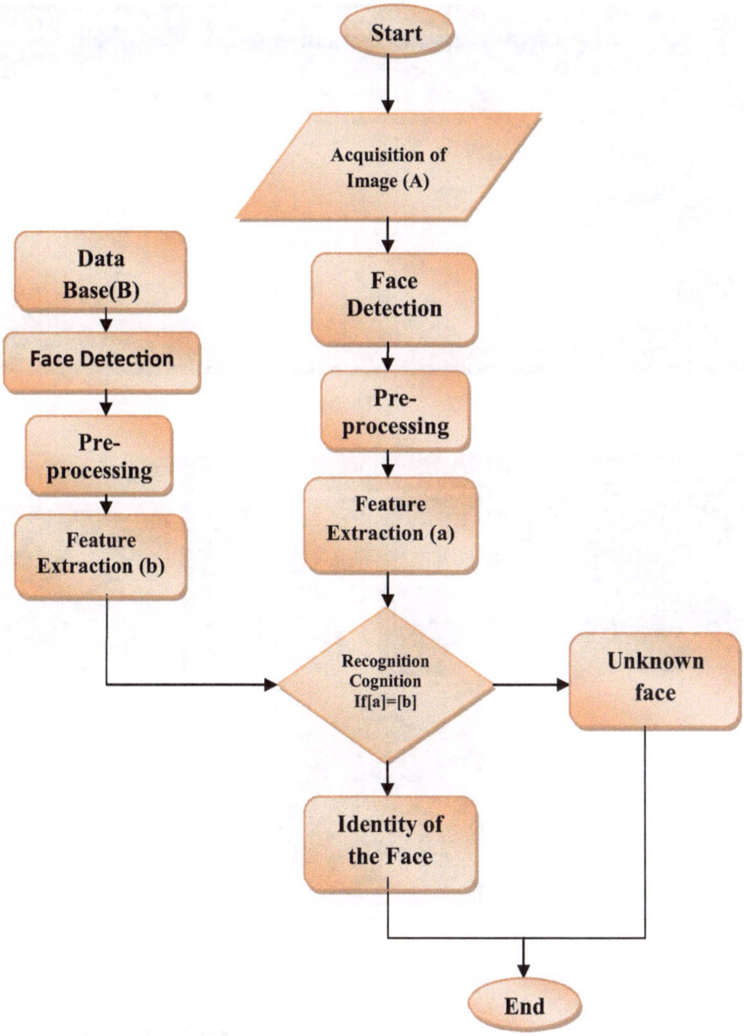

Fig. 11.1 Flow graph of the facial recognition system

searching for missing persons. This thesis focuses mainly on biometric recognition by hybrid fusion of 2D face feature and machine learning. It's demonstrated that the proposed technique beats all compared state-of-the-art and baseline algorithms, which illustrates the robustness of the proposed methods against the appearance variations of articulation, lighting and so forth. The proposed techniques ideally can inspire another new thinking and better approach to handle the face recognition issues.

The feature extraction has processed through 3 stages, i.e. of PCA, ICA and DWT. The PCA resulted in second stage global features, while ICA modified those features

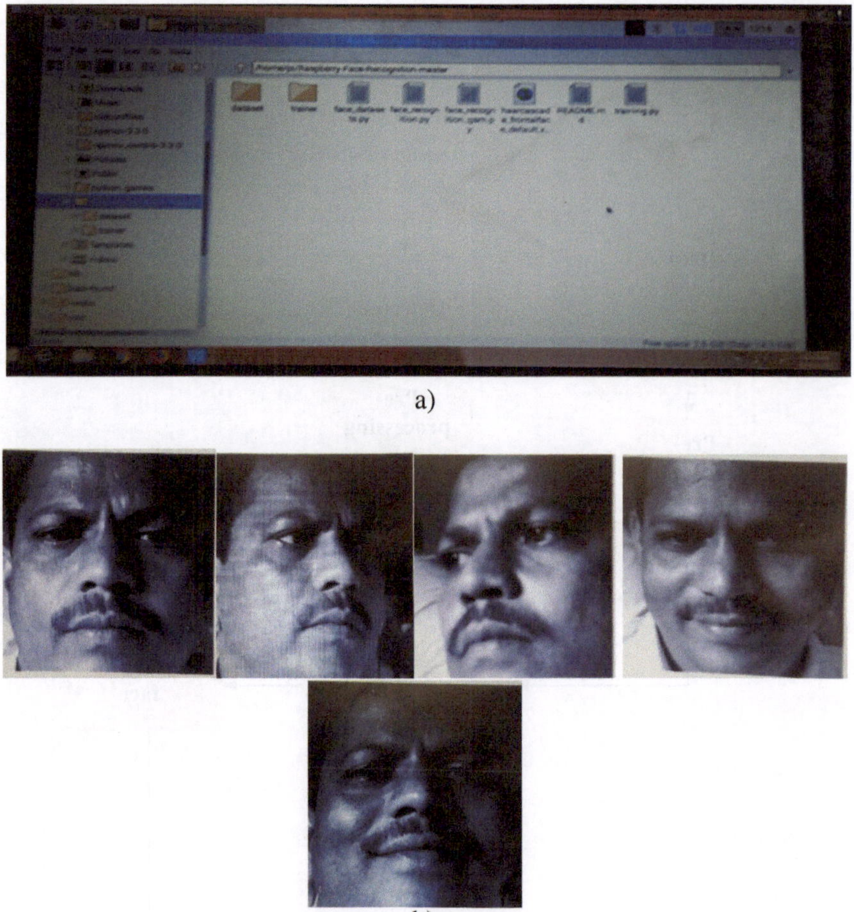

a)

b)

Fig. 11.2 **a**, **b** Training phase in the recognition process

Fig. 11.3 Output of testing image in real-time

and provided local spatially reduced features. For sufficient analysis, the DWT is employed further to ICA for feature extraction, and the final sets of features are classified for face recognition. In recognition, two experiments are parallel conducted; one for SVM and other for Neural Network. The simulation results are compared with the existing works, where PCA, DWT, SVM and other methods are compared for better analysis of the proposed system. It has been observed that the system did not perform optimally with SVM. In opposite to it, Neural Network performed exceptionally well and gave recognition rates up to 96%.

Further, a way to reduce the limitations of single-mode biometric systems is to develop multimodal biometrics. It has been studied the different types of combinations of possible modalities; can use the fusion level in a multimodal approach. The classifications of the characteristics, by the Euclidian difference of other IDs, are achieved by each model. Further, these scores (ID) are fused with the proposed fusion rule to claim the highest accuracy with less training rate. Fusion-based face recognition method gives the precision of around 99.45%. Fusion rule is created based on some hit and trails, which may not be Universal for entire datasets.

Hybrid face recognition, based on the fusion of local information, is proposed. In this texture extraction of image uses LBP + HOG, GLCM, Gabor wavelet decomposition methods separately. DWT has applied in whole train image. Further, the complete extracted features have fused with PCA. It represents faces, which can be roughly reconstructed from a small set of weights and a standard face image. These features are then subjected to a known classifier for its effectiveness in Neural Network.

The developed prototype has been thoroughly tested in Essex, ORL, Indian and YALE B datasets. This approach considered only one image for training from each class. The outcome claims better accuracy by Neural Network (99.40%) than PCA (96.99%). It is said that this research work has achieved a system that meets the objective, initially set up. Though till now, none of the face recognition algorithms has given 100% accurate results under all conditions.

References

1. Jain, A.K., Kumar, A. (2012). Biometric recognition: An overview. In *Second generation biometrics: The ethical, legal and social context* (pp. 49–79).
2. Inamdar, V., Rege, P., & Arya, M. (2010). Offline handwritten signature based blind biometric watermarking and authentication technique using biorthogonal wavelet transform. *International Journal of Computer Applications, 11*(1), 19–27.
3. Choi, J. Y., De Neve, W., Plataniotis, K. N., & Ro, Y. M. (2010). Collaborative face recognition for improved face annotation in personal photo collections shared on online social networks. *IEEE Transactions on Multimedia, 13*(1), 14–28.
4. de Freitas Pereira, T., Komulainen, J., Anjos, A., De Martino, J. M., Hadid, A., Pietikäinen, M., & Marcel, S. (2014). Face liveness detection using dynamic texture. *EURASIP Journal on Image and Video Processing, 1*, 1–15.

5. Hu, H., Ahn, G. J., & Jorgensen, J. (2012). Multiparty access control for online social networks: model and mechanisms. *IEEE Transactions on Knowledge and Data Engineering, 25*(7), 1614–1627.
6. Ding, C., & Tao, D. (2015). Robust face recognition via multimodal deep face representation. *IEEE Transactions on Multimedia, 17*(11), 2049–2058.
7. Unar, J. A., Woo, C. S., & Almas, A. (2014). A review of biometric technology along with trends and prospects. *Pattern Recognition, 47*(8), 2673–2688.
8. Pittman, M., & Reich, B. (2016). Social media and loneliness: Why an Instagram picture may be worth more than a thousand Twitter words. *Computers in Human Behavior, 62*, 155–167.
9. Tyagi, S., & Khanna, P. (2013). Face verification and identification using DCT-NNDA and SIFT with score-level fusion. *International Journal of Biomedical Engineering and Technology, 13*(2), 154–176.
10. Mohammed, A. A., Minhas, R., Wu, Q. J., & Sid-Ahmed, M. A. (2011). Human face recognition based on multidimensional PCA and extreme learning machine. *Pattern Recognition, 44*(10–11), 2588–2597.
11. Luo, M., & Zhang, K. (2014). A hybrid approach combining extreme learning machine and sparse representation for image classification. *Engineering Applications of Artificial Intelligence, 27*, 228–235.
12. Naik, G. R., & Kumar, D. K. (2011). An overview of independent component analysis and its applications. *Informatica, 35*(1).
13. Borgi, M. A., Labate, D., El'Arbi, M., & Amar, C. B. (2013). Shearlet network-based sparse coding augmented by facial texture features for face recognition. In *International conference on image analysis and processing* (pp. 611–620). Springer.
14. Goh, Y. Z., Teoh, A. B. J., & Goh, M. K. O. (2011). Wavelet local binary patterns fusion as illuminated facial image preprocessing for face verification. *Expert Systems with Applications, 38*(4), 3959–3972.
15. Lydia, E. L., Kannan, S., Rajest, S. S., & Satyanarayana, S. (2020). Correlative study and analysis for hidden patterns in text analytics unstructured data using supervised and unsupervised learning techniques. *International Journal of Cloud Computing (IJCC), 9*(2/3).
16. De Marsico, M., Nappi, M., Riccio, D., & Wechsler, H. (2012). Robust face recognition for uncontrolled pose and illumination changes. *IEEE Transactions on Systems, Man, and Cybernetics: Systems, 43*(1), 149–163.
17. Wei, J., Jian-Qi, Z., & Xiang, Z. (2011). Face recognition method based on support vector machine and particle swarm optimisation. *Expert Systems with Applications, 38*(4), 4390–4393.
18. Thai, L. H., Hai, T. S., & Thuy, N. T. (2012). Image classification using support vector machine and artificial neural network. *International Journal of Information Technology and Computer Science, 4*(5), 32–38.
19. Liau, H. F., & Isa, D. (2011). Feature selection for support vector machine-based face-iris multimodal biometric system. *Expert Systems with Applications, 38*(9), 11105–11111.
20. Tome, P., Fierrez, J., Vera-Rodriguez, R., & Nixon, M. S. (2014). Soft biometrics and their application in person recognition at a distance. *IEEE Transactions on Information Forensics and Security, 9*(3), 464–475.
21. Rao, A. N., Vijayapriya, P., Kowsalya, M., & Rajest, S. S. (2020) Computer tools for energy systems. In *International conference on communication, computing and electronics systems* (pp. 475–484). Springer.
22. Smeets, D., Claes, P., Vandermeulen, D., & Clement, J. G. (2010). Objective 3D face recognition: Evolution, approaches and challenges. *Forensic Science International, 201*(1–3), 125–132.
23. Jain, A. K., Klare, B., & Park, U. (2011). Face recognition: Some challenges in forensics. In *2011 IEEE International Conference on Automatic Face and Gesture Recognition (FG)* (pp. 726–733). IEEE, (2011).
24. Babu, N. N., Fathima, A. A., & Vaidehi, V. (2011). An efficient face recognition system using DWT-ICA features. In *2011 International conference on digital image computing: techniques and applications* (pp. 146–151). IEEE.

25. Kämäräinen, J. K., Hadid, A., & Pietikäinen, M. (2011). Local representation of facial features. In: *Handbook of face recognition* (pp. 79–108). Springer.
26. Wagner, A., Wright, J., Ganesh, A., Zhou, Z., Mobahi, H., & Ma, Y. (2011). Toward a practical face recognition system: Robust alignment and illumination by sparse representation. *IEEE Transactions on Pattern Analysis and Machine Intelligence, 34*(2), 372–386.
27. Satonkar Suhas, S., Kurhe Ajay, B., & Prakash Khanale, B. (2012). Face recognition using principal component analysis and linear discriminant analysis on the holistic approach in facial images database. *Int Organ Sci Res, 2*(12), 15–23.
28. Liao, S., Jain, A. K., & Li, S. Z. (2012). Partial face recognition: Alignment-free approach. *IEEE Transactions on Pattern Analysis and Machine Intelligence, 35*(5), 1193–1205.
29. Han, H., Klare, B. F., Bonnen, K., & Jain, A. K. (2012). Matching composite sketches to face photos: A component-based approach. *IEEE Transactions on Information Forensics and Security, 8*(1), 191–204.
30. Belhumeur, P. N., Jacobs, D. W., Kriegman, D. J., & Kumar, N. (2013). Localising parts of faces using a consensus of exemplars. *IEEE Transactions on Pattern Analysis and Machine Intelligence, 35*(12), 2930–2940.
31. Datta, D., Mishra, S., & Rajest, S. S. (2020). Quantification of tolerance limits of engineering system using uncertainty modeling for sustainable energy. *International Journal of Intelligent Networks, 1*(2020), 1–8. https://doi.org/10.1016/j.ijin.2020.05.006
32. Zhou, H., Mian, A., Wei, L., Creighton, D., Hossny, M., & Nahavandi, S. (2014). Recent advances on single modal and multimodal face recognition: a survey. *IEEE Transactions on Human-Machine Systems, 44*(6), 701–716.
33. Li, H., Lin, Z., Brandt, J., Shen, X., & Hua, G. (2014). Efficient boosted exemplar-based face detection. In *Proceedings of the IEEE conference on computer vision and pattern recognition* (pp. 1843–1850).
34. Posiyya, A., & Saikumar, K. (2020). Data management by a librarian for academic and research perceptions. *International Journal of Psychosocial Rehabilitation, 24*(8), 545–551.
35. Soumya, N., Kumar, K. S., Rao, K. R., Rooban, S., Kumar, P. S., & Kumar, G. N. S. (2019). 4-bit multiplier design using CMOS gates in electric VLSI. *International Journal of Recent Technology and Engineering (IJRTE).* ISSN 2277-3878.
36. Raju, K., Pilli, S. K., Kumar, G. S. S., Saikumar, K., & Jagan, B. O. L. (2019). Implementation of natural random forest machine learning methods on multispectral image compression. *Journal of Critical Reviews, 6*(5), 265–273.
37. Vijetha, A., Shashirekha, T., Raju, K., Rooban, S., & Saikumar, K. (2020). A robust fake currency detection model using ESVM machine learning technique. *Journal of Advanced Research in Dynamical and Control Systems, 12*(06), 170–179.
38. Saikumar, K., & Rajesh, V. (2020). Coronary blockage of the artery for heart diagnosis with DT artificial intelligence algorithm. *International Journal of Research in Pharmaceutical Sciences, 11*(1), 471–479.
39. Ding, C., & Tao, D. (2016). A comprehensive survey on pose-invariant face recognition. *ACM Transactions on Intelligent Systems and Technology (TIST), 7*(3), 1–42.
40. Al-asadi, T. A., & Obaid, A. J. (2016). Object Based Image Retrieval Using Enhanced SURF. *Asian Journal of Information Technology, 15*(16), 2756–2762.
41. Meshram, C., Ibrahim, R. W., Obaid, A. J., Meshram, S. G., Meshram, A., & Abd El-Latif, A. M. (2020) Fractional chaotic maps based short signature scheme under human-centered IoT environments. *Journal of Advanced Research.*
42. Obaid, A. J., Alghurabi, K. A., Albermany, S. A. K., & Sharma, S. (2021) Improving extreme learning machine accuracy utilizing genetic algorithm for intrusion detection purposes. In *Advances in intelligent systems and computing* (pp. 171–177). Springer.
43. Obaid, A.J. (2021). Critical research on the novel progressive, JOKER an opportunistic routing protocol technology for enhancing the network performance for multimedia communications. In *Research in intelligent and computing in engineering. Advances in intelligent systems and computing*, (vol. 1254, pp. 369–378). Springer.

44. Sharma, S., & Obaid, A. J. (2020). Mathematical modelling, analysis and design of fuzzy logic controller for the control of ventilation systems using MATLAB fuzzy logic toolbox. *Journal of Interdisciplinary Mathematics, 23*(4), 843–849.

Chapter 12
Machine Learning Based Online Handwritten Telugu Letters Recognition for Different Domains

B. Sankara Babu, Sunanda Nalajala, K. Sarada, V. Muniraju Naidu, Nagendar Yamsani, and K. Saikumar

Abstract Trending innovations utilized in numerous handwriting (HW) pattern detection of diverse individuals are machine learning and neural networks. The (CNN) handwritten characters of multiple entities are also quite challenging to identify and understand. Recognition of names linked to the Telugu film industry is part of the reorganization of trends in the study concept over the past few years. Neural networks (NN) play an essential role in the identification of HW character by Telugu. A digital machine's ability to obtain and comprehend intelligible HW data from records, photographs, touch screens and several other electronic devices etc. is HW recognition. These can be online or offline. In this case, online recognition involves translating visual pen tip gestures into a collection of originates used as input for the classification scheme where photographs of characters such as information were being used as unencrypted identification. This is a successful accomplishment for HW reorganization NN was accomplished, and the productivity up to 98.3% increased. Use the Convolutional Neural Network (CNN) to learn how to create an

B. Sankara Babu (✉)
Department of Computer Science and Engineering, Gokaraju Rangaraju Institute of Engineering and Technology, Hyderabad, India

S. Nalajala
Computer Science and Engineering, Koneru Lakshmaiah Education Foundation, Vaddeswaram, Andhra Pradesh, India
e-mail: sunanda@kluniversity.in

K. Sarada
Electrical and Electronics Engineering, KLEF, Vaddeswaram, India
e-mail: saradak@kluniversity.in

V. Muniraju Naidu
Department of Computer Science and Engineering, Narayana Engineering College (Autonomous), Nellore, India

N. Yamsani
Department of Computer Science and Engineering, SR Engineering College, Warangal, India

K. Saikumar
Electronics and Communication Engineering, Malla Reddy University, Hyderabad, India
e-mail: saikumarkayam4@ieee.org

P. Kumar et al. (eds.), *A Fusion of Artificial Intelligence and Internet of Things for Emerging Cyber Systems*, Intelligent Systems Reference Library 210,
https://doi.org/10.1007/978-3-030-76653-5_12

image recognition engine. CNN is a standard network where it is possible to train a computer to identify pictures based on picture patterns. It can be used once learned to recognize objects in the images.

Keywords Neural networks · Handwritten characters · Pattern reorganization · Convolutional Neural Network (CNN) · Recognition of characters (RC)

12.1 Introduction

Definitions of hand-designed characteristics involve densities of pixels over the image, curvature of personality, proportions & proportions, and vertically & horizontally rows. Vikram and et then used [MLPS] multilayer perceptron and achieved 98.3 overall accuracies on the handwriting Telugu personality data source.

Figure 12.1 represents the Telugu words' data collection, i.e. kaagunithalu & shay vothuluetc, both chosen from the data set accessible on XXX. At the final training samples, approximately 300 samples are evaluated as image templates.

Figure 12.2 of text includes single Unicode moral principles dependent upon that post-processing, which illustrates the data set definition of Unicode morals to be done.

This chapter gives a study and different methods for recognizing online handwritten characters' recognition in the Telugu language. An inevitable part of Optical Character Recognition (OCR) As OCR represents optical character acknowledgement, the question of identifying all sorts of different characters is solved through OCR technology. It is possible to understand and translate all handwritten and typed characters to a machine-readable, digital data format. And a very challenging field of pattern recognition is written by hand Character Recognition (HCR). For handwritten character recognition for Marathi, different approaches have been applied. Here we focus primarily on analyzing the execution, accuracy and comparison of some of the current web character recognition systems of Telugu screenplays, underfitting issue and why HMM & SVM remain efficient ways aimed at online character recognition. Telugu remains a north Indian language primarily spoken in the southern regions of India. It consumes nearly 160 million fluent languages, making it the world's largest fifteenth-most widely speech-language. We provided a complete directory of handwritten Tamil protagonists to start driving text classification for this template [1]. We

Fig. 12.1 Example data set

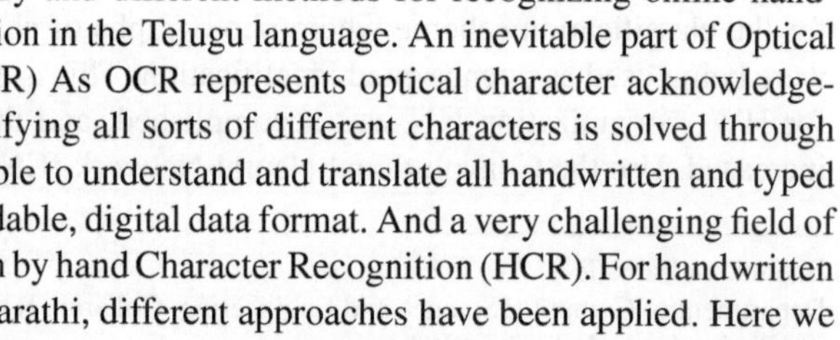

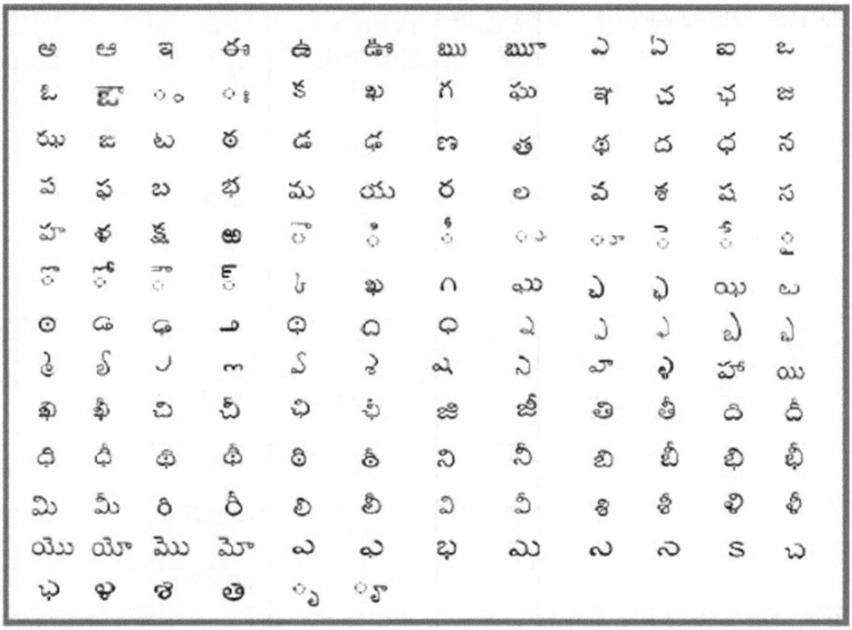

Fig. 12.2 The 150 data from available data sets

assume that all this is important as we have placed together with the Telugu orthography's most extensive set of vowels, consonant, vowel-consonant, & consonant-consonant sets. This work generates such a directory to typewritten real-world offline plot lines harvested from document images, making it the domain's most extensive and most varied data system. It explains the method of data collection, pre-processing steps, and the evaporation approach for obtaining individual Telugu characters.

The data set remains also made available for use by way of an experiment set towards evaluate text classification approaches then other maintenance works. This work too suggested a system of handwriting recognition of Telugu characters using Convolutional Neural Networks by way of a baseline classification model and a more efficient and comprehensive solution for Visual Attention Networks in Fig. 12.3. Finally, the new structure is contrasted with existing methods, and the outcomes are mentioned [2].

Fig. 12.3 Unicode values

Class Id	Telugu Symbol	Unicode
0	అ	0C05
1	ఆ	0C06
2	ఇ	0C07
3	ఈ	0C08
4	ఉ	0C09
5	ఊ	0C0A
6	ఋ	0C0B
7	ౠ	0C60
8	ఎ	0C0E
9	ఏ	0C0F
10	ఐ	0C10

12.2 Methodology

12.2.1 Requirements

In the HP dataset of Telugu characters, the model's assessment criterion would be transcription accuracy on the test photos. Accuracy is described as effectively anticipating Telugu character transcribed in the picture. The theory must accurately estimate not only the number of digits present. However, every one of those digits should likewise be accurately defined. I will test it against a specific model, such as transcribed Hindi/Tamil/Bengali character acknowledgement in both precision & tempo, as stated above, about benchmarking against past work performance in Fig. 12.4.

12.2.2 Data Investigation

The dataset was uploaded as from webpage of HP Labs India.

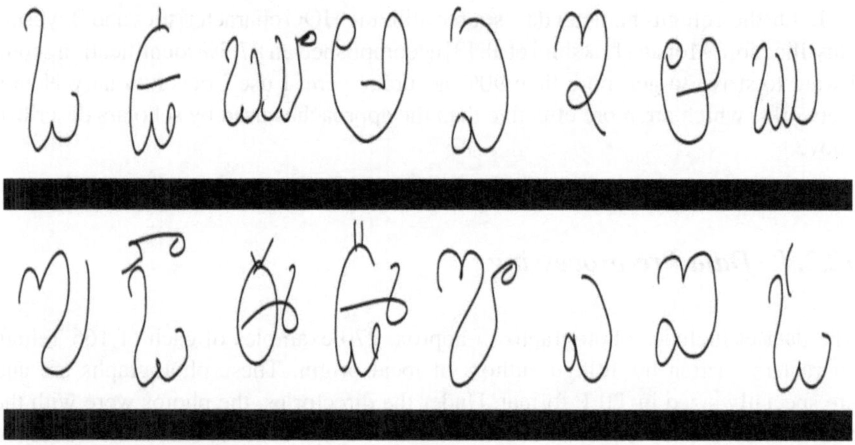

Fig. 12.4 Telugu letters vowels

12.2.3 Convolution Neural Network Algorithm

$$x_j^l = f(\sum_{i \in M_j} x_i^{l-1} k_{ij}^l + b_j^l) \tag{12.1}$$

$$x_j^l = f(\beta_j^l down(x_j^{l-1}) + b_j^l) \tag{12.2}$$

I plan to equate the observations to past projects. I would analyze the exactness/mean-squared mistake to see which one is the most accurate and compare my approach's pace utilizing CDNN and the past methods utilized in Fig. 12.5.

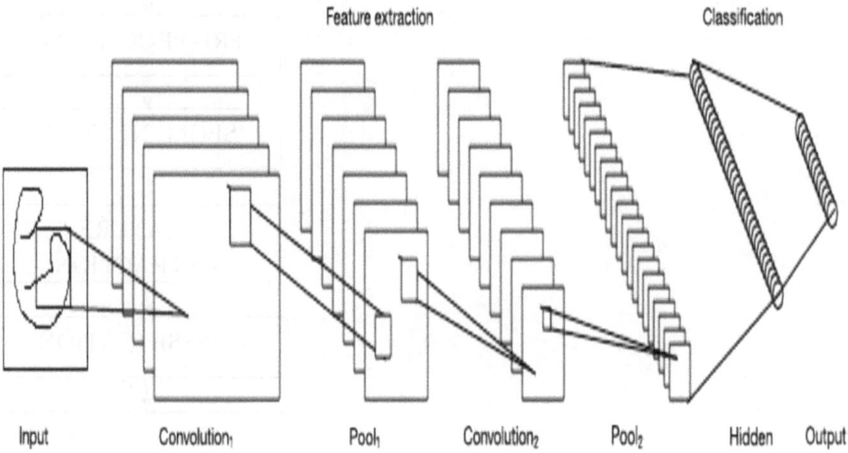

Fig. 12.5 CNN network

K. On the Telugu character data source utilizing HOG characteristics and Bayesian classification, Mohana Lakshmi et al. [3] accomplished an 87.5% identification score. I want to strive to get more than 90% accuracy here. I use Coevolutionary Neural Networks, which are more effective than the approaches used by scholars described above.

12.2.4 Data Pre-processing

The dataset includes photographs of approx. 270 examples of each of 166 Telugu characters written by Telugu authors of local origin. These photographs are and are specially sized in TIFF format. Under the directories, the photos were with the author's names and the picture document's name as the Unicode character. I modified the filename of the picture here as the essayist's name followed by the Unicode character with the goal that each picture gets a unique name for the image. And then all the photos of the same Unicode were grouped as their folder name under a folder of Unicode, so I could utilize the load files model in the method learn. datasets2. The initial elliptical pictures of a variable dimension are then resized into 80×80 square images and stored as JPG files [4] and in Fig. 12.6.

We evaluated the job done on Telugu linguistic scripts on Online written by hand Character Recognition (OHCR) in this work. Here, we briefly addressed multiple stages used to advance OHCR and work done on Telugu text recognition. This could remain used by way of a step in developing the OHCR Telugu message system. We

Fig. 12.6 Flow diagram

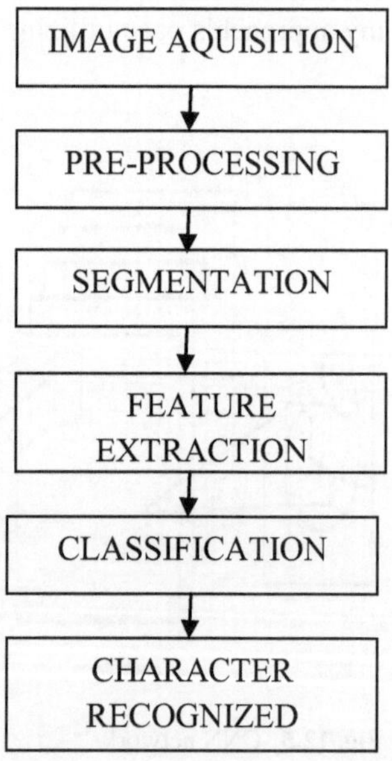

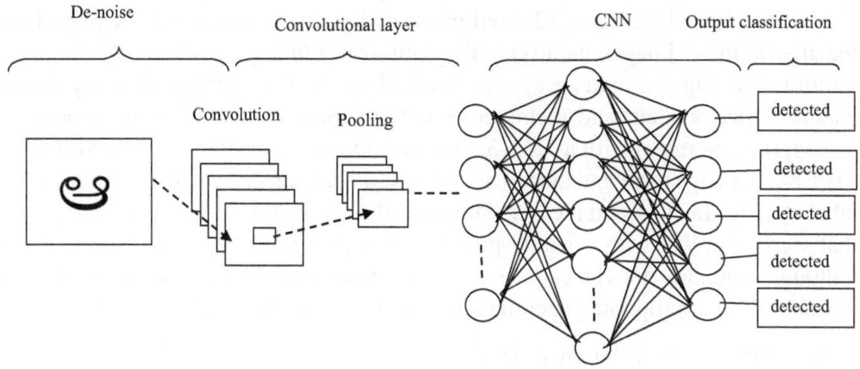

Fig. 12.7 HW character recognition

mentioned most purchase techniques such as HMM and SVM and during the evaluation of OHCR processes and troubles within these methodologies and issues in other methodologies, their accuracy, and their strategies of identification and approach. Due to the more significant commonality of actors and molecule protagonists in the Telugu language, we have found no complete, precise OCR systems. Due to overfitting problems, the significant issues in the data pre-processing concert and optimization stage. The scope for developing an OHCR scheme with greater accuracy in font, autonomous size and typewritten Telugu text is available. The figures, tables, and models given in this paper depend on the authors' examination of the improvement of character acknowledgement frameworks for Telugu's southern India mother tongue. As described in the methods and algorithms, the algorithm utilized to train the knowledge is deep learning neural network. A 4D Tensor is a contribution to the deep neural network. The 80 × 80 picture is reconfigured into a shaped 4D tensor (examples, channels, lines, sections) in Fig. 12.7, which is then gone through a heap of various layers and follows:

Input—4D form tensor (n, 80, 80, 1) where n is the quantity of pictures information, and 1 is the number of channels as the pictures are parallel. (Channel number = 3, if the videos remain RGB).

Layer 1—Coevolutionary Layer with 32 philtres then philtre size 3 × 3 as Max pooling layer by philtre size 2 × 2.

Layer 2—Coevolutionary Layer with 64 philtres and philtre size 3 × 3 as Max pooling layer by philtre size 2 × 2.

Layer 3—Coevolutionary Layer with 128 philtres and philtre size 3 × 3 and max-pooling with philtre size 2 × 2. Flatten) (——— This converts the output of.

Layer 4—Fully Connected layer (Dense) with 256 neurons, with a dropout normalization rate of 0.4 (0.4 possibilities that any specified component should be released throughout coaching).

Ultimately Linked layer (Dense) by138 neurons. Layer 5—RELU direct actuation is utilized here for convolution layers, and complete pooling layers and SOFTMAX enactment for the yield layer (Fully Linked Layer) is used.

Initially, I tried the 32 × 32 sized pictures. I got even less precision when I ran the algorithm, so I began modifying the optimizer. I first used Adam and attempted multiple learning rates and decay variations. Then, even though the efficiency did not change, I tried adadelta and attempted boundary tuning shown in 7. I experienced the dataset (i.e., pictures) with a 32 × 32, which was not noticeable. Then modified the scale of the image to 64 and then to 80. Then saw some repetitive pictures of different label names which caused meagre output for the algorithm. Then deleted the photos that were identical with separate product titles [5, 6]. Then I got 138 characters, although originally got 166 characters when I downloaded the data from the HP labs website. With appropriate generalization methods, I applied and explored:

- Structure in a coevolutionary layer
- Dropout in a related layer
- A role for Data augmentation
- Batch standardization.

Applying these approaches did not offer a significant output increase. Then I got 88% test accuracy and 91% preparing precision with three convolutional layers and Adam as an analyzer [7]. However, after adjusting the optimizer to SGD with a learning rate of 0.1 and introducing another co-evolutionary sheet, training accuracy improved to 94% and test accuracy to 91%.

12.2.5 Model Evaluation and Validation

On the Telugu character dataset of 20 epochs, I got 91% test accuracy and 95% training accuracy. If we raise the number of ages, then the accuracy would further improve. When preparation improved, the loss declined from 4.03 to 0.4. The preparation exactness is not precisely the approval precision for the initial not many ages and improved test accuracy after several generations. In the approval, bends were given.

The precision gained by the model is greater than the previously stated benchmark. It's possible that including other epochs will help improve precision. The recorded goal is 90% of test accuracy [8]. But the model achieved 91% accuracy and 95% training accuracy by operating 20 epochs. The precision can be further improved by introducing a few more ages. The relation between the accuracy achieved by the CNN model and different models used to identify handwritten Telugu characters is seen in Fig. 12.8.

The distinction amid the salt-and-pepper commotion then spots in the individual character is challenging. A more comprehensive strategy is needed to combat these issues. X is the horizontal orientation and Y is the axis along with the duration of the scan. The suggested method to collect the individual characters follows the following steps: Executing the sweep's Otsu limit followed by Connected Component Review to secure all connected segments inside a region's upper and lower point [9]. Locate the dark dab box by separating the locale fixated on the cases in the image's best

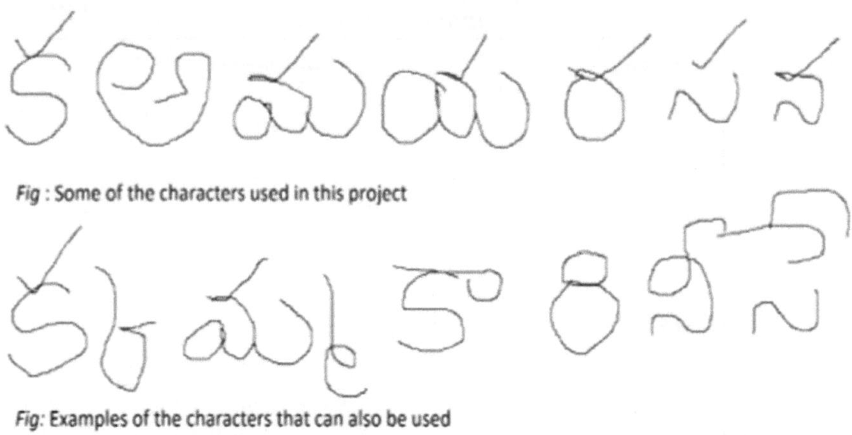

Fig. 12.8 Telugu letters

500 pixels. Afterwards, it utilizes the spot's X-facilitate to get the assortment of pages. The subsequent bundles are consolidated depending on distances between their vertices the X way or whether the crates remain covered. This is rehashed aimed at a set number of cycles, or until not all the more joining is doable [10, 11]. The distance limits are chosen, so the long columns' headers showed in turning out to be one section or a bundle. They will be the most critical boxes in the casing. The bigger containers (which are lines) are gathered, and the leftover boxes are part by these line headers into bunches [12]. The two packages with the least difference between their centres are combined if a collection does not have ten boxes in it. Until precisely, ten boxes remain; this step is repeated [13]. The characters in both sets remain sorted thru the y-coordinate at the bottom right and organized by a unique mark depending on the page number, row & column in which they are placed. In the following subsections, each of the primary measures is extended in Fig. 12.9.

On a subset of both the dataset obtained through the procedure detailed in the preceding part, all channels are educated, validated and checked. There were two experiments carried out. With just Telugu consonant clusters (36 classes), one was, with both vowels (52 lessons), the other was. With such an 81:9:10 split of the overall data across all considered characters, the train, validation and scientific history were developed. To achieve that splits and implement equal checking of the templates, sampling design was introduced. To get an equivalent amount of photos per individual, the train collection was then arbitrarily oversampled. Both models have been programmed by an early stop procedure that stops training if the lack of validity over a specified number of iterations does not decrease [14]. Training to avoid overfitting was done with on-the-run transfer learning in Fig. 12.10.

Identification: By far, the final main stage in the method of identifying handwriting characters is classification. In this stage, different techniques or models remain used to map other groups of the feature vectors and thus define the symbols or words expressed by the features in Table 12.1. The classifier to be used is determined

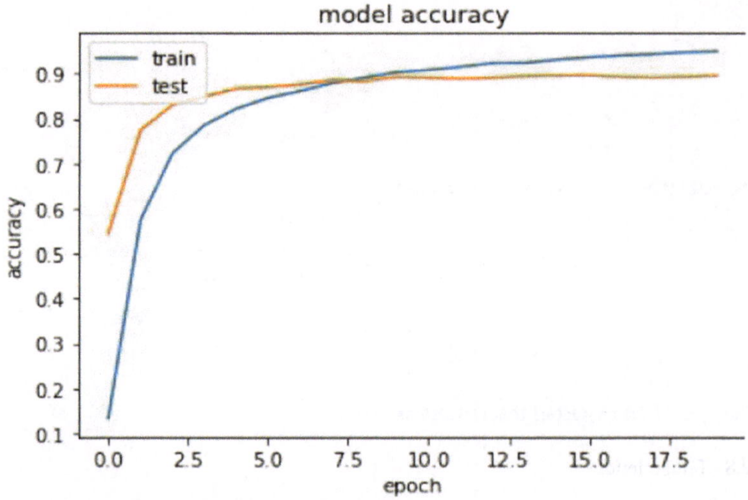

Fig. 12.9 Telugu letters model accuracy

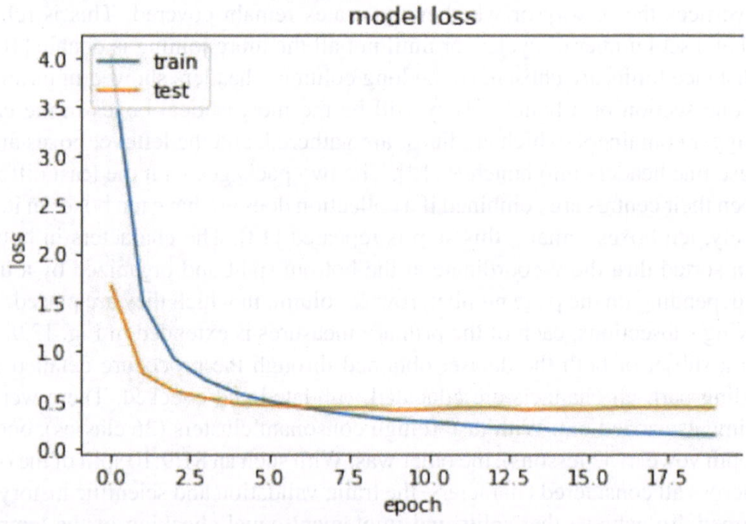

Fig. 12.10 Telugu letters model loss

based on different variables, taking account the issues of the world environment in Fig. 12.11. A mixture of the algorithm is also often used for classification because it is more powerful than having a choice to keep [15].

A CNN close in form towards the baseline model is used for the focus network. The awareness process falls after the convolution and max-pooling surfaces [16, 17]. It is introduced as another coevolutionary layer in this design that takes the past data.

Table 12.1 Results analysis

Method	Accuracy
CNN (Proposed Method)	91%
[1] Zoning feature	78%
[2] Adaptive and static zoning methods	88.8% for 50 characters
[3] Two dimensional fast Fourier transform and support vector machine	71%
[4] Multi-layer perceptron's (MLP)	85%
[5] Bayesian classifier	87.5%

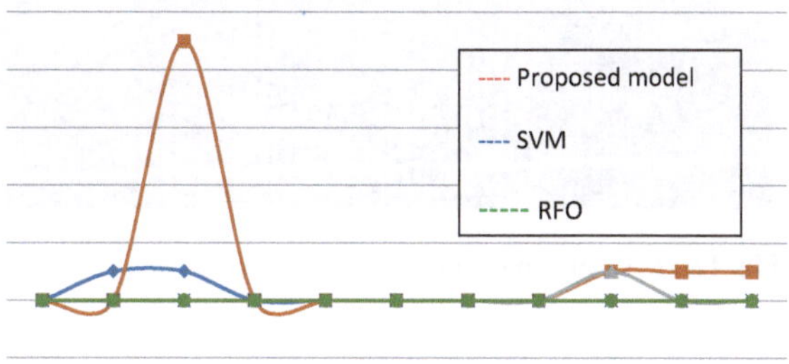

Fig. 12.11 Comparison of results

Still, it consumes a 1 × 1 kernel or the same number of philtres by way of the last layer, and a function of Schiff bases derived to send the attention vector [18].

An insightful component duplication between the yield of the underlying layers (up to the produced" of focus), then the focus vector provides a snapshot of interest [19]. Different components (4 × 4 × 128 to 4 × 512) are then transformed, then the axes remain translated as time stages to measurements per time phase. This serves by way of contribution to a bi-directional Long Short-Term Memory (LSTM) organization that functions in this design as the transmitter shown in Fig. 12.12 and Table 12.2 [20].

By no way studies on manuscript identification of Telugu characters a modern frontier. Field study goes back further than a decade ago [21, 22]. The strategies for addressing the issue are rather complicated and thus impossible to evaluate. In comparison, in terms of the number of observations and groups, the datasets they use are somewhat different. The variety per class-where a class remains an instance of a specific Telugu individual. Therefore, as those accessible for English OCR questions, there is a solid requirement for a genuine benchmark dataset [23]. The difficulty in creating a comprehensive database for any activity relating to handwriting is to capture the heterogeneity in writers that represent the total population of writers. This need, under the same circumstances, a vast number of authors to create the same characters. This is done in this work by making form-like sheets of boxes per character. In their natural handwriting, the writers fill out this form. The exact process in which this sheet-level dataset remains produced is defined elsewhere [24–28]. Now comes the challenge of removing from the pictures the actual characters. Owing to a

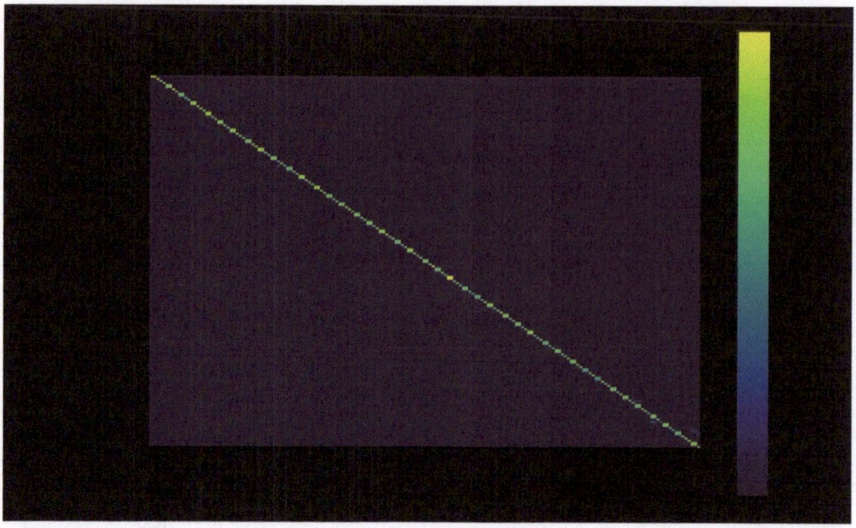

Fig. 12.12 Confusion matrix analysis

Table 12.2 Test versus training

Original	Tested	Original	Tested
47	47	110	110
46	46	134	134
60	60	67	67
68	68	59	59
121	121	65	65
15	15	119	119
119	119	72	72

human error then the scanning process, the issue is the intrinsic noise in the images. This contributes to several complex concerns explored in subsequent sections, along with the procedures adopted to mitigate their impact [29, 30].

Finally, pathways to identification are discussed. Using Coevolutionary Neural Networks (CNN) also auto-encoder systems by an array classifier are the approaches suggested for this mission. Besides, with modest performance, the extraction of features by versatile and static drafting techniques or 2D Fourier Transforms also Support Vector Machines (SVM) & Quadratic Discriminate Classifiers has been tested. Through factual pruning dependent on use or focusing on either a subset of characters (like only phonemes, for instance), all these techniques use smaller classes. In this thesis, the design examined is that of a Visual Attention Network that utilizes a CNN pair with a consideration framework and a Recurrent Neural Network (RNN) for encoding towards concentrate minutely on some observable traits. The primary emphasis of this article, although, is the production of datasets. Because of its space restriction, the complete value cannot be clarified in more depth.

12.3 Conclusion

The real labels and the expected marks of the initial 50 pictures, aside from 2 or 3 out of 50 images, indicate that the model predicts nearly the right labels. The Confusion Matrix is used to verify neural network recognition accuracy. Displays the uncertainty grid plot used to distinguish transcribed Telugu Characters for the model. Deep coevolutionary neural networks are well known to be very effective at classifying picture images. Many studies have been carried out to identify handwritten characters utilizing coevolutionary neural organizations for English letters in order, numbers, Chinese, Arabic, and some Indian dialects, such as Hindi. Devanagari, etc. Still, there is far less contribution to the identification of Telugu language characters. Due to so limited research, the material accessible on the internet is not so strong. Initially, I didn't note it and encountered several repetitive data issues. I noticed that the data was the issue, so I removed the repetitive ones and explained it. Then I noticed that tuning the methodology was hard. I tried numerous optimizers with varying learning speeds and tried a model adjusting number of layers and number of philtres and hilter size. At long last, I wound up with a model that gives 91 per cent precision that met my hopes of having about 85 to 90 per cent.

12.3.1 Future Work

This application is most prominently used in future language detection schemes; earlier stage models cannot exactly work for the character recognition process. Therefore, this type of deep learning and machine learning model with API applications gives humans better experience at unknown languages.

References

1. Sastry, P. N., Lakshmi, T. R. V., Rao, N. V. K., Rajinikanth, T. V., Wahab, A. (2014). Telugu handwritten character recognition using zoning features. In *2014 International conference on IT convergence and security (ICITCS)*.
2. Prasad, S. D., & Kanduri, Y. (2016) Telugu handwritten character recognition using adaptive and static zoning methods. In: *2016 IEEE students' technology symposium (TechSym)*. Kharagpur.
3. Dara, R., Panduga, U. (2015). Telugu handwritten isolated characters recognition using two dimensional fast Fourier transform and support vector machine. *International Journal of Computer Applications*.

4. C. Vikram, C. ShobaBindu, C. Sasikala, "Handwritten Character Recognition for Telugu Scripts Using Multi LayerPerceptrons (MLP)", 2013 International Journal of Advanced Research in Computer Engineering & Technology, 2013.
5. Mohana Lakshmi, K., Venkatesh, K., Sunaina, G., Sravani, D., & Dayaka, P. Hand written Telugu character recognition using Bayesian classifier. In: P. Vijayaraghavan, & M. Sra (Eds.), Handwritten Tamil recognition using a convolutional neural network. https://web.media.mit.edu/~sra/tamil_cnn.pdf.
6. Pal, U., Sharma, N., Wakabayashi, T., & Kimura, F. (2008). Handwritten character recognition of popular South Indian scripts. In D. Doermann, & S. Jaeger (Eds.) *Arabic and Chinese handwriting recognition. SACH 2006. Lecture notes in computer science* (vol. 4768). Springer. https://doi.org/10.1007/978-3-540-78199-8.
7. Varalakshmi, A., Negi, A., & Krishna, S. (2012). Dataset generation and feature extraction for Telugu hand-written recognition. *International Journal of Computer Science and Telecommunication, 3*(3), 57–59.
8. Soman, S. T., Nandigam, A., Chakravarthy, V. S. (2013) An efficient multiclassifier system based on convolutional neural network for offline handwritten Telugu character recognition. In: 2013 National conference on communications (NCC) (pp. 1–5). New Delhi, India. https://doi.org/10.1109/NCC.2013.6488008.
9. Dewan, S., & Chakravarthy, S. (2012) A system for offline character recognition using auto-encoder networks. In T. Huang, Z. Zeng, C. Li, C. S. Leung (eds.) *Neural information processing. ICONIP 2012. Lecture notes in computer science* (vol. 7666). Springer. https://doi.org/10.1007/978-3-642-34478-7.
10. Prameela, N., Anjusha, P., Karthik, R. (2017). Offline Telugu handwritten characters recognition using optical character recognition. In *2017 International conference of electronics, communication and aerospace technology (ICECA)* (vol. 2, pp. 223–226). https://doi.org/10.1109/iceca.2017.8212801.
11. Prasad, S. D., & Kanduri, Y. (2016). Telugu handwritten character recognition using adaptive and static zoning methods. In *2016 IEEE students technology symposium (TechSym)* (pp. 299–304). https://doi.org/10.1109/techsym.2016.7872700.
12. Dara, R., & Panduga, U. (2015). Telugu handwritten isolated characters recognition using two dimensional fast Fourier transform and support vector machine. https://doi.org/10.5120/20330-0820.
13. Salazar, J., Kirchhoff, K., Huang, Z. (2019). Self-attention networks for connectionist temporal classification in speech recognition. In: *ICASSP 2019—2019 IEEE International conference on acoustics, speech and signal processing (ICASSP)* (pp. 7115–7119). https://doi.org/10.1109/ICASSP.2019.8682539.
14. Xu, K., Ba, J., Kiros, R., Cho, K., Courville, A., Salakhudinov, R., Zemel, R., & Bengio, Y. (2015). Show, attend and tell: Neural image caption generation with visual attention. In *International conference on machine learning* (pp. 2048–2057).
15. Srivani, I., Prasad, G. S. V., & Ratnam, D. V. (2019). A deep learning-based approach to forecast ionospheric delays for GPS signals. *IEEE Geoscience and Remote Sensing Letters, 16*(8), 1180–1184.
16. Mannepalli, K., Sastry, P. N., & Suman, M. (2016). MFCC-GMM based accent recognition system for Telugu speech signals. *International Journal of Speech Technology, 19*(1), 87–93.
17. Sivakumar, S., Nayak, S. R., Vidyanandini, S., Kumar, J. A., & Palai, G. (2018). An empirical study of supervised learning methods for breast cancer diseases. *Optik, 175*, 105–114.
18. Narasinga Rao, M. R., Prasad, V. V., Teja, P. S., Zindavali, M., & Reddy, O. P. (2018). A survey on prevention of overfitting in convolution neural networks using machine learning techniques. *International Journal of Engineering and Technology (UAE), 7*(2.32), 177–180.
19. Santoso, L. W., Singh, B., Rajest, S. S., Regin, R., & Kadhim, K. H. (2021). A genetic programming approach to binary classification problem. *EAI Endorsed Transactions on Energy, 8*(31), 1–8. https://doi.org/10.4108/eai.13-7-2018.165523
20. Abdy Sayyed, M. A. H., Gupta, R., & Tanyimboh, T. (2019). Combined flow and pressure deficit-based penalty in GA for optimal design of water distribution network. *ISH Journal of Hydraulic Engineering.*

21. Agrawal, R., Jana, D., Upadhyay, R. K., & Rao, V. S. H. (2018). Dynamic relationship between the mutual interference and gestation delays of a hybrid tritrophic food chain model. *ANZIAM Journal, 59*(3), 370–401.
22. Rajest, S. S., & Suresh, P. (2018). The "Four Cs" education for 21st century's learners. *Research Guru Online Journal of Multidisciplinary Subjects (RGOJMS), XII*(I), 888–900.
23. Gupta, J., Singla, M. K., Nijhawan, P., Ganguli, S., & Rajest S. S. (2020). An IoT-based controller realization for PV system monitoring and control. In A. Haldorai, A. Ramu, S. Khan (Eds.), *Business intelligence for enterprise internet of things. EAI/Springer innovations in communication and computing*. Springer.
24. Ahmed, S. M., Madhuri, G., Reddy, M. S. & Condoor, S. S. (2018). Skill development in freshmen by adopting project based learning-"introduction to engineering" course. *Journal of Engineering Education Transformations, 12*(Special Issue 6).
25. Ali, M. Y., & Ahmed, S. M. (2017). A delay-sensitive unwired net to improve quality transmission. In: *Proceedings 2017 International conference on recent trends in electrical, electronics and computing technologies, ICRTEECT 2017* (pp. 17).
26. Meshram, C., Ibrahim, R. W., Obaid, A. J., Meshram, S. G., Meshram, A., & Abd El-Latif, A. M. (2020). Fractional chaotic maps based short signature scheme under human-centered IoT environments. *Journal of Advanced Research.*
27. Obaid, A. J., Alghurabi, K. A., Albermany, S. A. K., & Sharma, S. (2021). Improving extreme learning machine accuracy utilizing genetic algorithm for intrusion detection purposes. In *Advances in intelligent systems and computing* (pp. 171–177). Springer
28. Obaid, A.J. (2021). Critical research on the novel progressive, JOKER an opportunistic routing protocol technology for enhancing the network performance for multimedia communications. In *Research in intelligent and computing in engineering. Advances in intelligent systems and computing* (vol. 1254, pp. 369–378). Springer Singapore.
29. Sharma, S., & Obaid, A. J. (2020). Mathematical modelling, analysis and design of fuzzy logic controller for the control of ventilation systems using MATLAB fuzzy logic toolbox. *Journal of Interdisciplinary Mathematics, 23*(4), 843–849.
30. Al-Yasiri, D. S. Q., & Obaid, A. J. (2018). A new approach for object detection, recognition and retrieving in painting images. *Journal of Advance Research in Dynamic and Control System, 10*(2), 2345–2359.

Chapter 13
Mutation Testing and Web Applications—A Test Driven Development Approach for Web Applications Built with Java Script

S. Suguna Mallika and D. Rajya Lakshmi

Abstract Defect Prevention refers to the process of ensuring that no defects occur in the first place rather than later detecting defects and correcting them. A test-driven development approach naturally contributes to defect prevention along the software development process pipeline. A generic test suite that suits the testing needs of any web application is needed. The test suite should serve a regression test suite's purpose to ensure an application's general health providing a confident user experience. The authors presented a generic set for checking web apps, mutation operators. In this work, the operators' comprehensive suite is presented as an exhaustive regression test suite to prevent web application defects. Small and medium business enterprises can bank on the regression test suite proposed for customer retention while saving on their testing budgets. The test suite promises a robust web app persistent against any kind of malignant usage.

Keywords Mutation testing · Web application testing · Mutation operators · Defect prevention

13.1 Introduction

Organizations are increasingly relying on web applications to manage their operations, interdepartmental transactions, project management, etc. Web applications' flexibility and remote monitoring and management facilities have made them sought after solutions for small and medium and large companies [2]. The business impact created due to web applications' advent ranges from creating a global opportunity for businesses to ease information. From the reinvention of business models to a customer-centric services offering, web applications made a groundbreaking impact.

S. Suguna Mallika (✉)
CSE, CVR College of Engineering, Hyderabad, Telangana, India

Part-time Research Scholar, JNTUK, Kakinada, Andhra Pradesh, India

D. Rajya Lakshmi
Professor, CSE, JNTU UCEN, Narsaraopet, Guntur District, Andhra Pradesh, India

P. Kumar et al. (eds.), *A Fusion of Artificial Intelligence and Internet of Things for Emerging Cyber Systems*, Intelligent Systems Reference Library 210,
https://doi.org/10.1007/978-3-030-76653-5_13

The traditional brick and mortar business establishment model is owned by an individual/few of them, from interactive games to online polls, blogs to online forums, social media. The applications to online reservations of bus, train, flights, eCommerce/online shopping cart applications. Web application development's content management systems offer a scope and span across almost every society targeted at different age groups.

However, the expedition in developing web applications is a direct cause of concern for the consumers given the information security threats. Despite the availability of commercial off the shelf components (COTS), most developers develop websites from scratch. However, they may make mistakes that can stop the web page visitors from turning into paying customers. Some of the most common mistakes made by web developers often overlooked by them, result in a deep crisis once the web application gets into deployment mode [1]. Mutation testing is an efficient testing procedure for exposing some of the potential defects of software. Mutation testing could be divided into three types: Strong mutation, weak mutation, & firm mutation. While robust mutation testing emphasizes changing the source program with lots of changes one at a time, soft mutation testing makes fewer changes to the source program. Firm mutation goes midway between substantial Modification and soft mutation testing. However, owing to its lengthy and exhausting test suites, not many testers and developers prefer it. Mutation testing deploys a set of operators on the application that can test. These operators are meant to replace source code, and the application runs against test cases. If the application does not deliver output that was supposed to be with the unmutated code, the test case is a pass. If the application gives the result as expected of an unmutated source code, then the mutation has did not affect the application.

It could be perceived in two ways, either the input sample size needs to be enhanced, or there needs to be an investigation backward with what could be wrong in the code's functioning. Several debates centred around whether mutation testing is scalable is practical with real-world software that the industry is producing aggressively. But by prioritizing a set of mutants coupled with a Test-driven development approach for web applications might result in healthier, safer, and secure web applications for organizations to help retain the existing customer base. Vandana et al. [29] summarized the various web application vulnerabilities and attention points for testers to ensure healthy web applications. With the test-driven development picking up in the latest software development processes, an attempt to recommend a mutation operator suite for a web app's test-driven development is made here. The web app would be incorporated with desired security and performance features during the development phase to avoid the ice cream cone anti-patterns in testing.

There are different categories of faults possible with web applications due to their complex heterogeneous construction and deployment [23]. Some of the classified defects pertain to web app specific defects, incorrect session management, wrong multilanguage support, faulty form construction, cross-browser portability problems, missing plugins, authentication problems, the inappropriate cookie value, updation of database failures, link conversion problems, GUI faults, improper session management, multi-tier architecture problems, operating system compatibility, and multi-window faults [4, 11]. The listed defects are summarized. Existing systems or earlier literature in mutation testing focused mostly on replacing operators or link transitions between pages of a web application [3]. However, not much work done in the mutation testing area related to authentication, incorrect cookies, failures about database updation, and session management handled incorrectly. The authors worked in the areas above for inclusiveness of non-functional attributes. The authors proposed 19 operators for mutation testing of web apps and a tool MUTWEB developed. Exclusively to test servlet-based web apps to be tested against these 19 operators. The classification of operators was done based on the non-functional attributes performance and security and presented.

Presently we will see every day pass-software JavaScript problems and a way to restore them. This recollects records for using program dev instruments to find and fasten troubles, using Polyfills and libraries to paintings around issues, getting present-day JavaScript highlights operating in more mounted packages, and that is only the tip of the iceberg.

An exceptional deal of exploration has to be achieved to recognize the principle drivers of programming new agencies' disappointments. One of the essential explanations in the back of such disappointments was low pleasant confirmation at some point of the product advancement measure.

Programming trying out is an essential part of the product advancement lifecycle. Without it, you can omit usefulness problems or significant convenience blemishes that wind up baffling your quit clients. In any case, all experiments made equivalent. Composing to high-quality, useful experiments is also as wide as checking out your packages. Truth instructed helpless investigations can result in a testing cycle that is minimum more effective than now not checking out by any manner.

Figure 13.1 presents all the various web vulnerabilities which need attention when testing a web application. It clearly explains the brief analysis of mutation schemes, which are the most prominent in data mining.

13.2 Existing Work

Jeff Offutt and Upsorn proposed a test suite of 15 operators focusing mostly on challenges like user's ability to control web applications, Identifying web resources with URL's, communication depending on HTTP requests, communication via exchanges of data. New control connections, state management both on the client-side and server-side. However, none of them focused on session management, database update failures, cookie management, cross-site scripting, etc.

Shahriar and Zulkernine [28] proposed three types of security attacks using cross-site scripting and proposed 11 mutation operators to test cross-site scripting attacks. There are three types of cross-site scripting attacks possible: stored, reflected, then Document object model-based. When unsanitized information is retrieved from the database having some javascript or HTML code to generate some dynamic HTML content, such attacks are called stored XSS attacks. Reflected attacks are those attacks

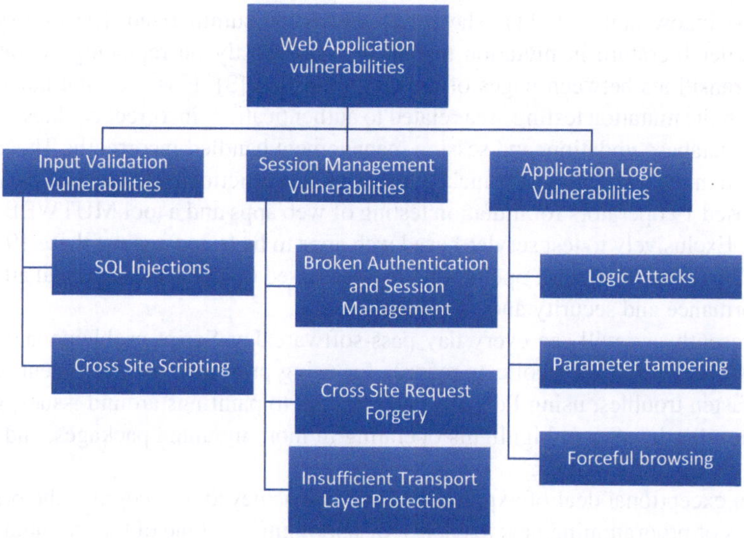

Fig. 13.1 Summary of web vulnerabilities presented by Vandana et.al

in which the application server returns the code supplied by a user. DOM-based attacks occur when javascript code processes inputs based on DOM objects. An operator ADES proposed to replace a write function calls arguments with escape function calls. Another operator RESC meant to replace escape function calls with XSSV code in javascript code [27].

The third operator RWWE uses the eval function to replace a write or a writeln part. The fourth operator RIHA uses DOM text nodes to return the properties of the Inner HTML. The MARF operator proposed works on replacing the search string with a counter replacement string. The sixth operator AHSC uses an HTML equivalent to replace each character in an accepted string argument. RHSC injects XSSV's in place of HTML special chars function calls. AHEN operator uses PHP variables to generate HTML contents [31].

RHEN works on removing the HTML entities function calls. MALT operator works on strip tags function calls where additional tags are injected in the tag argument. RSTT eliminates calls to the function strip tags to apply XSSV attacks on applications. The work of Hossain et al. is briefly presented in Table 13.1.

Table 13.1 Summary of cross-site scripting operators proposed by Hossain et al.

S. no.	Name of operator	Description	Key principle
1	ADES	Add escape function calls	Replace a Write-calling feature arguments by Leakage task calls
2	RESC	Leakage task calls removed	Replace escape function calls with XSSV code in javascript code
3	RWWE	Write function is replaced with Eval	Uses Eval function to return a write or a writeln function
4	RIHA	Inner HTML property by text node addition in DOM replaced	Uses DOM text nodes to replace the properties of the Inner HTML
5	MARF	Arguments of replace function calls are modified	RETURNS the search string with a counter replacement string
6	AHSC	Add HTML special chars function calls	Uses an HTML equivalent to replace each character in an accepted string argument
7	RHSC	Remove HTML special chars function calls	Injects XSSV's in place of HTML special chars function calls
8	AHEN	Add html entities function calls	Uses PHP variables to generate HTML contents
9	RHEN	Remove function calls to HTML entities	Removes the HTML entities function calls
10	MALT	Modifying parameter strip tags function calls that can be permitted	Alters strip tags function calls where extra labels injected in the acceptable label Discussion statement
11	RSTT	Remove function calls to strip tags	Removes calls to strip-tags of function to permit. Some arbitrary HTML label in the created performance through PHP variables

13.3 Mutation Operators and Their Importance

13.3.1 Comprehensive Mutation Test Suite

Most of the developers tend to implement small client validations for providing authenticated access to the users. But the risk of built-in browser tools is often overseen by the web application developers as in authentication. A lack of foresight in a scalable server for hosting databases creates havoc after the web application gets deployed. SQL injection attacks are paramount when it comes to accessing the database from simple execution of authentication queries. Most developers give authenticated access to the user trying to log in without establishing the logged user's right context. For example, it is easy for any user who logged into the system to access the application's inside irrespective of the application's actual, registered user due to the inbuilt browser tools. Malicious users can log in to the web application using specially crafted strings to bypass the authentication process.

SQL Injection attacks are the primary kind of attacks that can be done on a database with potential vulnerabilities. BAR and AAR is a mutation operator exclusively designed to check whether the authentication is provided appropriately to a genuine user [18]. The user's mutant code can bypass the whole login process and give unauthenticated access to an evil user. If this is not ascertained, then database access to unscrupulous users is possible, resulting in compromised security and privacy. The application users might warrant themselves from using the web application, affecting customer retention. If the query is a simple client-side script, it accesses the internal tables stored in the database. If input data is not filtered and placed into an inner database query, there is every risk that the site is vulnerable [12].

Cross-site scripting lets a malicious user run a script on a web page using the search field or comment fields or guest books/to steal other users' data of the web applications [19, 24]. If the search field executes a script, then it is evident that it is vulnerable to attacks. Cookies can be stolen, phishing (redirection to a similar-looking page) can be done, page ranks could increase by placing links on frames, spreading worms, etc., are just some of the spiteful things that can be done to the web app [13].

The authors presented in the earlier works are classified based on the categories of web application faults shown in Fig. 13.2. These operators proposed to help the developers prevent any defects from rolling into web apps development, which could adversely affect the non-functional requirements like availability, security, etc. Some of the predominant non-functional requirements that significantly impact a web application's functioning are Ease of Use, Security, Reliability, Minimum Down Time, Availability, etc. [17]. A simple classification of the proposed operators based on non-functional requirements affecting a web app is presented in Table 13.2. Cookies are simple text files containing information about the application state maintained by web browsers on the user's machine. They aim to improve the application's performance while accessing the web servers, thereby giving the user a "feelgood" experience, as shown in Table 13.2.

13.4 Mutation Testing Tools

Many tools have been developed towards develop the automation of mutation testing [10, 14]. Some of the existing tools include PIT [5], Cosmic Ray [6], Jumble [7], Selenium [15], Ruby on Rails, Mutpy, EATOOS [9], MUTWEB [25, 26], reAjax [21], Veriweb [20], WEBVIZOR [22], WATIRetc. The authors also developed a tool, MUTWEB, for automated testing of their proposed mutation operator suite. For

Fig. 13.2 Web applications faults classification

testing the proposed operators, an automated testing tool MUTWEB developed by authors used to analyze operators' performance on the listed experiments. MUTWEB accommodates exclusively operators designed for testing web applications in the earlier works. PIT [5] is an open-source trying out apparatus created through Coles. So, it has four ranges to be unique, freak age, test determination, embedding freak, and figuring out freak within the execution of the device. It employs operators centred around conditions, like removing them, manipulating them, and is good with giving the coverage metrics for the test suite [16]. Jester is a time-consuming mutation testing tool as it consumes a lot of time to deliver the interpreted result. Jumble is a bytecode level tool that mutates java class [8].

13.5 Results and Discussion

http://github.com/nanpj hosts a series of open source web apps utilized by Upsorn and Offutt aimed at performing their proposed mutation operators. The same is currently used in the present work to evaluate the authors' usefulness of the mentioned operators. A total of 10 applications are taken into consideration in the current position, and a careful hand seeding of mutants is done to evaluate operators' efficacy here. The selection of apps is made based on their development using servlets. The apps

Table 13.2 Classification of proposed writers' web mutation operators constructed on non functional requirements

S. no.	Operator's name	Explanation	Group	non functional requirement affected
1	DSID	This operator supports in the identifying. If a web submission's personal page portion remains available in spite the user signing to expired applications,	Incorrect Conference Administration	Security
2	DACD	Remove the process of cookie repercussions With this operator, the working and usefulness of a web application was studied	Incorrect Cookie Administration	Performance
3	DHBR	Boolean HTTP Substitution -On going session authentication will logically face problems by removing the Boolean meaning	Incorrect Session Management	Security
4	DFIR	In order to analyze the working of the servlets amid the interchange, Forward Involved the inter-changed keyword mechanism can helps the application efficiency	Incorrect Session Management	Security
5	DRDUR	Request Dispatcher URL Replacement—Modification of the URL in the request dispatcher method to verify web submission rendering	Incorrect Conference Administration	Security
6	DCD	Close Removal form—conn. Close () approach where the creator unknowingly loses out as coding allows resource lines such as database access to be maintained, which could impede the output of a web service	Incorrect Conference Administration	Performance
7	DSSR	Entity name can be used to Substitution—set Attribute is validate the appearance of a web application	Incorrect Conference Administration	Security
8	DGSR	Replacement of session name—URL session names are manipulated using some arbitrary values to check and see if unauthorized access to the web app can be obtained	Incorrect Session Management	Security
9	DCDM	Modification of the Cookie Process—this model is changed and a value that is arbitrary is replaced to check the actions of the web client	Incorrect Cookie Administration	Performance
10	BAR	Basic Authentication Substitution—tries to manipulate the server's internal demand to authenticate the customer to obtain web server entry	Incorrect Conference Administration	Security
11	AAR	Advanced Authentication Substitution—The backend bugs of the mobile interface can verified this operator easily	Incorrect Session Management	Security

(continued)

Table 13.2 (continued)

S. no.	Operator's name	Explanation	Group	non functional requirement affected
12	XSSC	In order to insure that that the attempt is halted, the web site cross verification and intrusion are very danger to using the search query given by the web service	Cross-Site Scripting Vulnerability	Security
13	DRUR	URL modification is performed using a script like below: RequestDispatcherrd = request.getRequestDispatcher("Welcome.html");	Incorrect Session Management	Security
14	DSGD	Web app's get Quality task is removed	Incorrect Conference Administration	Security
15	DPR	User interfaces that are based on JDBC are updated with its port numbers	Database Updation Failures	Security
16	DPD	JDBC dependent web apps with ip addresses are deleted	Database Updation Failures	Security
17	DDNR	Database names for JDBC related web apps are substituted with different names and the default password protection is verified	Database Updation Failures	Security
18	DLDR	The same driver used in the back—end database relation is substituted by a new driver	Database Updation Failures	Security
19	DSSD	Servlet-based web applications are set to uninstall the Attribute function	Incorrect Conference Administration	Security

are referred to as Experiments represented as and refers to experiments. Web Stutter, Study InfoSys, computer GPA, BS Voting, Conversion, H.L. Voting, KS Voting, faults needing, Irregular String, and Check24Online are ten programs selected for experimentation. Web Stutter offers a recurrence of rehashed phrases for a given statistic. KS Voting, BS Voting, and HL Voting are understudy online hustings programs that permit an understudy to solidify a poll in opposition to specific understudies' votes. Given a small college management application were for the students enrolled credit hours and grades are given, PC GPA distributes an understudy GPA. Change manages to change one estimation over to some other on the internet. Fault seeding is an app that maintains all faulty versions of the software. Random String enables a user to choose strings randomly with or without string replacement. Check24online is an online card game that presents output based on the entry of the user.

A portion of the highlights of the examinations under Test like information base availability, treats, meetings, confirmation, and so on are considered for approval in opposition to the proposed administrators. A plotting of mutants generated versus mutants killed based on performance operators is presented in Fig. 13.3. it is evident that many applications fail to implement some performance operators. There can be performance degradations in the web application on the fly. The analysis reveals that the number of persistent mutants is more in number with the proposed operators, as shown in Fig. 13.4. Hence it is highly recommended to use the proposed mutation operators for web application development to ensure performance and security. The persistent mutants reveal that not all web applications developed to view the required essential security and performance features. Using this mutation testing site as a reference point, to begin with, web application development would become easy, bug-free, and better.

Table 13.3 summarizes the experiments under research with the interest of a range of code lines in every net utility below assessment. The absolute quantity of segments each web app encompassed.

Table 13.4 summarizes all the mutants generated against each experiment and the number of mutants killed. Mainly focuses on the performance testing operators of web apps. Killed mutant signifies that the test case is intact. Table 13.5 presents the security-based mutation operators for web apps, the number of generated mutants against each web app and mutants killed are presented. However, it is observed that some operators cannot be applied to certain web applications as those features might not be implemented in those web applications. Then those operators become irrelevant from a testing perspective. Plotting the generated mutants against the killed mutants for performance-based operators and security-based operators is presented in Table 13.6.

Summary of mutants killed and generated experiment wise are analyzed in Tables 13.7 and 13.8.

A M score is utilized towards identifies the usefulness of a test suite and hence helps use to infer whether the test suite is adequate or not [30]. Projects the mutation operator suite's average mutation score against each experiment and overall mutation score against all the experiments is 0.9666, which indicates the quality of test data used towards performing mutation testing of web applications. Mutation verification

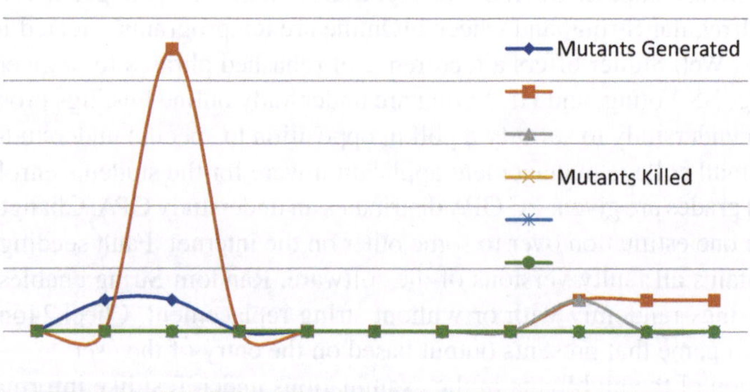

Fig. 13.3 Plotting of mutants generated and killed based on performance operators

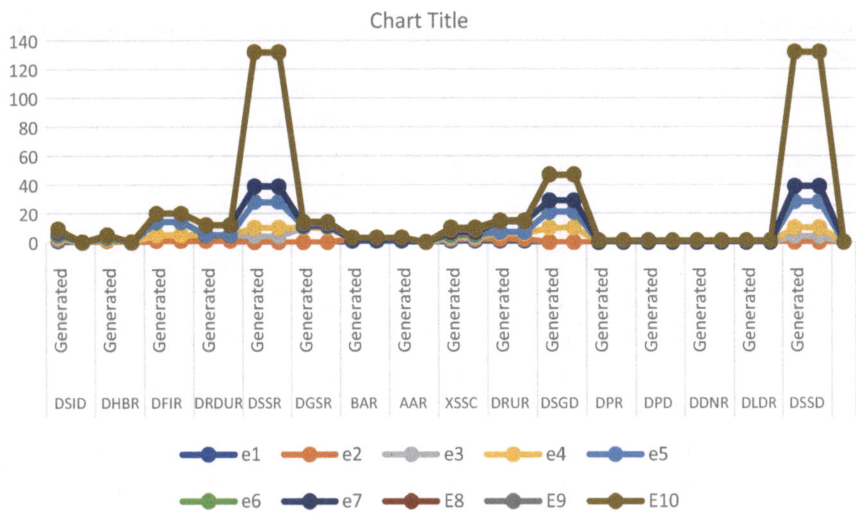

Fig. 13.4 Security operators mutants generated versus killed

Table 13.3 Experiments list and size of experiments

S. no.	Subjects	Components	LOC
1	webStutter	3	12
2	StudInfoSys	3	1766
3	computeGPA	2	581
4	BSVoting	11	930
5	Conversion	1	388
6	H.L. Voting	12	939
7	KSVoting	7	1024
8	Faultseeding	5	1541
9	RandomString	1	285
10	Check24Online	1	1619

could be used for a Test-driven development (TDD) approach using an optimized operator suite for better performance and web applications security [32–35].

13.6 Conclusion

The proposed mutation operators are adept at identifying potential futuristic shortcomings of the application under Test like performance and confidentiality of the web extension. The operators are successfully pointing the future security holes in the experimented subjects. They are proven to be high utility test cases that can help

Table 13.4 Summary of mutants generated and killed affecting the performance of a web application

	Mutants generated			Mutants killed		
S. no.	DACD	DCD	DCDM	DACD	DCD	DCDM
e1	1	0	0	0	0	0
e2	1	9	0	0	0	0
e3	0	0	0	0	0	0
e4	0	0	0	0	0	0
e5	0	0	0	0	0	0
e6	0	0	0	0	0	0
e7	0	0	0	0	0	0
e8	1	1	1	0	0	0
e9	0	1	0	0	0	0
e10	0	1	0	0	0	0

a tester and a developer, thereby setting a high standard for developing web apps. The DACD, DCD, and DCDM operators caution the developers of the impending problems that the web app can create if not handled rightly. The operators analyzed here mostly address servlets based web applications.

13.6.1 Future Enhancements

There is a shortage of operators in python and other programming languages based web applications. There is a need to integrate the mutation testing suite into a CI/CD (Continuous Integration/Continuous Deployment Pipeline) to help developers come up with a bug-free code, to begin with. The MUTWEB tool needs to be enhanced further to perform automated mutation analysis of the applied operators on the subject web applications. The usefulness and importance of each proposed mutation operator need to computed using a quantitative measure, thereby prioritizing the operators' needs.

Table 13.5 Mutants generated operator wise experiment wise for security operators

	DSID	DHBR	DFIR	DRDUR	DSSR	DGSR	BAR	AAR	XSSC	DRUR	DSGD	DPR	DPD	DDNR	DLDR	DSSD	Total mutants
e1	1	1	1	1	0	0	1	1	1	1	0	0	0	0	0	0	9
e2	1	0	0	0	0	0	1	1	1	1	0	0	0	0	0	0	5
e3	1	1	4	4	4	10	0	0	1	4	10	0	0	0	0	4	43
e4	1	0	0	0	6	0	0	0	1	0	0	0	0	0	0	6	14
e5	1	1	9	0	18	1	0	0	1	1	11	0	0	0	0	18	61
e6	1	1	6	7	11	1	0	0	1	7	8	0	0	0	0	11	54
e7	1	1	0	0	0	0	0	0	1	0	0	0	0	0	0	0	3
e8	1	0	0	0	93	1	1	1	1	0	18	1	1	1	1	93	213
e9	0	0	0	0	0	0	0	0	1	0	0	0	0	0	0	0	1

Table 13.6 Mutants killed operator wise experiment wise for security operators

	DSID	DHBR	DFIR	DRDUR	DSSR	DGSR	BAR	AAR	xssc	DRUR	DSGD	DPR	DPD	DDNR	DLDR	DSSD	Total mutants
e1	0	0	1	1	0	0	1	0	1	1	0	0	0	0	0	0	5
e2	0	0	0	0	0	0	1	0	1	1	0	0	0	0	0	0	3
e3	0	0	4	4	4	10	0	0	1	4	10	0	0	0	0	4	41
e4	0	0	0	0	6	0	0	0	1	0	0	0	0	0	0	6	13
e5	1	1	9	0	IS	1	0	0	1	1	11	0	0	0	0	18	61
e6	0	0	6	7	11	1	0	0	1	7	8	0	0	0	0	11	54
e7	0	0	0	0	0	0	0	0	1	0	0	0	0	0	0	0	3
e8	0	0	0	0	93	1	1	0	1	0	18	1	1	1	1	93	213
e9	0	0	0	0	0	0	0	0	1	0	0	0	0	0	0	0	1
e10	0	0	0	0	0	1	0	0	1	1	0	0	0	0	0	0	4

Table 13.7 Summary of mutants killed versus generated experiment wise

S. no.	Experiment no.	Mutants	Equivalent	Killed
1	e1	9	2	5
2	e2	5	2	3
3	e3	43	1	41
4	e4	14	1	13
5	e5	61	0	61
6	e6	54	0	54
7	e7	3	0	3
8	e8	213	1	210
9	e9	1	0	1
10	e10	4	0	4

Table 13.8 Mutation score experiment wise

S. no.	Experiment no.	Mutation score
1	e1	0.71
2	e2	1
3	e3	0.976
4	e4	1
5	e5	1
6	e6	1
7	e7	1
8	e8	0.990
9	e9	1
10	e10	1
Avg. mutation score		0.9666

References

1. Schuler, D., & Andreas Zeller, J. (2009). Avalanche: Efficient mutation testing for Java. In *ESEC-FSE*, 24–28 August. ACM
2. Mirshokraie, S., Mesbah, A., & Pattabiraman, K. (2015). Guided mutation testing for JavaScript web applications. *IEEE Transactions on Software Engineering, 41*(5).
3. Praphamontripong, U. (2012). Web mutation testing. In*IEEE fifth international conference on software testing verification and validation.*
4. Lakshmi, D. R., & Mallika, S. S. (2017). A review on web application testing and its current research directions. *International Journal of Electrical and Computer Engineering, 7*(4), 2132.
5. PIT. http://pitest.org/. Last accessed on October 1, 2019.
6. Cosmic Ray. https://github.com/sixty-north/cosmic-ray. Last accessed as on October 1, 2019.
7. Sourceforge, "Jumble". (2007). http://jumble.sourceforge.net/. Last accessed as on October 1, 2019.
8. Smith, B. H., & Williams, L. (2007). An empirical evaluation of the Mujava mutation operators. In *Proceedings of the 3rd workshop on Mutation Analysis (MUTATION '07), published with*

the proceedings of the 2nd testing. Academic and Industrial Conference Practice and Research Techniques (TAIC PART' 07). IEEE Computer Society (Vol. 10–14, pp. 193–202).
9. Mallika, S. S. (2013). EATOOS-testing tool for unit testing of object oriented software. International Journal of Computer Applications, (0975–8887) 80(4), 6–10.
10. Augsornsri, P., & Suwannasart, T. (2014). An integration testing coverage tool for object-oriented software. In: *International conference on information science and applications*. IEEE. https://doi.org/10.1109/icisa.2014.6847360
11. Mutpy. https://bitbucket.org/khalas/mutpy. Last accessed as on October 1, 2019.
12. Kundu, S. (2012, March). Web testing: Tool, challenges and methods. *IJCSI International Journal of Computer Science Issues, 9*(2), 3.
13. Arora, A., & Sinha, M. (2012, February). Web application testing: A review on techniques, tools and state of art. *International Journal of Scientific & Engineering Research, 3*(2).
14. Girgis, M. R., Mahmoud, T. M., Abdullatif, B. A., & Zaki, A. M. (2014, August). An automated web application testing system. *International Journal of Computer Applications (0975–8887),* 99(7).
15. Gogna, N. (2014, August). Study of browser based automated test tools WATIR and Selenium. *International Journal of Information and Education Technology, 4*(4).
16. Sharma, M., & Angmo, R. (2014). Web based automation testing and tools. *(IJCSIT) International Journal of Computer Science and Information Technologies, 5*(1), 908–912.
17. Križani, J., Grguri, A., Mošmondor, M., & Lazarevski, P. (2010). Load testing and performance monitoring tools in use with AJAX based web applications. In *MIPRO 2010*, 24–28 May.
18. Dukes, L., Yuan, X., & Akowuah, F. (2014). A case study on web application security testing with tools and manual testing. 978-1-4799-0053-4/13 2013. IEEE.
19. Mesbah, A., & Prasad, M. R. (2011). Automated cross-browser compatibility testing. In ICSE' 11, May 21–28.
20. http://www.cs.utah.edu/~juliana/pub/veriweb-www2002.pdf as on April 8, 2017.
21. Santoso, L. W., Singh, B., Suman Rajest, S., Regin, R., & Kadhim, K. H. (2021). A genetic programming approach to binary classification problem. *EAI Endorsed Transactions on Energy, 8*(31), 1–8. https://doi.org/10.4108/eai.13-7-2018.165523
22. Sprenkle, S., Esquivel, H., Hazelwood, B., & Pollock, L. (2015). *WEBVIZOR: A visualization tool for applying automated oracles and analyzing test results of web applications.*
23. Dallmeier, V., Pohl, B., Burger, M., Mirold, M., & Zeller, A. (2014). WebMate: Web application test generation in the real world. In *ICSTW' 14 Proceedings of the 2014 IEEE international conference on software testing, verification, and validation workshops* (pp. 413–418), March 31–April 04.
24. Shahriar, H., & Zulkernine, M. (2009). MUTEC: Mutation-based testing of cross 28 site scripting. In *ICSE Workshop*. IEEE.
25. Suguna Mallika, S., & Rajya Lakshmi, D. (2019). MUTWEB—A testing tool for performing mutation testing of Java and Servlet based web applications. *International Journal of Innovative Technology and Exploring Engineering (IJITEE), 8*(12), 5406–5413. ISSN: 2278-3075.
26. Offutt, J., & Lee, Ss. D. (1994, May). An empirical evaluation of weak mutation. *IEEE Transactions on Software Engineering, 20*(5), 337–344.
27. Offutt, J., Pan, J., Tewary, K., & Zhang, T. (1996, February). An experimental evaluation of data flow and mutation testing. *Software-Practice and Experience, 26*(2), 165–176.
28. Shahriar, H., & Zulkernine, M. (2009). MUTEC: Mutation-based testing of cross site scripting. In*ICSE Workshop on Software Engineering for Secure Systems* (pp. 47–53). https://doi.org/10.1109/IWSESS.2009.5068458
29. Dwivedi, V., Yadav, H., & Jain, A. (2014). Web application vulnerabilities: A survey. *International Journal of Computer Applications, 108*, 0975–8887.
30. Nica, S., Wotawa, F., & Ramler, R. (2011). Is mutation testing scalable for real-world software projects? In *The third international conference on advances in system testing and validation lifecycle*, 978-1-61208-168-7.
31. Mansour, N., & Houri, M. (2005). "Testing web applications", information and software technology (pp. 31–42). https://doi.org/10.1016/j.infsof.2005.02.007

32. Kumar, D. P., Kumar, K. S., & Asadi, S. S. (2018). Evaluation of customer perception towards online advertisement technology an empirical study from Vijayawada. *International Journal of Mechanical Engineering and Technology, 9*(2), 136–150.
33. Al-Yasiri, D. S. Q., & Obaid, A. J. (2018). A new approach for object detection, recognition and retrieving in painting images. *Journal of Advance Research in Dynamic and Control System, 10*(2), 2345–2359.
34. Al-asadi, T. A., & Obaid, A. J. (2016). Object detection and recognition by using enhanced speeded up robust feature. *International Journal of Computer Science and Network Security (IJCSNS), 16*(4), 66–71.
35. Al-asadi, T. A., Obaid, A. J., Hidayat, R., & Ramli, A. A. (2017). A survey on web mining techniques and applications. *International Journal on Advanced Science Engineering and Information Technology, 7*(4), 1178–1184.

Chapter 14
Recommending Products Based on Visual Similarity Using Machine Learning

Umair Ali Khan, Fayaz A. Memon, M. Amir Bhutto, and Adnan A. Arain

Abstract A product recommendation system is aimed at providing an improved shopping experience to the users, thereby increasing the revenues. The text-based product search on most of the online stores is based on the past search history of the user and/or the other characteristics associated to the products, such as the annotated labels, price range, category, description, size, color, and other attributes. Though, this traditional method has adequate performance, however, it is prone to the issues pertaining to improper annotation of the metadata related to the products. Due to the recent advancements in machine learning, product recommendation using machine learning is getting increasing attention. Over the last one decade, tremendous progress in machine learning and computing has paved the way for efficient product recommendation. In this book chapter, we present a critical analysis of product recommendation using the traditional methods. Specifically, the book chapter explores the prospect of product recommendation based on visual similarity using machine learning which is rather a new concept in this domain and is getting increasing attention in research community. The chapter also explores the underlying challenges and discusses the future directives.

14.1 Introduction

A product recommendation system is aimed at providing an improved shopping experience to the users, thereby increasing the revenues. The product search on most of the online stores is based on either the past search history of the user or the other characteristics associated to the products, such as the annotated labels, price range, category, description, size, color, and other attributes. Though, this traditional method has been known to provide satisfactory results, however, it is prone to the

U. A. Khan (✉) · F. A. Memon · M. A. Bhutto · A. A. Arain
Universiti Teknologi Malaysia, Johor Bahru, Malaysia
e-mail: umair.khan@quest.edu.pk

A. A. Arain
e-mail: adnan.ahmed03@quest.edu.pk

P. Kumar et al. (eds.), *A Fusion of Artificial Intelligence and Internet of Things for Emerging Cyber Systems*, Intelligent Systems Reference Library 210,
https://doi.org/10.1007/978-3-030-76653-5_14

issues pertaining to improper annotation of the metadata related to the products. The same issue arises in online search of images, videos, and other type of data. The research on manual metadata generation has shown that the human annotated data is imprecise and lacks consistency [1].

As depicted in Fig. 14.1, the existing product recommendation systems can be broadly classified into three categories: (i) collaborative recommendation systems [2], (ii) content-based recommendation systems [3], and hybrid systems. In collaborative recommendation, the shopping or product browsing actions of a user are used to predict the choices of another user. For example, if one user browses the items of LCD monitor, but ultimately purchases a printer, the collaborative technique will attempt to find a correlation between these two products. By getting data from more users and confirming this association, it will eventually begin to influence which products are being recommended. Collaborative filtering can be further classified into model- or memory based. Model based systems require a model for user-item interaction. Whereas, memory based systems rely on similarities between users or items in terms of observed interactions. An obvious drawback of collaborative filtering is that it requires past history of users' actions to build an association among the similar products. In addition to this, they require huge data and also pose challenge in terms of new products and users.

In content-based product recommendation system, the items liked or purchased by a user in the past are used to recommend the similar items. This technique requires matching the attributes of a user profile in which preferences and interests are stored, with the attributes of a content object, in order to recommend to the user new interesting items [4]. The hand-crafted features of products used in this technique require expertise, are prone to known issues, and have limited ability to expand on the user's existing interests. A hybrid recommendation system combines the features of both the systems for product recommendation.

It is evident that in the aforementioned systems, a past history of product associations and/or users' pre-stored preferences are required. Hence, the performance of these systems improves over time, rather than giving immediate results. Therefore, a

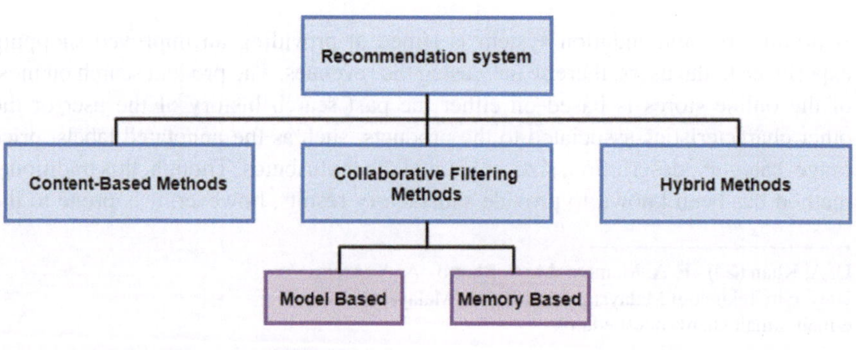

Fig. 14.1 Classification of recommendation system

recommendation system which can immediately recommend the similar products to a user without requiring his past usage or preference history could be more efficient.

In contrast to these approaches, a recommendation system based on the visual similarity of products could be more productive. At the same time, such system could be immensely effective in searching/browsing the products based on the similarity of visual features of a given image of a product. For example, one can search a particular chair of his choice by passing its visual features (query image) to the recommendation system to obtain the desired recommendations. This visual similarity will further help to find the matching of related products. For example, one can find a compatible table matching with a particular chair based on the visual similarity. Such a system falls into the category of content-based filtering.

The development of such a system requires extracting the visual image features of a product of interest and matching them with the features of the whole database of products to find the objects with the best match. That said, the diversity of product shapes could make the mathematical formulation of hand-crafted features [5] an extremely challenging and inefficient task. At the same time, the issues pertaining to the hand-crafted features offer yet another dedicated challenge.

The recent advancements in machine learning have paved the way for efficient pattern matching and learning. Especially, the emerging field of deep learning [6] has proved to be immensely effective to learn complex patterns in a given data. Another appealing feature of deep learning is its efficacy to learn the underlying patterns in a data without requiring a- priori mathematical model. Hence, deep learning could be efficiently used to find the visual similarity of products in a recommendation system.

In this chapter, we first critically analyze the existing progress of machine learning for product recommendation. Subsequently, we explore the potential of deep learning and the corresponding challenges for product recommendation based on visual similarity of products. The survey presented in this chapter stands apart from other related studies in the sense that we first provide a comprehensive review on the existing product recommendation systems and then compare their performance with the one using machine learning and, in particular, deep learning. Apart from that, we also highlight the potential challenges and research directives in this domain which could be useful for the researchers starting to work on this topic.

14.2 Related Work

A number of techniques have been introduced for product recommendation systems. The authors in [7] use a product taxonomy and customer classification to identify customers' shopping behavior, thereby recommending the relevant products. The technique proposed in [8] uses multi-attribute decision making to find the utility values of products in the same product class. These values are used to recommend the similar products. Another technique proposed in [9] exploits the consumers' current needs obtained from the system-user interactions to develop a fuzzy-based recommendation system for consumer electronics to retrieve optimal products. In

[10], the authors build a customer profile model to recommend similar products by using information such as clicks, basket interactions, purchases, and interest fields.

The technique proposed in [11] derives product ratings to apply collaborative recommendation to online transaction data even when no explicit rating information is available. Another technique proposed in [12] detects users' purchase intents from their microblogs in near real-time and makes product recommendation based on matching the users' demographic information extracted from their public profiles with product demographics learned from microblogs and online reviews.

The product recommendations in [13] are based on intelligence contained in processing elements and subjective and/or objective product information received from consumers or input to the systems as part of their initial setup. The output of the system comprises sets of products that they predict the consumer will prefer and/or perform well for the problem or concern identified by the consumer. Similarly, the technique proposed in [14] obtains users' feedback to score products and assign them certain ranks. The ranks are finally used to get an aggregated final users' ranking.

In a different approach [15], the authors propose a technique to extract product adopter mentions from online reviews. This information is categorized to form different demographic user groups. The aggregated demographic information of many product adopters can be used to characterize both products and users, which can be incorporated into a recommendation method using weighted regularized matrix factorization.

In a more recent approach [16], machine learning is used to combine the benefits of memorization and generalization for product recommendation systems. In a similar way, microblogging and social media information is also used for product recommendation systems. The technique proposed in [17] uses the linked users across social networking sites and e-commerce websites (users who have social networking accounts and have made purchases on e-commerce websites) as a bridge to map users' social networking features to another feature representation for product recommendation.

A more recent approach [18] uses a product configuration process which is tailored to the characteristics of the particular decision makers. Exploiting the potential of deep learning, the technique proposed in [19] uses a specific deep neural network to extract the content features of the items. A collaborative filtering model, which models and utilizes temporal dynamics of user preferences and item features, is modified to take the content features into prediction of ratings for cold start items. In [20], the authors use collaborative filtering approaches to address the issue of sparsity and scalability using dimensionality reduction and ontology techniques. The ontology is used to improve the accuracy of the recommendations in the collaborative filtering part.

14.3 Visual Similarity Based Product Recommendation Using Machine Learning

Sometime the user wants to search for a product based on a query image. The need for this type of search arises in the following cases.

- The appropriate descriptive information of the product is not available. For example, the user sees a product somewhere, likes it and wants to buy it for himself. However, he does not know the precise name of the product, manufacturer and other specifications. The least he can do is to take a photo of the product and randomly search it on online shopping stores.
- The service provider or the online vendor does not have the accurate or complete metadata about the product. This restricts the user to find his item of choice easily as he does not have descriptive information of the product.
- The user may want to find items which are visually similar to the item of his choice irrespective of the model, manufacturer, price range and other features.
- In case of a "*cold start*", a problem where no search or preference history is available, it is very difficult for an online vendor to recommend products.

The existing product recommendation systems cannot server purpose in the above-mentioned scenarios. Only a machine learning based product recommendation system, capable of matching the visual features of the products, can be effective in these cases. Since a content-based machine learning algorithm, which is intended to exploit the visual similarity of the products, is required to be trained on the visual features. The extraction and/or modeling of visual features using the traditional approaches is not trivial and robust. But deep learning solves this problem in the sense that it does not require any engineered features to learn the product similarity. Hence, the current research on product recommendation based on visual similarity (and other aspects) has been directed on deep learning. This is the very reason why the traditional machine learning algorithms such as support vector machines, shallower neural networks, Naïve Bayes, random forests, decision trees, etc. have no longer been effective to address the ever-increasing diversity of products and sheer amount of data generated on daily basis.

The recent advancements in machine learning, in particular deep learning, and the availability of data and required computing power have accelerated research in this domain. The relevant literature demonstrates several techniques of product recommendation based on visual similarity. The authors in [21] propose a technique which determines a user's preferences by initial recommendations of products and refining its recommendations. The same technique is used by some movie streaming websites such as *Netflix* which captures a user's preferences by initial recommendations of diverse movie genres. However, this technique is strictly dependent on the diversity and completeness of the initial recommendations.

The technique proposed in [22] constructs an item-item visual similarity matrix by computing visual similarity between each two products. However, this technique also requires a user-item matrix to integrate with the visual similarity matrix which limits the performance of this technique in case of new users or incomplete ratings given by the users.

Shankar et al. [23] more effectively address this problem by proposing a deep learning based network, called *VisNet*, for quantitative estimation of visual similarity between two images containing fashion items. Their proposed system has been implemented on a large online store and is reported to deliver an impressive accuracy. However, after the emergence of more efficient and complex deep learning architectures, the efficacy of this technique can be further improved. In a similar but more recent technique [24], the authors propose a two-stage deep learning framework to recommend fashion images based on a user-input image.

In a different technique [25], the authors use Generative Adversarial Networks (GANs) to retrieve a ranked list of shoe images similar to queried shoe image. GANs are originally used to generate visually realistic fake images. However, this technique also demonstrates the potential of GANs for an entirely different task of product recommendation. However, since the proposed system was only trained for shoe images, its scalability is yet to be ascertained.

14.4 Challenges

The development of a product recommendation system based on visual similarity offers several challenges including, but not limited to, the following:

1. The training dataset comprising catalog images captured under controlled conditions should cover different variations between images of the same item such as complex backgrounds, extreme perspective variations, partial views and poor lighting [23].
2. The visual similarity based product recommendation system must encompass the higher and multiple level of abstractions in the same way as human brain does.
3. The system must be able to exploit transfer learning to incorporate new products with minimal effort and expertise.
4. The system should be able to cover the enormous diversity of products and their variants.

14.5 Research Directives

Following potential research directions are highlighted in this domain.

1. Visual similarity based product recommendation may be integrated with collaborative filtering to give a better user experience. In case where the user data is available, the fusion of these two techniques is expected to deliver better performance.

2. For images with cluttered background or unnecessary parts, the object of interest first needs to be segmented before sending the query image to a recommender engine. This will improve the recommendation accuracy and alleviate the amount of data to be analyzed.
3. In case the query image contains multiple objects, the policies to consider these objects together or separate objects may be devised.
4. A ranking of various clusters of similar products can be generated which may be modified based on user's feedback.

14.6 Conclusion

Product recommendations based on visual similarity does not require user's data and eliminates the problem of cold start. Due to the immense potential of deep learning, this technique is finding increasing acceptability and is soon to replace the traditional recommendation systems. Alternatively, the good features of content-based and collaborative filtering may be combined to give an even better performance.

In this chapter, we first described the need of a recommendation system based on visual similarity and highlighted the potential of machine learning, in particular deep learning, for this task. We discussed the traditional approaches of recommendation systems with their categories. We also provided a critical analysis of the existing techniques of recommendation systems based on visual similarity. At the end, we presented some potential challenges in this domain and future directives.

References

1. Khan, U. A., Ejaz, N., Martínez-del-Amor, M. A., & Sparenberg, H. (2017). Movies tags extraction using deep learning. In *14th IEEE international conference on advanced video and signal based surveillance (AVSS)*, Lecce (pp. 1–6).
2. Errico, J. H., Sezan, M. I., Borden, G. R., Feather, G. A., & Grover, M. G. (2015). *U.S. Patent No. 8949899*. U.S. Patent and Trademark Office.
3. Lops, P., De Gemmis, M., & Semeraro, G. (2011). Content-based recommender systems: State of the art and trends. In *Recommender systems handbook* (pp. 73–105). Springer.
4. Lops, P., de Gemmis, M., & Semeraro, G. (2011). Content-based recommender systems: State of the art and trends. In F. Ricci, L. Rokach, B. Shapira, & P. Kantor (Eds.), *Recommender systems handbook*. Springer.
5. Xia, R., et al. (2014). *Supervised hashing for image retrieval via image representation learning* (Vol. 1, No. 2014). AAAI.
6. LeCun, Y., Bengio, Y., & Hinton, G. (2015). Deep learning. *Nature, 521*(7553), 436.
7. Hung, L. (2005). A personalized recommendation system based on product taxonomy for one-to-one marketing online. *Expert Systems with Applications, 29.2*, 383–392.
8. Choi, S. H., Kang, S., & Jeon, Y. J. (2006). Personalized recommendation system based on product specification values. *Expert Systems with Applications 31.3*, 607–616.
9. Cao, Y., & Li, Y. (2007).An intelligent fuzzy-based recommendation system for consumer electronic products. *Expert Systems with Applications, 33.1*, 230–240.

10. Park, Y.-J., & Chang, K.-N. (2009). Individual and group behavior-based customer profile model for personalized product recommendation. *Expert Systems with Applications, 36.2*, 1932–1939.
11. Choi, K., et al. (2012). A hybrid online-product recommendation system: Combining implicit rating-based collaborative filtering and sequential pattern analysis. *Electronic Commerce Research and Applications, 11.4*, 309–317.
12. Zhao, X. W., et al. (2014). We know what you want to buy: A demographic-based system for product recommendation on microblogs. In *Proceedings of the 20th ACM SIGKDD international conference on Knowledge discovery and data mining*. ACM.
13. Shaya, S. A., et al. (2014, March 4). Intelligent performance-based product recommendation system. U.S. Patent No. 8666844.
14. Sohail, S. S., Siddiqui, J., & Ali, R. (2015). User feedback based evaluation of a product recommendation system using rank aggregation method. In E. S. El-Alfy, S. Thampi, H. Takagi, S. Piramuthu, T. Hanne (Eds.), *Advances in intelligent informatics. Advances in intelligent systems and computing* (Vol. 320). Springer.
15. Wang, J., et al. (2015). Leveraging product adopter information from online reviews for product recommendation. In *ICWSM*.
16. Cheng, H.-T., et al. (2016). Wide & deep learning for recommender systems. In Proceedings of the 1st workshop on deep learning for recommender systems. ACM.
17. Zhao, W. X., et al. (2016). Connecting social media to e-commerce: Cold-start product recommendation using microblogging information. *IEEE Transactions on Knowledge and Data Engineering, 28.5*, 1147–1159.
18. Scholz, M., et al. (2017). A configuration-based recommender system for supporting e-commerce decisions. *European Journal of Operational Research 259.1*, 205–215.
19. Wei, J., et al. (2017). Collaborative filtering and deep learning based recommendation system for cold start items. *Expert Systems with Applications, 69*, 29–39.
20. Nilashi, M., Ibrahim, O., & Bagherifard, K. (2018). A recommender system based on collaborative filtering using ontology and dimensionality reduction techniques. *Expert Systems with Applications, 92*, 507–520.
21. Hsiao, J., & Li, L. (2014). On visual similarity based interactive product recommendation for online shopping. In *2014 IEEE international conference on image processing (ICIP)*, Paris (pp. 3038–3041). https://doi.org/10.1109/ICIP.2014.7025614
22. Guo, C., Zhang, M., Liu, Y., & Ma, S. (2016). A picture is worth a thousand words: Introducing visual similarity into recommendation. In *2016 seventh international conference on intelligent control and information processing (ICICIP)*, Siem Reap (pp. 153–160). https://doi.org/10.1109/ICICIP.2016.7885893.
23. Shankar, D., Narumanchi, S., Ananya, H. A., Kompalli, P., & Chaudhury, K. (2017). Deep learning based large scale visual recommendation and search for e-commerce. arXiv preprint arXiv: 1703.02344.
24. Tuinhof, H., Pirker, C., & Haltmeier, M. (2018). Image-based fashion product recommendation with deep learning. In *International conference on machine learning, optimization, and data science* (pp. 472–481). Springer.
25. Ay, B., Aydın, G., Koyun, Z., & Demir, M. (2019). A visual similarity recommendation system using generative adversarial networks. In *2019 international conference on deep learning and machine learning in emerging applications (Deep-ML)*, Istanbul, Turkey (pp. 44–48). https://doi.org/10.1109/Deep-ML.2019.00017

Chapter 15
AIIOT: Emerging IoT with AI Technologies

Sandeep Mathur, Tanvi Singla, Kunwar Bharat, and Ajay Rana

Abstract The term internet of things (IoT) refers to the interconnection of various smart devices and objects across the world. In recent times IoT technology has seen an immense increase in its usage as the government around the world is getting automated. The services are getting advanced day by day and affecting the general population. It is dependent on Big data analysis, such as smart governance, automated grid system, automated medical services, and many more. Its usage is getting extended to the services of every mainframe function from defense technology to heavy grade industries. IoT is in the implementation stage. Its impact is more than technological advancement. In the current research work, the following sections will illustrate the introduction of IoT, IoT functioning, important components of IoT, the architecture of IoT, application of IoT in general life, smart governances, smart cities with various automated services which are increasing day by day. The biggest example can be referred to as fast tags services which include automated service of doing payment on the national highway of India. Further, the role of artificial intelligence in IoT has been described and it is followed by the most alarming topic in IoT that is intrusion detection in IoT.

Keywords Internet of Things · Automation · Artificial intelligence · Intrusion detection

---

S. Mathur (✉) · T. Singla · K. Bharat · A. Rana
Amity Institute of Information Technology, Amity University, Noida, Uttar Pradesh, India
e-mail: smathur@amity.edu

T. Singla
e-mail: tanvi.singla@student.amity.edu

K. Bharat
e-mail: kunwar.singh8@student.amity.edu

A. Rana
e-mail: ajay_rana@amity.edu

P. Kumar et al. (eds.), *A Fusion of Artificial Intelligence and Internet of Things for Emerging Cyber Systems*, Intelligent Systems Reference Library 210,
https://doi.org/10.1007/978-3-030-76653-5_15

15.1 Introduction

IoT is defined as a network of physical objects-devices, instruments, vehicles, buildings, and alternative things that are equipped with circuits, wires, physical science, software, sensors, and property tools that change these objects to gather and exchange information. To make these physical objects independent of human help for running their functioning and ability the physical object is equipped with sensors, communication, networking, and information processing technology. It can perform various IoT functions like data transfers, home automation, robotics, and many more. To date, varied technologies have been concerned in IoT, like wireless sensing element networks (WSNs), barcodes, intelligent sensing, radio-frequency identification (RFID), Near Field Communication (NFCs), cloud computing Etc.

15.1.1 Architecture of IoT

The basic and most important requirement of an IoT system is a proper connection for its continuous automated function for performing various tasks, the connection must make sure that physical objects used in IoT are connected for this process a proper architecture is required and that IoT architect must assurance of IoT operations, which connects the physical and virtual worlds. The architecture of IoT will involve elements such as networking, communications, processing, etc. as shown in Fig. 15.1.

15.1.1.1 Service-Oriented IoT

Things in IoT will have interactions in real-time. This will force IoT to be adaptive because efficient adaptability will provide easiness in the interaction between the devices undisturbed communication services among them. So, in other terms, IoT must have a decentralized and heterogeneous system. One of the most important features of IoT is to treat a complex system in the form of a subsystem and define its functionality more simply. Each subject device in the system can be reused and maintained individually so it can be reused whenever they are required.

15.1.1.2 Sensing Layer

The way IoT has described itself as an extended smart interconnected network, in which smart things are connected steadily and can be accessed from anywhere. The term smart system in IoT is functional on sensors and tags. This system of automated objects is in place for automated sensing of the ecosystem in which it allows them to interchange the data in that present ecosystem, so things can be uniquely characterized, and surrounding ecosystem can be used for various application [2]. Every

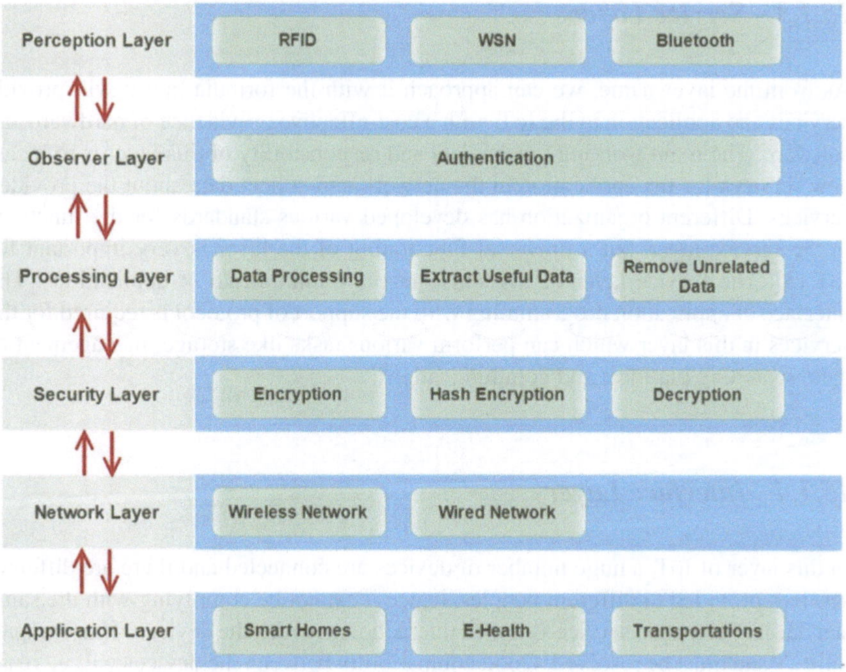

Fig. 15.1 Architecture layers of IoT [1]

sensor is a sensing layer that has a digital identity, and it can be found in the digital domain of the sensing layer. UUID Universal Unique identification is used to give a unique identity to every object. The identifiers contain names and addresses.

15.1.2 Network Layer

The main role of the network layer is to perform tasks in a subsystem manner. In the network layer, everything is connected, and it gives them an allowance for the awareness of the surroundings. In network layer data is shared between connected things and it is crucial to automated processing management in IoT as shown in Fig. 15.1. For continuous sharing of data by smart objects require a strong connected network with full automation, so every object should be able to switch and perform their roles when required.

15.1.3 Service Layer

As from the layer name, we can approach it with the formula that it will provide services and application to the IoT, with a cost-effective providence of hardware and software. The main working mechanism and responsibility of this layer is to locate new services for the application in the network and collect data about the provided services. Different organization has developed various standards for the functions in the service layer, but a universal functioning of the layer is very important for IoT [3]. The service layer of IoT must have a minimum no of applications. The interface of application programming with the support of protocol is required for the services in this layer which can perform various tasks like storage, management of data, searching platform and communications.

15.1.4 Interface Layer

In this layer of IoT, a huge number of devices are connected and there are different services provided to different peoples, hence it cannot be complying with the same standards. For a good service-oriented interaction between the devices. That compatibility issue must be resolved. Good compatibility between the devices will increase the efficiency of the exchange of data, communication, and event processing. To simplify the management and interconnection between the devices a strong effective mechanism is required. In other words, the interface layer can resolve this issue and act as the front end of the IoT.

15.2 Advantages of IoT

15.2.1 Efficiency of Decision Making

The biggest advantage of IoT in the industrial sector is the efficiency of decision-making Example—IoT solutions into the traditional ERP (Enterprise resource management) and CRM system is to monitor them and collect data, the received data from the software can help the corporate to improve their reaction time so further it can transform into improved diagnostics, higher revenue, and better customer satisfaction [4].

15.2.2 Good Maintenance or Predictive Maintenance

IoT technology provides real-time data so you do not have to depend on primitive data for planning maintenance schedule for your functioning machine.

15.2.3 Timesaving

By using IoT on daily basis on the industrial level it saves time by performing many automation tasks which on the other hand if done by humans then the process will extend and take more time.

15.2.4 Analysation of Data

The main mechanism of IoT is to collect data and analyse it on a certain level as the technology is evolving day by day the efficiency of collecting data by IoT is also increasing and its effects and improve several factors related to IoT like performance and reliability.

15.2.5 Safety and Security of a Manufacturing Plant

The big data analysis of IoT can improve the overall workers and employee security in the plant [5], by monitoring key performance indicators (KPI's) health and safety like rate of illness, the no of injuries, near misses' accidents, vehicles accidents, and property damage or loss during daily operations.

15.3 Challenges in IoT

The Internet of things (IoT) has been one of the biggest developments that hit the world in the last some decades. It was a necessary development in today's world where we are using the Internet to run our devices and connect with others. Today's ideal world has the potential to change life as we know it. Implementing IoT architecture is not going to be easy. According to some researches, the implementation of IoT is going to be complicated or not so easy. For developers, IoT is one of the major challenges that need to be understood and applied properly. The consequences of poor implementation can be drastic. The following points given below are to prevent IoT from getting hampered:-

I. **Scalability**—The implementation of IoT at a larger scale is very important for its proper and efficient functioning.
II. **Security**—The biggest challenge for IoT developers since its initiation has been its security. Connected devices with each other can leave them open to spiteful imposition and hacking. Security in IOT infrastructure concerns not just developers, but also corporations and government. Hacking into such infrastructure has already created many problems, such as the attack on Ukraine's power grids or the 2009–10 attack that destroy Iran's nuclear centrifuges. Corporations open to such attacks can see paralyzing losses. But as IoT becomes more and more implanted in our life, it is not just the corporate information and government installation that are under attack. Hacking of devices like automated cars can ever cause loss of life. we have today, the solution also must be multi-tiered. We need gateways to gather sensor data and using a cryptographic layer for authentication. This will ensure that devices cannot be easily hacked, and malicious commands relayed. We need a stronger password framework since it has proven it to be the most effective first layer of defence [6]. Fortunately, we have many developers working on developing security software to create an ironclad framework for the burgeoning IoT infrastructure. This includes working on better sensors at gateways, secure software development and using the cloud to update and manage security.
III. **Connectivity**—We are facing the major difficulty of connectivity. The most effective data gathering is typically through server-ride software. However, poor connectivity can hamper this function. this is true for certain devices that are used in remote locations. The solutions can be the delay in uploading the data. All information can be stored and then uploaded when the device connects to the web.
IV. **Adaptability**—Adaptability is one of the biggest issues in evolving IoT technology, among the industries this. will require a change at certain levels in the way organizations design and argument their industrial technique and process. IoT systems must be extensible and adjustable through software or services that integrate the overall solutions. Changing any technology and developing it into IoT is a very complex process it requires regular hardware and software updates with maintenance and many more things which will increase the complexity in setting up the IoT technology.
V. **Cloud computing**—For any innovation that cycles or stores data, information security and protection issues are its primary goal. In like manner, for Cloud-IoT, the issues identified with security and protection are the best concerns since it is straightforwardly connected with the trust of clients. Previously, there had been hacking and falsification cases with both presumed and non-rumoured organizations on the lookout, which brought about extraordinary misfortunes. Security danger related with IoT is substantially more disturbing because for an example, dissimilar to security dangers in financial area which may bring incredible monetary misfortune, any danger to medical services IoT applications associated with the Cloud is perilous.

15.4 Literature Review

As we have earlier discussed the architecture of IoT, some of its advantages and challenges also. We have mentioned below some of the papers which were published under the same domain. Table 15.1 mentioned below shows the summary of previous research done on the Internet of things, artificial intelligence and their applications including intrusion detection and many other topics related to IoT.

Table 15.1 Reviews on various previous work on IoT

Author's name	Outcomes	Future work	Reference number
Misra, D., Das, G., & Das, D.	This paper introduces the opportunities of IoT, application fields and analyse the challenges of IoT	The future of IoT will be expected to be unified, unsealed, and extensive. IoT has become an unavoidable trend of development of information industry	[6]
Ashish Ghosh, Debasrita Chakraborty, Anwesha law	Artificial intelligence technology may be led to control human life in many aspects and we will depend on them for everything	We may have a smart cyber revolution and automation without unrestrained employment	[7]
Artur Miguel Arsenio, Fernando Nabais	IoT focused on observing current approaches for all these intelligent things connected and communicating	Man, in the middle is used in attacking to penetrate any system and we can use it in securing IoT	[8]
Okwori Anthony	This paper represents an overview of intrusion detection in IoT as well as detail knowledge of various threats	Detection system will identify suspicious event and let the system administrator or security officer decide whether to start an investigation	[9]
Pradyumma Gokhale, omkar Bhat, Sagar Bhat	It describes the upcoming technologies of the internet were the physical things could be acquired and identified through the internet	Every available thing is getting smarter and many new technologies will emerge in upcoming years	[10]
Ah-lian Kor, Colin Pattinon, Max Yanovsky, Vyacheslav kharchenko	This paper describes about smart homes, smart health and care, smart quality of life	It involves the development of an advanced and flawless integrated IoT enabled ecosystem and services to support smart livings	[11]

15.5 Applications of IoT

15.5.1 Smart Governance

Smart governance is an extended version of e-government both built on the Internet of things. In the current scenario, the smart government can be categorized as new generation and extended smart government. The Internet of things has modernized the various things in the twenty-first century and played a key role in making and maintaining smart cities. Smart cities imply in establishing an imperishable technological process for managing cities from utility control, electricity housing to transport control. The main goal behind developing this technology is to improve the quality of life while reducing the cost. Despite some challenges, smart government initiatives have made in Dubai, Australia, Singapore, and India with positive results. Governments around the world have realized the immense benefits of IoT, and they have started allocating heavy budgets for the development of smart government technology with IoT [12]. The e-government function on the automation with process and support of administration paper-free offices, data maintenance, information recovery, interdepartmental connections, and workflow mechanized. Smart government, on the other hand, provides the public with open and transparent data. Smart government is a combination of digital government and smart cities and we can divide smart government into two parts, first is called extension smart government which introduced to an extension of traditional e-government. The second is the next generation smart government which is the collation of government 2.0 and smart cities (Fig. 15.2).

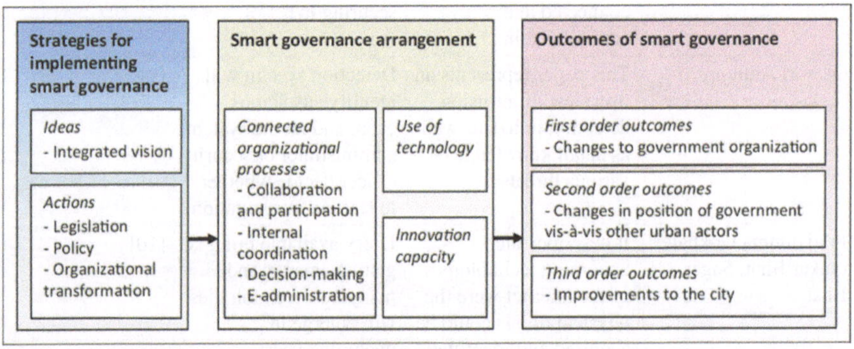

Fig. 15.2 Model of smart governance [13]

15.5.2 Smart Homes with IoT

IoT is a set of terms used for all the technology that allows the connection of the device to the internet and this system functionality is dependent on the collection of data and processing it for various automated IoT functions that can be used for a certain scenario when the respective situation arrives. Let us take an example of an automated bread toaster machine that can be programmed for automatic off and on the function when the machine completes its task of toasting the bread. Now imagine the same situation with all the devices present in the house from a washing machine, television to the home theatre even the doors and windows are on the automated function where you don't have to go to every device for shutting it down manually you can do it via voice command or with an AAP, now just imagine furthermore forward where you don't have to even give command also everything is analysed on the activity of your daily life via collecting data from various automated sensors installed in the home and they perform the tasks by analysing the same data like closing doors, windows, lighting, almost everything in the house and that is where the future of IoT with a smart home lies. A set of systems of light that is connected to an automated system and its functioning work on various criteria like daylight brightness, automatically switched on as sunsets, detecting motions at midnight and switching on the light, and there can be many others functions which can be included in this automated lighting functions [14]. Doors in the future can be smarter, just think they open and close when you enter and exit, and it can be programmable for various other functions including security functions also example just by your voice command all the doors in the home are on lockdown and this can be made possible by facial recognition, smart apps, and voice recognition. These days thermostats are controlled via various apps but as the IoT is progressing in the future we will not we are needing the apps for that. The thermostat will be able to recognize you near home and then check the home and external temperature and set the preferred temperature for you. It may monitor your actions and adjust the temperature accordingly like when you are exercising or showering. Now combine all the above system in one home connected to a smart device, there are smart sockets presents in a home that turns off and on automatically, smart alarm, voice assistants can read the newspaper and the whole routine which includes several home devices like lights, home appliances, thermostats, and other devices are controlled shutters in morning opening by themselves and many more other things IoT is smart home is the future (Fig. 15.3).

15.5.3 Fastags with IoT

Fastags device comes under the national electronic toll collection (NETC) program and it is designed to catch radio frequency identification (RFID) technology to make toll payments directly while the vehicle is in motion. This tag will be fixed on the windscreen of the vehicle and gives a luxury to the customers to make toll payments

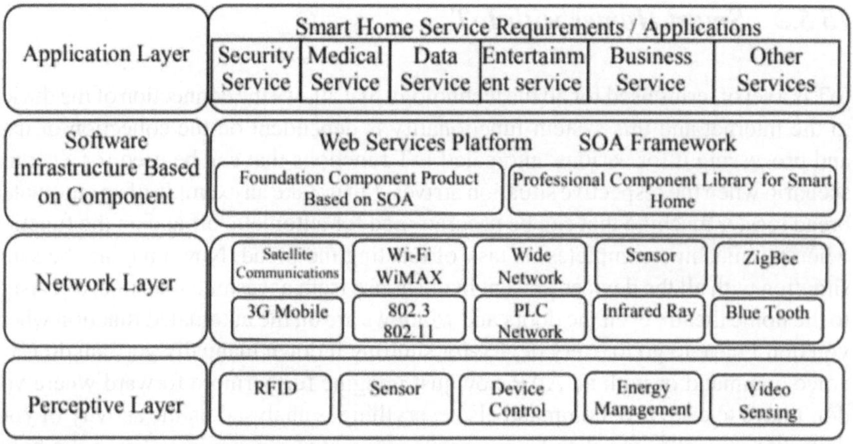

Fig. 15.3 The architecture of smart home application based on IoT and component technologies [15]

directly from their account which is linked to the fastags payment portal which is connected to a gateway payment services. This system was initially brought in the picture as a pilot project in 2014 on the stretch of golden quadrilateral between Ahmedabad and Mumbai. This was implemented in the Delhi Mumbai region of the quadrilateral. This fastags technology is an absolute example of IoT. Just think in this manner that the tag with radio sensor of RFID and it is a very minute technology in this aspect of fastags, then it relates to a payment gateway which is linked directly to your bank account or the wallet of the fastags services the next step is it will automatically deduct the required tax according to the distance covered by an individual from the last toll. This whole process involves so many automation services like capturing of the tags then the details are processed related to the captured tags which have all the information related to the vehicle, tags, model of the car then all details are confirmed by the NETC mapper which checks and validates the status of the tag after that host will calculate the fare and a debit request will be initiated to NETC system after the validation of the tag, now this process request will transfer to the issuer bank for deducting the money from the account of the customer now the host from the issuer will send an SMS to the tag holder of deducting the required fare from the account, and lastly, the host will send the notification to the person in toll plaza of payment confirmation. This whole process required no output or input from human beings it was automated as everything is connected through the internet and data processing is the key role for its proper functioning and this process is less time consuming with a healthy rate of efficiency. But it is a new technology, and it requires improvement at certain levels and those improvements are done daily as this new technology evolves day by day.

Following components are shown in Fig. 15.4 mention below-

1. Auditing module at the reader layer.

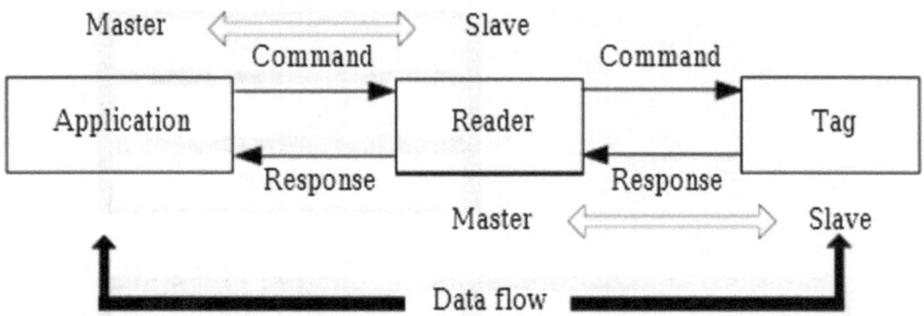

Fig. 15.4 RFID system model [16]

2. Detection module at the middleware layer.
3. Action module.

15.5.4 Connected Cars

Sometimes we tend to visualize a scene degree passing associate degree exceedingly in a very fantasy flick where AN actor is driving an automobile by voice command. That fiction is turning into true these days with the blessing of IoT. The thought of a connected automobile is once AN automobile or the opposite vehicle area unit usually controlled by a smartphone or the opposite device that is connected with the net.

- This manner of application contains plenty of sensors to look at ofttimes this standing of the car.
- Analyses if any impediment is presently at the front of the car and take immediate steps to avoid any accident.
- It should be a self-learning automobile with the employment of AI. With time, AN automobile becomes a great deal of advanced and sensible.
- Predicts any machinery failure before it happens, that creates it a great deal of wholly different than a daily automobile.
- Users can begin the engine by on voice command from a smartphone before driving.
- The report of NHTSA says that it's safer to use AN automatic automobile rather than self-driven.
- Potential to manage home devices from automobile (Fig. 15.5).

15.5.5 Agricultural Drone

Monitoring a huge crop field is not simple. The agricultural drone is associate degree aerial vehicle that will survey the massive field and reveal problems. Birds-eye scan

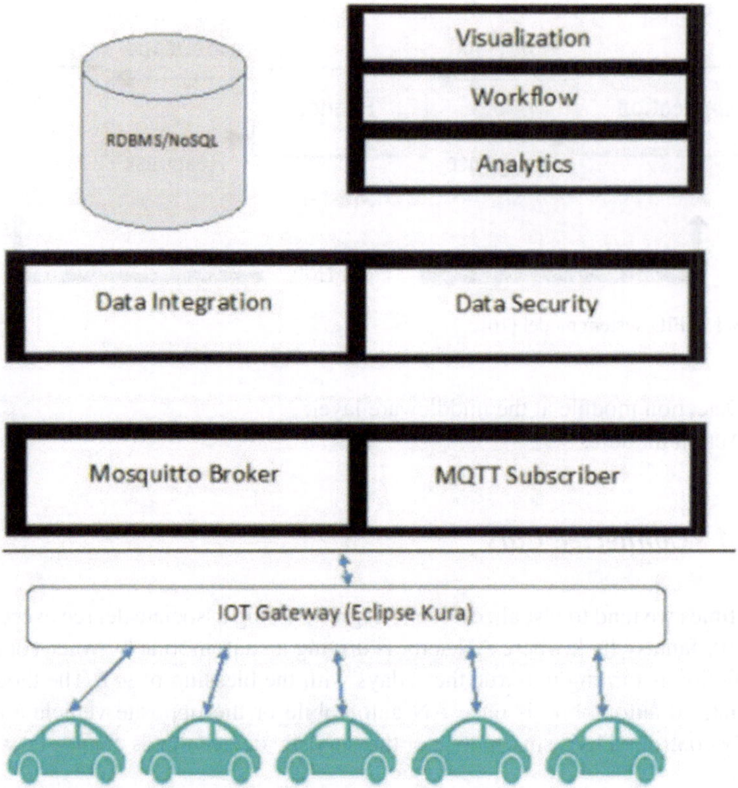

Fig. 15.5 Architecture diagram of 'Connected car' ecosystem [17]

from the sky with sensors provides a clear and effective image. It will keep a decent impact on agricultural business by saving time and proper land management choices.

- Valuate field analysis, crop health, crop growth, and tormentor attack.
- Analysis goes on administrate planting by drone technology.
- It will spray fertilizer and pesticides, that's so much effective than regular spraying systems.
- With correct image method algorithms, a drone can notice the plant life attacks.
- Excessive water flow on the crop fields can ruin the fields. To forestall that status device from a\u00a0\u00a0\u00a0\u00a0drone can notice further status during a district.
- New job likelihood as agricultural drone pilot is returning (Fig. 15.6).

Fig. 15.6 Agricultural drone system [18]

15.5.6 Smart Grid

The previous grid system sends electricity from power plants to homes and industries. However, the good grid consists of electrical instrumentality, electronic network, automation, and power provide. In a different way, the good grid is two-way communication between client want and utility. The good grid makes responses in keeping with the requirement of the client. It brings the chance to extend the potency and output of a rustic.

- Sensing the cable good grid ensures correct electrical provide.
- Self-repairing characteristics of the good grid mean it will mechanically reroute electricity throughout an influence failure.
- The transmission level of electricity is simpler than a typical grid system.
- Reduces electricity prices of the patron by suggesting them low-priority electronic devices.
- During serious load, this IoT application suggests users cut back power consumption (Fig. 15.7).

15.5.7 IoT in Military Field

There are several fields in the military in which we are using IoT devices. For example—The environment of the Military can be put under observation to give help to war fighters. Regardless of whether there is no war circumstance the basic natural condition can cause misfortune to military units. Illustration of Siachen glacier is well known to the Indian landmass, where war fighters need to face basic natural

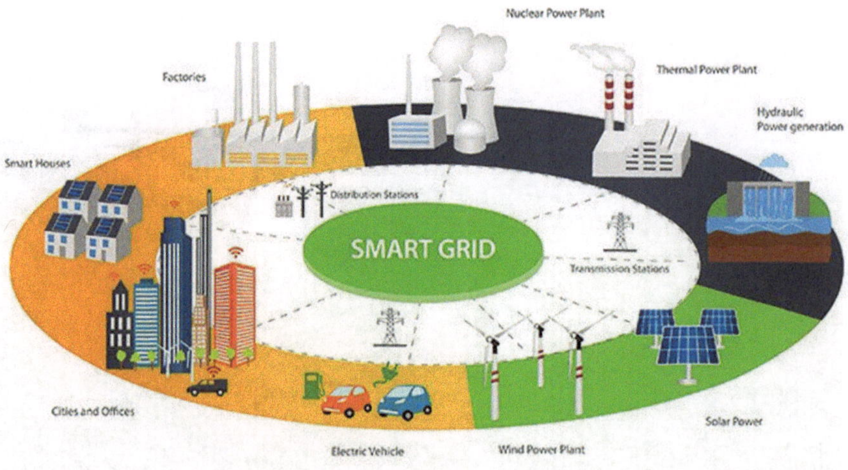

Fig. 15.7 Smart grid system [19]

circumstance for the expense of their lives. Using IoT, we have developed a wearable gadget. This gadget has sensors and trackers to remove data as to the condition and wellbeing status of military soldiers. Fighters can get to this strategic information by means of show. A similar data can likewise be made accessible to military central command to investigate the circumstance and make a suitable choice dependent on it.

15.6 Intrusion Detection in IoT

Let us take an example of IoT connection, IoT connects with the internet through the internet protocol version 6 (IPv6). It also has a six low power wireless personal area network (6LowPan). Despite having various security features like encryption and authentication the number of threat and attacks on IoT services are increasing day by day. Wireless sensor networks (WSN), 6LoWPAN network or WSN are connected to the un-secure and un-trusted internet services and sources. These vulnerabilities have created a threat and open target for the physical interface, wireless protocols and user interface of IoT [20]. Various penetration attacks can be done by hackers by using the same resources that are used in the functioning of IoT. Therefore, intrusion detection is the most important system for the protection of internals as well as externals IoT components. A realistic approach can be described in four parts first is a prevention unit, the detection unit for identifying components for security breaches and an investigation unit can also be installed for studying the breaches and making prevention methods for future intrusions. The focus of intrusion detection should be on identifying unauthorized access in IoT networks and it can also inspect various hardware and software tools that can detect machine and user actions detect well known malicious network activity and detect network intrusion.

(i) **Common intrusion prone services of IoT**

1. Home network routers.
2. Automated health devices.
3. Grid system services with IoT.
4. Connected car's system.
5. Supervisory control and data acquisition system (SCADA).

(ii) **Vulnerabilities on the internet of things network**

1. Insecure network interface.
2. Insufficient authentication/authorization.
3. Lack of transport encryption /integrity verification.
4. Improper security configuration and /insecure software and firmware.
5. Poor physical security.

(iii) **Types of attacks on the internet of things**

1. Botnets.
2. Man, in the middle internet of things attack.
3. Data and identify theft internet of things and network.
4. Social engineering.
5. Denial of service attack.

The Intrusion detection in IoT network is possible in the following ways:

(i) **Misuse detection**: The whole database of attack is stored here, and it matches the current scenario with the past attacks in the network, which has already been stored form the studies of the past attacks. Its efficiency is directly proportional to the information in the database.
(ii) **Anomaly detection intrusion system**: This intrusion system must start the admin point it defines and controls the flow of network traffic load. In the form of the software, the anomaly detection system will check the inconsistency and compare it with the normal state that which constantly function normally if there is any inconsistency in the network it will alert the admin support.
(iii) **Network-based intrusion detection system**: In this detection system, every packet is examined properly which is transmitted through the network. It can detect the harmful packets across the networks which are ignored by the firewall security. A network-based id's sensor has two interfaces with one in a manageable unit. It communicates with the sensor through the management interface. The observing interface is linked to the network section that is being observed as shown in Fig. 15.4. It observes each packet that crosses through the traffic and apply known attack signatures to every frame to detect malicious activity in the traffic. If it detects a match against any signature it alerts the admin console.

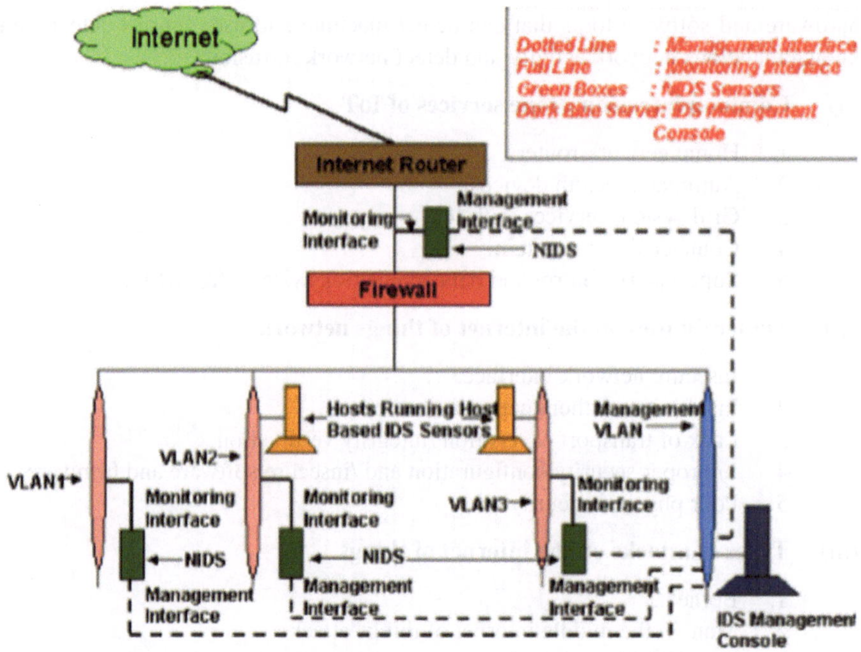

Fig. 15.8 Deployment of ids sensor and management console in a network [9]

(iv) **Submissive and reactive intrusion detection system**: In submissive intrusion detection system mainframe work is to observe potential security breach and if the malicious activity is detected by the sensor then it will send an alert to the console of the owner. In reactive intrusion detection, the focus is on the suspicious activity in the firewall then it will reprogram all the system in the firewall which will log off the suspected user then the sensor will scan the whole system for the suspected user foe expelling the user out of the system (Fig. 15.8).

15.7 Role of Artificial Intelligence in IoT

Internet of things is all about connecting physical objects around the internet and they run totally on sensors and communicative devices without any intervention of human need. As in 2020, only the world-wide number of IoT connected devices is around 17.6 billion [21]. So, what will happen when we combine artificial intelligence with IoT artificial intelligence brings so many advanced functions and features which use the collected data of IoT at so many different levels by using machine learning datasets and many more tools and domains of artificial intelligence.

15.7.1 Artificial Intelligence in Industrial IoT

Getting analytics by combining IoT and artificial intelligence can predict the equipment failure in industries, and this will provide a boost start for early maintenance procedures. It can predict the damages and help the company to boost its economy, infrastructure and many more, the machine learning domain gives the industry a possibility to identify patterns that lead to machine failures.

15.7.2 Increasing Operational Efficiency

Machine learning combined with IoT just not increase operational efficiency, machine learning can forecast operational conditions in advance, and it can predict the ecosystem with a neutral environment for efficient function of the machine. Machine learning has a unique ability to detect patterns in constant data flow which are barely visible to human eyes.

15.7.3 Enhancing Risk Management

Artificial Intelligence and machine learning combined with data set of IoT Government and private organizations can detect risk in their provided services and automated response to those risks can be added as a service feature by private organizations and several government domains. Cameras are placed in various public places by the government for safety purposes if we apply machine learning combined with IoT to detect several risks at those places like an accident happens automated response bill alerts the nearby authorities and nearest hospitals after detecting the risk factors from the accident [22]. Law enforcement agencies around the world have a huge database of crime that happened in the country in the past, machine learning combined with IoT can be used to study those crimes and make predictions for possible crimes in the future. IoT is all about sensors, tags, storage, management of data capability of high in processing while staying connected with the internet. Functions change the ecosystem in which they exist they do so together data and provide services accordingly. If we enable IoT with artificial intelligence, then it will make the Iron Man's powerful Jarvis AI a reality.

15.7.4 Voice Assistant

The famous examples of IoT combined with artificial intelligence is voice assistant. Voice assistant services respond to the owner's command and the performance is dependent on the third-party applications combined with cloud-based services. They can perform various tasks like answering basic queries, booking cabs, reservation for movies and restaurants, controlling home with automation services using relay protocol light switching on and off light controlling smart doors and smart Windows.

15.7.5 Cyber-Physical System

CPS is an advanced engineered system that is built on high-end computational algorithms and physical networks. Today's CPS system is built on the foundation of smart objects' unification infrastructure communicational framework. These interconnected substructures like smart cities, smart governance, automated electric grid system and many other smart devices which are connected for proper function of these connections they are needed to be flexible and less expensive in implementation [23].

15.7.6 Artificial Intelligence and IoT CPS

The IoT CPS scenario deals with the Hunk amount of data. Data act as a Centre command in the big world of IoT. For intelligence data dependence decisions taken by smart objects in the IoT system define the relation of IoT-CPS and artificial intelligence [24]. It will make every smart object independent to make some decisions by having some intelligence for local processing. Innodata dependent decisions more amount of data should be utilized but should be feasible time when it is in use for the services. Automated machine learning system how to process play amount of data must have the following conditions. 1. Data preparation capabilities 2. Learning basic and advanced algorithms 3. Adaptive processes with automation. 4. Scalability. 5. Enable modelling and real-time decision making. The IoT-CPS involves functions that interact with a complex environment and this is a challenging scenario and these systems must have advanced assets to aware and automatically asses the coming faults and failures and these possibilities and scenarios of IoT-Cps will take our planet towards a bright future (Fig. 15.9).

15.7.7 Cognitive AI and IoT CPS System

IoT is not just a subsystem of wireless networks, sensors, accumulators or collecting hunks of data sensed by smart objects present in the range of the sensor. IoT is a futuristic approach that features to connect every smart object in the world [26]. The mechanism of these connected devices is justified by Programmable computing and constant flow of data which have a constant progression and follow certain rules.

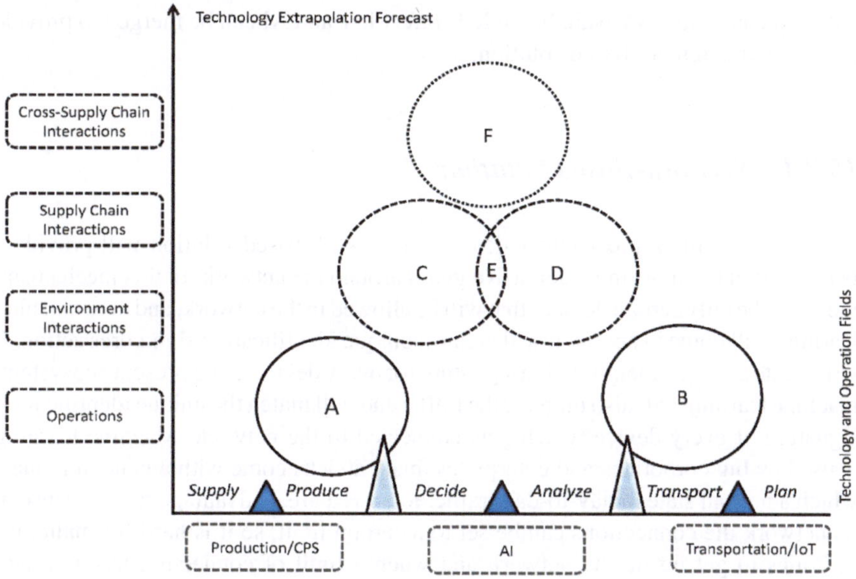

Fig. 15.9 Technology extrapolation forecast for the fields CPS, AI and IoT [25]

But on the other hand, having a psychological framework is a good option because they extract the hidden relations inside the ecosystem of IoT. The IoT CPS combined with artificial intelligence I can manage 80% of the world's data, that research called unstructured data as shown in Fig. 15.5 for examples tweets, journals, images etc. [27]. Cognitive in artificial intelligence with IoT-CPS works on the mechanism of extracting the data from the understanding of the ecosystem with the help of artificial intelligence which trains the data packets and acquires them with a pre-knowledge for better functions and efficiency. Hence this system can be taken as an extension of IoT with the potential of self-learning.

15.8 AI Using IoT Devices for Security Additions

Today's IoT has indulged itself in our daily life from the health care system to infrastructure to interconnected smart cities and that's where security in IoT arrived, this does not mean that IoT is not a good initiative. Yes, there are many flaws in IoT but with the help of several artificial intelligence domains, it can be improved. IoT at smart homes has not the same security as the generic computers have. For one thing managing so many devices is a nightmare. This is where artificial intelligence comes in the scene as the many domains of artificial intelligence can manage flaws of security in IoT. AI especially adept at finding patterns when it is fed a large amount of data and data is always available in IoT. Infect IoT and AI can be merged to provide the following domain-based solution.

15.8.1 Network-Based Solution

Instead of installing pre-security devices, a network-based solution will provide a better option by imposing a defensive guard around the network. In this mechanism, there will be only certain devices that will be allowed in the network, and here machine learning will come in action, it will create a unique identification signatures action. It will create a unique identification signature for every device in the present ecosystem, machine learning will also monitor the traffic and will match the unique identification signature of every device which gets connected to the network. Machine learning is used by big corporates and enterprises many attacks come with a general request which a normal state in day-to-day traffic. IoT has a great advantage here because in its network the connections can be set to a certain limit, so it is hard for malicious activities to get inside the network and when a limit of connection has a certain number then it is very easy to differentiate the changes between normal behaviour and suspected behaviour of the connected device. Also, the traffic monitoring services of machine learning can be used to monitor the interconnections between the devices to find the malware that might have to get past the outer perimeter of the network and can identify the devices which are compromised [28].

15.8.2 Device-Based Solution

IoT devices face a very alarming issue of power processing and storage capacity to facilitate the security functions and store huge database of signature, attacks, malicious activity to protect them against the threats. Machine learning can help by setting certain exit point security for IoT. The performance-based solution can be developed over the signature-based solution and it is less resource-demanding and can run without bogging down small processors.

15.9 Conclusion

The way IoT has grown in past years has affected every aspect of human lives from early 2000. No one could have imagined that in just 20 years around 20.1 billion devices with an investment of 6 trillion dollars worldwide would be connected with several types of sensors and devices which make it independent of human hands for them to up and running as the IoT technology is evolving day by day it is providing solutions which seems to be impossible in the early days of technological revolutions,

IoT is all about data small data big data from the microscopic level to the macroscopic level data is the main component in IoT. Although every developing technology has its flaws and advantages IoT has big security concerns related to its certain services from home automation to smart governance every service is vulnerable at some point for hackers and from its malfunction. But as the IoT is used by big corporations and enterprises its security has some vast improvements with the help of several domains of artificial intelligence like machine learning. IoT is the future as the other technology will evolve as the things will evolve more data analytics data processing everything which is connected to data even the smaller technology. Also, it will completely be dependent on IoT. We have discussed, IoT technology should be more used in the military of an individual country. This will open gates for advancement in IoT technology from its smaller to bigger components. IoT technology has the very much potential to provide us with a better lifestyle at the appropriate cost.

15.10 Future Work

IoT is all about sensors, accumulators, collecting tons of data that are sensed by IoT devices in ecosystems and environments. We have used IoT in smart governance, smart cities, smart cars, personal cloud storage, and voice assistant. IoT has the potential to do much better and can be used in the welfare of several other important assets. For Example combining the IoT and Machine learning domains of Artificial Intelligence to build a threat assessment system, and use this whole system for protecting important government assets, buildings, several army bases. Although combining their security mechanisms with IoT will ensure fewer human casualties and an early warning system that will prevent attacks in sensitive areas. IoT systems can be used in several disaster management events to ensure minimum damage and early warning systems will help the authorities to save more lives. The same security features which are used to protect the IoT system, when combined with IoT and AI then it is enough capabilities to secure defends and several grid systems on which the country is running. In a battlefield and live combat soldier which are equipped with combat suits, helmets, and weapon system all the equipment can be embedded with advance sensors which can analyse and filter the channels of data and can watch over soldiers biometric functions like iris, facial expressions, heart rate and later this analysed data can be used as various scenarios for combat practices and from the same data analyst and engineers can provide soldiers with advance equipment like connecting the drone control to the automated system embedded on soldiers suit and this can facilitate several advantages in combat like providing automated recon before the forces have arrived on the scene can mark the location of enemy combatants providing the types of weapons enemy has with the help of previously-stored combat scenarios and IoT embedded AI can predict the several outcomes which will directly or indirectly help the soldiers to not only survive lethal combat it will also increase the efficiency of stealth so high-value targets can be captured and higher threats can be neutralized behind the enemy lines as well as inside the country. The army base stations are an integral part of a country's security system as it has a tactical advantage in warfare. IoT embedded devices will act as an extra protection layer to the base. Fully automated screening systems, security cameras communicated in the network of the base stations that having the embedded sensors which will include thermal to infrared features can be used for detecting threats and unusual objects around the base. The whole security system of the base stations is related to the machine learning domain of artificial intelligence which will train the automated system by acting as a central command that will increase the efficiency of the security system around the base. It will decrease the human resources but will ensure the safety of the base stations. IoT will also help in the smart management of the base stations in terms of electricity, water, solar power, etc. IoT will help to increase the capacity and the output of the military bases and result in the smart functioning of the base stations to ensure the safety of the personal army officials inside the base stations.

References

1. Burhan, M., Rehman, R. A., Khan, B., & Kim, B. S. (2018). IoT elements, layered architectures and security issues: A comprehensive survey. *Sensors (Basel, Switzerland), 18*(9), 2796.
2. Hassan, Q. F., & Madani, S. A. (Eds.). (2017). *Internet of things: Challenges, advances, and applications*. CRC Press.
3. Allsbrook, A. (2018). *Four artificial intelligence challenges facing the industrial IoT*. Clear Blade, Inc.
4. Fletcher, D. (2015). Internet of Things. In M. Blowers (Ed.), Evolution of cyber technologies and operations to 2035. In *Advances in information security* (Vol. 63). Springer.
5. Lee, E., & Seshia, S. (2011). Introduction to embedded systems—A cyber-physical systems approach.
6. Misra, D., Das, G., & Das, D. (2018). Review on Internet of Things (IoT): Making the world smart. In R. Bera, S. Sarkar, S. Chakraborty (Eds.), *Advances in Communication, Devices and Networking. Lecture Notes in Electrical Engineering* (Vol. 462). Springer.
7. Ghosh, A., Chakraborty, D., & Law, A. (2018). Artificial intelligence in Internet of things. *CAAI Transactions on Intelligence Technology, 3*(4), 208–218.
8. Arsénio, A. M., Serra, H., Francisco, R., Nabais, F., Andrade, J., Serrano, E. (2014). Internet of intelligent things: Bringing artificial intelligence into things and communication networks.
9. Okpe, O. A., John, O. A., & Emmanuel, S. (2018). Intrusion detection in internet of things (IOT). *International Journal of Advanced Research in Computer Science, 9*(1).
10. Gokhale, P., Bhat, O, & Bhat, S. (2018). *Introduction to IOT*.
11. Kor, A. L., Pattinson, C., Yanovsky, M., & Kharchenko, V. (2018). IoT-enabled smart living. In M. Dastbaz, H. Arabnia, B. Akhgar (Eds.), *Technology for smart futures*. Springer.
12. Baheti, R., & Gill, H. (2011). Cyber-physical systems. *The Impact of Control Technology, 12*(1), 161–166.
13. Bolívar, M. P. R., Meijer, A. (2015). Smart governance: Using a literature review and empirical analysis to build a research model. *Social Science Computer Review*.
14. Câmara, D., & Nikaein, N. (Eds.). (2016). *Wireless public safety networks 2: A systematic approach*. Elsevier.
15. Abbasi, M. A., Memon, Z. A., Memon, J., Syed, T. Q., & Alshboul, R. (2017). Addressing the future data management challenges in IoT: A proposed framework. *International Journal of Advanced Computer Science and Applications, 8*(5), 197–207.

16. Jia, X., Feng, Q., Fan, T., & Lei, Q. (2012). RFID technology and its applications in Internet of Things (IoT). In *2012 2nd international conference on consumer electronics, communications and networks*.
17. Dhall, R., & Solanki, V. (2017). An IoT based predictive connected car maintenance approach. *International Journal of Interactive Multimedia and Artificial Intelligence, 4*(3).
18. Jensen, J. (2019, April 26). Agriculture drones: Drone use in agriculture and current job prospects. *UAV Coach*. https://uavcoach.com/agricultural-drones/
19. https://tse4.mm.bing.net/th?id=OIPu.ZX7DoTwI1vnQgVXjvj4e-AHaEQ&pid=Api&P=0&w=265&h=153
20. Gershenfeld, N. (2014). *When things start to think: Integrating digital technology into the fabric of our lives*. Henry Holt and Company.
21. AlEnezi, A., AlMeraj, Z., & Manuel, P. (2018, April). Challenges of IoT based smart-government development. In *2018 21st Saudi Computer Society National Computer Conference (NCC)* (pp. 1–6). IEEE.
22. Alam, F., Mehmood, R., Katib, I., Albogami, N. N., & Albeshri, A. (2017). Data fusion and IoT for smart ubiquitous environments: A survey. *IEEE Access, 5*, 9533–9554.
23. Fortino, G., & Trunfio, P. (Eds.). (2014). *Internet of things based on smart objects: Technology, middleware and applications*. Springer Science & Business Media.
24. Yoskovitz, S., & Shaul, G. (2019). *U.S. Patent No. 10330647*. U.S. Patent and Trademark Office.
25. Abomhara, M., & Køien, G. M. (2014, May). Security and privacy in the Internet of Things: Current status and open issues. In *2014 international conference on privacy and security in mobile systems (PRISMS)* (pp. 1–8). IEEE.
26. Jeong, S. (2015). *The internet of garbage*. Forbes Media.
27. Kaplan, J. (2016). *Artificial intelligence: What everyone needs to know*. Oxford University Press.
28. Soomro, S., Miraz, M. H., Prasanth, A., & Abdullah, M. (2018). *Artificial intelligence enabled IoT: Traffic congestion reduction in smart cities*.

Chapter 16
Smart HR Competencies and Their Applications in Industry 4.0

Yashonidhi Srivastava, Souvik Ganguli, S. Suman Rajest, and R. Regin

Abstract With the advent of artificial intelligence and IoT linkages, human resource management (HRM) is undergoing radical changes and moving onto the new coinage called smart HR. The features of HR in industries are intentional, along with the need to shift to a smarter domain. With regard to industry dynamics, as well as working conditions and the overall organizational structure, the advantages of the technologies implemented and the methodologies suggested have been presented. Also addressed are the industrial shifts relating to the evolving economy and other developments brought about by the introduction of novel systems and technologies. The research also took on the influence of the introduction of new working techniques such as smart work (SW), structural equation modeling (SEM), and smart city projects (SCPs) in the field of smart HR. Several steps taken for the development and growth of the sector were also highlighted, along with recommendations that could be taken up to further boost the experience of work and organization. For a deeper understanding of the principles, a lot of studies were also included, including supply chain systems and cyber-physical systems. Besides, the identification of domains needing further study was also highlighted. In addition, to get the best performance, strategic alliances, line management, and organizational design were addressed too. Focus on the digitization of human resource management was also discussed on the changes of the organizations from e-HRM to d-HRM.

Keywords Smart HR · Artificial intelligence · Internet of Things (IoT) · Industry 4.0 · Blockchain technology · Cyber-physical systems (CPS)

Y. Srivastava · S. Ganguli (✉)
Department of Electrical and Instrumentation Engineering, Thapar Institute of Engineering and Technology, Patiala, Punjab 147004, India
e-mail: souvik.ganguli@thapar.edu

S. Suman Rajest
Vels Institute of Science, Technology & Advanced Studies (VISTAS), Chennai, Tamil Nadu, India

R. Regin
Department of Information Technology, Adhiyamaan College of Engineering, Hosur, Tamil Nadu, India

P. Kumar et al. (eds.), *A Fusion of Artificial Intelligence and Internet of Things for Emerging Cyber Systems*, Intelligent Systems Reference Library 210,
https://doi.org/10.1007/978-3-030-76653-5_16

16.1 Introduction

Incorporations of technologies like big data, analytics, and artificial intelligence (AI) in the management encouraged the HR domain to adopt new methodologies and systems and dwell into smart HR. Some companies had been operating on big data from the very beginning and henceforth did not face difficulties in including added technologies to enhance their firm. However, many organizations faced concerns with the adoption of novel techniques. Manifold strategies regarding organizational order and training along with frameworks were proposed to ease their way into industry 4.0. Management strategies involving Analytics 3.0 were approved to enhance the organizational structure and to throw light on the relevant skill requirements [1].

The technology of web scraping was put to work to extract information from online portals regarding job profiles in the growing industry. The skill sets required to work for the industry constantly moving towards the inclusion of big data were also discussed in [2]. The responsibilities of the HR domain in the degrading economy were also examined featuring the development of trusty teams encouraging personal and professional growth [3]. Distinctive importance on the sectors of production, enhancement of industrial practices, and research have turned out to be the key to succeeding in the manufacturing sector [4]. The relationship between supply chain management (SCM) and HRM was explored along with the transmutations of industry 4.0. Business trends and factors affecting the addition were analyzed and comments regarding further research were made [5].

Smart work (SW) was ratified increasingly to obtain spatial flexibility and to provide optimum working conditions for the employees. SW models for higher productivity, innovation, and flourishing performance were also examined. The need for HR practices to adapt according to the growing industry was highlighted through the conduction of several interviews and surveys [6, 7]. Discussions were carried out regarding the SEM model for deciphering the salaries of the employee along with their promotion and loyalty to the organization [8]. For remodeling businesses to come to terms with the changes of industry 4.0, DSCC and CDSCI models were discussed keeping stakeholders, novel technologies, and working abilities in the loop. The digitalization process enhanced by the inclusion of novel technologies like big data, AI, cloud computing, and analytics was also elaborated. Sensor-based systems were greatly embraced in the industry [9–11].

Possible areas of research and development were discovered in the field of management procedures and the technologies being adopted in the recent past. The relationship between HRM and the workers was also discussed and its consequent effects on the organization's performance were described [12]. The proceedings in the manufacturing industries post the enhancements brought about by the industrial revolution were included along with the discussion on a Croatian model made for innovation. The objective was to reduce the gap between the functionalities of the industry [13]. The dependency of the industry on supply chain systems was brought up and suggestions were made regarding their inclusion in upcoming projects. Frameworks and other methodologies were also proposed to move towards a prosperous future

[14]. The issues of monitoring the vast amount of data accumulated by the inclusion of technologies like big data were discussed along with provisions to manage, analyze, and understand it [15].

Strategic partnerships were promoted in the HR domain by strengthening planning, organization design, and development along with the enhancement of line management and rotation and training of the HR domain [16]. The transformation of e-HRM to d-HRM was elaborated along with the necessities for the same [17]. To improve the recruitment process of the companies, a method based on blockchain called BcRMS was introduced considering factors of training companies, educational institutions, laws and customs, and the previous companies [18]. The increased usage of video conferencing platforms like zoom was discoursed along with a description of the challenges faced during their usage. The need for extensive research in the field was explained followed by the methodologies that exist for its implementation [19, 20].

Next-generation wireless technology, along with data processing, supports notably high data rates and smart handheld devices, providing a big challenge for transferring these contents from server to client. In a SDN setup, the proposed Improved Database Synchronization (IDBSync) mechanism accelerated the synchronization of data between the server side and the mobile database [21]. This technology can improve upon the smart HR competencies considerably. The use of big data applications, as well as its phenomenon, has increased by leaps and bounds during this age of information technology. This has been reviewed in [22] which can be made use of in the development of smart HR policies to aid industry 4.0.

The rest of the chapter is organized in the following way. In Sect. 16.2, the management practices of smart HR and their impact on the industries are discussed. Section 16.3 emphasizes on the utilization of the smart HR for the growth and development of the industries. Section 16.4 deliberates on the cohesion of smart HR with industry 4.0. Further, the enhancement and the digitalization of the human resource management are taken up in Sect. 16.5. Section 16.6 draws the important inferences with some directions of future work ahead.

16.2 Smart HR 4.0 Advantages, Learnings and Transformations

This section of the chapter discusses the functionalities of HR in industries along with the need to move towards a smarter domain. The advantages of the technologies adopted and of the methodologies proposed have been presented regarding the industry trends as well as the working conditions and the overall organizational structure. The learning from several surveys and interviews has been briefed and the lessons learned have been introduced in the form of new techniques and systems. The industrial transformations corresponding to the changing economy and other changes brought about with the inclusion of novel frameworks and technologies are also discoursed.

16.2.1 Management Practices

There are a few corporations like Google, Facebook, LinkedIn, and eBay that have been operating on big data since the very beginning. With the boom of the big data industry, these companies did not have to face the challenges of incorporating big data with their traditional data resources. Firms had to either switch or find ways to incorporate big data with their pre-existing IT infrastructures and models. For companies like Google or Facebook, established IT frameworks never endured for them hence there was no requirement of a possible switch. The hurdles for these companies were to have their big data analytics coexist with other frameworks, data, and analytics. To understand the coexistence of the analytics system, several companies were interviewed and the existence of big data accompanied by other data and their environment was studied. To make that a feasible system, Hadoop clusters were put to work along with IBM mainframes. The integration brought about a novel management strategy called Analytics 3.0. The thought process of companies regarding big data with emphasis on their organizational structure and the skill requirements for relevant occupations has been discussed. It was concluded that big firms were moving towards a merging data economy where traditional analytics and big data were no longer being separated but were being combined for improved outcomes. Progressing towards the adoption of a new model resulted in transformations of the structure of the organization along with the leadership and skill requirements. Industries adopting the novel framework combing sectors of data including weblogs, images and videos, social media data, documents along with reporting, OLAP, and modeling have witnessed tremendous transformations and have received prominent value in return [1]. Figure 16.1 shows the presence of the big data along with the data repository.

The work in [23] consisted of solutions about technologies to make the industry more 'smart' accompanied by organizational solutions. The proposed solutions were found to be implemented by utilizing logistics which has been increasingly adopted in the age of globalization. The solutions target individual companies in addition to national economies and also catered to global economies. With the ever-growing industry, the level of competitiveness was noticed not only in companies but in the supply chains as well. To grow with the growing economy, companies had to constantly introduce novel solutions, methodologies, and systems to keep progressing. Therefore, having solutions was at times not enough as the organizations required complementary implementations as well. The issues regarding the implementation of logistic solutions in the IT domain have further been discussed.

The significance of smart human resources 4.0 along with its functionalities as a catalyst in the proceedings of the disruption of the human resource sector was discussed. Examples from Credit Suisse were adopted to further elaborate on the

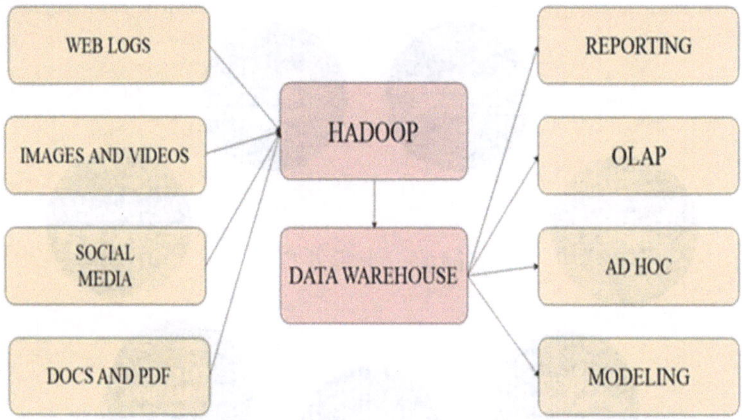

Fig. 16.1 Coexistence of big data with the data repository

importance of smart HR 4.0 and the need for its utilization in the industry. Credit Suisse is known to have reduced employee attrition by utilization of the methodologies on people analytics. A conceptual system known as the Smart HR 4.0 framework has been adopted to successfully illustrate the disruption of the procedures of talent onboarding along with talent off-boarding and the development process. Leadership style as well as organizational structures along with the application of emerging technologies discoursed. Big data has been applied integrally to overcome the challenges and come up with strategies to deal with the transformations in the industry. Technologies like the internet of things and artificial intelligence were also utilized along with big data to automate the proceedings in the HR domain. Leadership style along with organizational structure overviews resulted in massive growth for organizations and was hence included in the framework discussed [24]. The framework of the smart HR for the industry 4.0 is thus presented in Fig. 16.2.

The introduction of big data analytics has promoted several impactful transformations in the human resource industry. With the initiation of novel technologies in the eternally transcending domain, the job roles and skill requirement along with relevant job profiles for the position remained unclear. To provide clarity regarding the skill requirement, a study was conducted focusing on big data professions. Several distinct natures of jobs and skills were studied by examining real-world job posts that were put up online on job hunting platforms. The examination process began by pointing out the four big sectors of big data having potential jobs. It then proceeded onto the recognition of nine groups of big data-oriented skill sets that were in tremendous demand by the corporations. Each group was then classified based on the competence required for the occupations regarding big data. The classifications created have also been represented diagrammatically for better understanding. The fields of business analysis, developer, data science, and engineering have been explored and the relevant job profiles have been listed. A semi-automated, analytical method was hence proposed combining various technologies like machine learning and big

Fig. 16.2 A Smart HR based conceptual framework for industry 4.0

data analytics to classify job roles and relevant skill set requirements. The system proposed is novel, completely replicable, and works on several algorithms accompanied by valuable inputs from the experts. The technology of web scraping was implemented to create orders among jobs and skillsets from a considerable amount of job posts on online platforms. The data hence generated could be adopted by leaders and HR managers to come up with strategies to develop the accurate skills required for the execution of big data in the best possible way. The methodology introduced also facilitated the establishment of a common directory that could be employed by recruiters and teachers to improve the proceedings of demand and supply in the industry [2]. The job sectors of the big data analysis have been identified in Fig. 16.3.

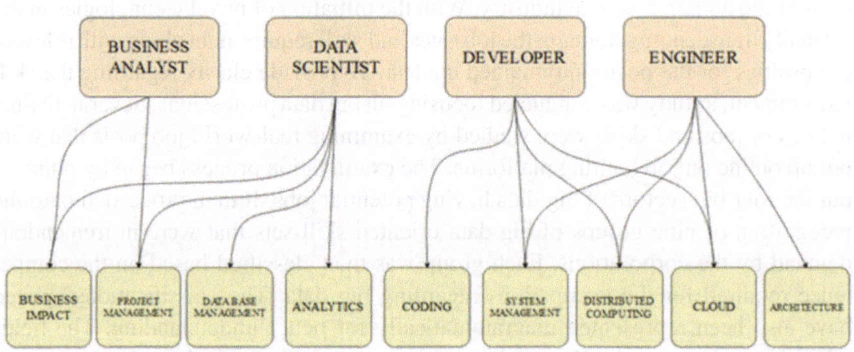

Fig. 16.3 Diagrammatic representation of big data job sectors

Management methodologies and practices to promote innovation as well as learning have discoursed to keep up with the advancements of industry 4.0. The paper provided novel perspectives and insights on techniques and systems that can promote the growth of the industry. The importance of Smart HR in industry 4.0 was discussed along with highlighting the need to embrace and adopt new technologies and learn necessary skills. A conceptual framework was included to emphasize the transformations in the role of HR 4.0 in a degrading economy. Practices of developing trusty teams with the best personal benefits and an emphasis on individual growth and development were discussed to strengthen retention maneuvering. Discussions based on ensuring the success and survival rate of organizations in the ever-growing industry were also carried out along with a stress on the significance of adopting new technologies for the heightened versions of the HR sector. The study provided relevant information on industrial connections and organizational behaviors concerning exercising managers along with researchers and scholars who aim to bring about innovations to the industry [3].

16.2.2 Impact of Smart HR on the Industry

To scrutinize industry 4.0, German companies and their strategies were examined. It was found that Germany had always taken the lead in the manufacturing sector crediting to the specialization in the fields of production, adoption of novel technologies to improve the industrial procedures, and research. Moreover, Germany also has an influential foundation of IT provisions and automation engineering which boosts their likelihood of a commanding performance as the industry transcends into industry 4.0. The introduction of industry $.0 also seemed to have feasible solutions for major issues like depreciating resources, metropolitan production, and demographic transitions. The availability of resources and their transportation to sustain productivity discoursed. The conception of flexible work practices was also discussed to facilitate the combination of the worker's jobs with their private lives which could result in better outcomes and extended productive hours at work to furthermore conclude into a balanced lifestyle.

The probable impacts on Human Resource Management (HRM) as a resultant of Industry 4.0 were discussed in terms of the skills needed, job profiles, and relevant qualifications. The consequent repercussions on supply chain management (SCM) also discoursed. The relationship between SCM and HRM focusing on Industry 4.0 was hence presented. To explore Industry 4.0 and study the consequences, it has had on the industry, available information was organized and then interpreted to decipher the recent trends and their impacts on the businesses. Based on the trends, recommendations were also made to facilitate further research. Co-citation software was adopted, utilizing the database of Web of Science to visualize the terminologies that had originated and were further used for the classification of data to be analyzed. The theme of the literature was based around Industry 4.0 and the incorporation of new technologies like big data and artificial intelligence to make the procedures more

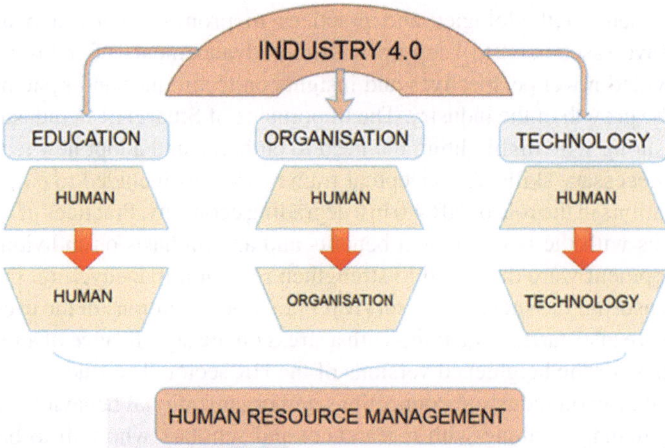

Fig. 16.4 Structure perceiving information corpus

efficient. The indulgence of industry 4.0 in the fields of education, organizations, and tie-ups with technological advancements were discussed. The flow of knowledge in these sectors concerning human resource management was likewise presented and has been portrayed diagrammatically in Fig. 16.4 for sounder comprehensions [4].

The studies concluded that the developed countries were doing exceptionally well in the domains of research production while Asian countries were lagging. Four major sectors were brought up namely educational transformations, employability, infrastructural resources, and work-related proposals. The categories were made keeping in mind the working conditions of the laborers as well as the environment and the skills necessary for the job. Furthermore, to successfully create an interface between humans and machines, industrial improvements were suggested and appropriate changes were proposed. The implications of HRM in regards to SCM were encompassed by socio-technical systems having fields of education administering with human competences, assimilation of SCM in terms of organizational competencies, and the management of data that came under the dominion of technical competences. The review presented could enormously benefit SCM managers in terms of developing strategies for the organization to promote training as well as for the advancement of new skills [5]. Figure 16.5 gives an overview of the socio-technical systems required to put in the improvements in HRM for the supply chain management system.

The impacts and consequences of balanced agile project management system (APM) were studied in terms of the mediating role of market orientations in addition to the strategies adopted by telecommunication providers. The analysis was carried out through a questionnaire designed for the telecommunication providers; additionally, the data was provided by the concerned executive members. The data hence received was used to examine the proposed model following the confirmatory analysis method. Fundamental equation modeling was brought into consideration

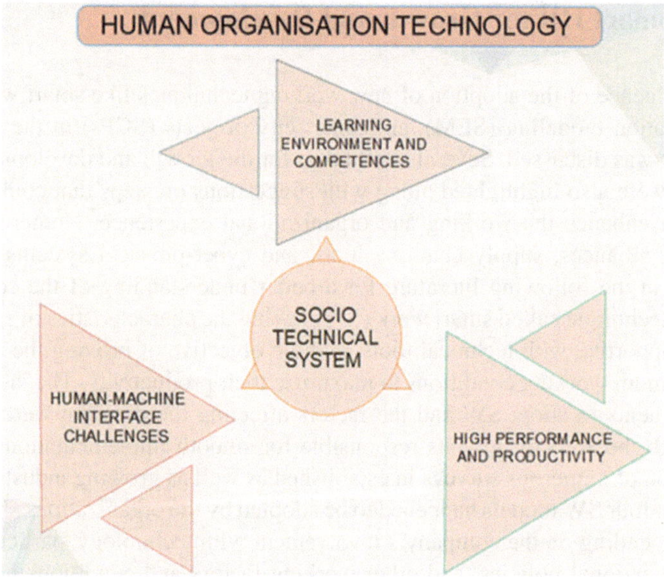

Fig. 16.5 Socio-technical systems to incorporate the suggestions of HRM for SCM

to successfully investigate causal relationships. It was found that the market was the link between balanced APM and imperative agility. Strategic agilities likewise directed out to be the nexus between the market and the organization's accomplishments. The investigation also suggested the extension of familiarity about project management by encouraging balance controlling. Business associate switching was also addressed to improve inter partner relations as well as for rendering immense value for clients [25].

The benefits and the need for sustainable management was highlighted accompanied by the roles and responsibilities of HR individuals to shoulder the idea of sustainable management ahead for more organizations to adopt. The transformations with stakeholders and their outlook on the managerial approaches have strived about the obligations of HR professionals for spreading sustainable strategies internationally. Examples have also been included for permissive applications in the industries. It was concluded that with the rapid resource drainage powered by the growing industries, there is a need for leaders who value and understand the importance of stimulating towards sustainable management. The acknowledgment of organizations agreeing to incorporate strategies to make their approaches more sustainable has been included. The significance of HR professionals in this domain has also been discussed by the inclusion of relevant examples and proposed strategies [26].

16.3 Smart HR for Growth and Development

In the influence of the adoption of new working techniques like smart work, structural equation modeling (SEM), and smart city projects (SCPs) in the domain of smart HR was discussed. Several steps taken for the growth and development of the industry were also highlighted along with suggestions on steps that could be taken to further enhance the working and organizational experience. Numerous studies including alliances, supply chain systems, and cyber-physical systems were also included in the following literature for a better understanding of the concepts. A working technique called smart work (SW) having the characteristics of spatial flexibility, supportive with technical tools, and the objective of proving the employees with optimum working conditions to maximize their productivity. The judgments of establishments to adopt SW and the factors affecting the decision were discussed along with the key constituents responsible for smooth implementations. Post the conduction of numerous surveys in established as well as growing industries, it was found that four SW models happened to be adopted by the organizations. The models varied depending on the company's involvement with technology, particularities of the organizational policies, and other working factors and conditions upon which the firm functioned. At least two elements were found in each of the companies that chose to move ahead with the SW mode. In the cases of having three possible elements in the SW process, high levels of labor productivity were experienced by the firm. Suggestions regarding the appropriation of SW practices for innovation and the flourishment in the employee's performance were additionally included [7].

A conceptual model to encompass employee behaviors in terms of the dynamic organizational statures was introduced for the crucial sector of big data in the HR domain. The proposed model took into consideration a large set of data that got accumulated from public sectors which was subsequently run through structural modeling. The structural equation modeling (SEM) helped in pinning down employee salaries, promotions in the firm, and organizational loyalty. The various features associated with job satisfaction is therefore provided in Fig. 16.6.

The work also included the significance of the approach described and illustrated the inclusion of big data postulates, implementational propositions, and other management requirements in the proposed structure. Employee behavior was also targeted in the process. In conclusion, the recommended approaches and frameworks were discussed under several parameters and suggestions for future research in the area were made [8].

The managerial aspect of multinational enterprises (MNEs) was studied to understand their process of managing HR in smart city projects (SCPs). Several case studies were conducted along with interviews of managers of MNEs who had been an integral part of SCPs. It was discovered that the approach of partnerships as well as temporal separation was utilized to maximize the benefits in projects. Several methodologies were applied exclusively for the management of social relations. The HR practices also varied based on the kind of alliance developed. Hence, it became

Fig. 16.6 A conceptual model for job satisfaction

all the more essential for HRM to support organizational abilities through collaborations and partnerships. Acumens on elevating the role and impacts of MNEs in the domain of SCPs have additionally been reviewed for the development of sustainable and budgetary cities [27].

The introduction of Industry 4.0 had brought about a need for remodeling of business models that were being attended to in companies. To redesign the systems, a framework for the digital supply chain system (DSCC) was proposed and its abilities were highlighted. The concept of a Critical digital supply chain integration (CDSCI) was also discussed emphasizing novel technologies for communication, stakeholders, and the capabilities of the workers. The study conducted resulted in the classification of six enabler technologies namely, big data analytics, blockchain, AI and workers, cloud computing, cyber-physical systems, and the internet of things. The study was directed towards proposing a framework for businesses to adopt and for managers to utilize in the process of digitalization. Valuable penetrations have been included that will be valuable for further research in this field [9]. The digital integration of the supply chain is thus drawn in Fig. 16.7.

The literature focused on decorating the digitalized mindsets of HR practitioners and elaborated on its importance in this era. The professional image of HR pupils was also described to be impactful from a managerial approach. The effects of digital practices in enhancing the relations between line managers and members of the HR department were also discussed in terms of producing solutions and efficient working [10].

Sensor-based practices, as well as cognitive assistance-based systems, have been tremendously appreciated and embraced in industry 4.0. Surveys conducted at organizations like HIS, IFR, and Deloitte were used to find significant pieces of information on the consequences of Industry 4.0, challenges faced, and the advancements being incorporated to make the system more effective. The practice of structural equation modeling was adopted to study the data collected in the process [11].

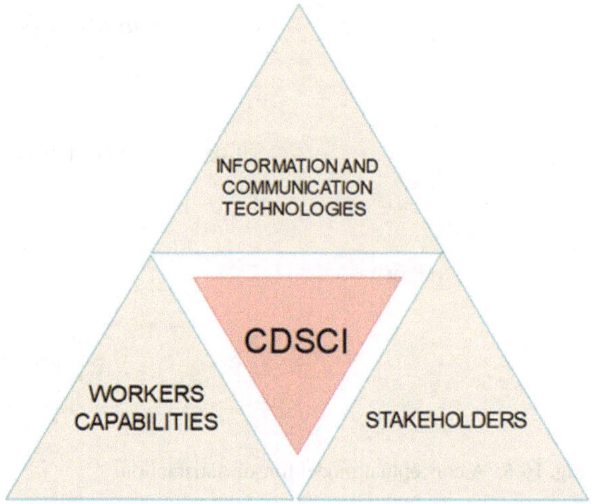

Fig. 16.7 Critical digital integration of the supply chain

The significance of Human Resource Management and its necessity in Industry 4.0 was highlighted along with its role in the growth and development of organizations dwelling towards a transition phase. The need for HR functionalities to adapt to the changing technological applications was also discussed emphasizing its function in extension, value creation, and productivity of the firm. The systems of Tech Mahindra as well as their digitalization process were illustrated to highlight the benefiting approaches in the epoch of Industry 4.0. The most beneficial approach for organizations turned out to be the enhancement of research and learning along with steps being taken to adopt technologies and learn necessary skills. It had made the functioning of the organization more efficient, productive, and economic [6].

16.4 Implementations and Scope of Research in Industry 4.0

The implementation of the framework and methods adopted post the conduction of surveys and interviews were discussed. The structure as well as the process of implementation were discussed with emphasis on the organizational structure and the most feasible approach meant for the firm. Domains requiring more research and studies were identified and suggestions were made regarding further work necessary for the field.

16.4.1 Research Scope and Agendas

The human resource management sector has experienced a rise in the research and studies conducted over the last two decades on their administration and strategies. Stages of research and development were identified in this chapter and expected pointers regarding research, management processes, and adopted methodologies were brought up. The relationship between human resource management and the administration of the workers as well as the organization's performance is a question that remained unanswered even after several years of research. Therefore, a need for additional work and research to be done in this domain was felt. Over the following decade, several journals and their findings have been discussed with emphasis on the associations between human resource management and performance. Despite the research done, sufficient reasoning was not found to explain the association described, however, over the next few years, the studies conducted resulted in invaluable knowledge and information for the provision of answers [12].

The objective of the paper was to improve the understanding of the manufacturing industry by emphasizing the progression of research in this field and on the promotion of innovation in organizations. The indulgence of novel manufacturing technologies, ICT integration in the domain, and the involvement of new conceptual ideas for production have been discussed. To carry forward the research process for industry 4.0, several online questionnaires were put to use along with the interviews from directors and CEOs of manufacturing industries for further insight on the subject. The objective was to obtain knowledge from as many resources as possible for a better understanding of the proceedings in the industry. It was found that several organizations were still operating on the technologies that were introduced in the manufacturing process about five decades ago. The technologies being utilized in the industry have changed drastically since the industrial revolution. The adoption of novel instruments and technologies has also been depicted diagrammatically in the following text. Such industries need to take up prominent measures to transform themselves with the aspiration of being encompassed in industry 4.0. The development of the Croatian model of Innovative smart Enterprise was proposed to bridge the gap between the functionalities of the industries and to prosper to new limits [13]. The various phases of the industry transformation are shown in Fig. 16.8.

To resolve the challenges faced at the global level in the manufacturing industry, theoretical as well as empirical research was carried out. The research has helped be supportive of the transforming industry and has also been a pillar for the organizations undergoing particular transmutations facilitated by the adoption of novel ideas and approaches. Previous work done in the field has also been included which has been appropriated for defining research agendas and targets for the upcoming future. The work done was analyzed to come to conclusions concerning the fields that required additional studies and research. The agenda to be adopted included the formulation of common terminologies, defined levels of implementation, classified phases of expansion for remodeling into industry 4.0, and a better understanding of the economy and its trends. The agenda also highlighted the importance of analyzing

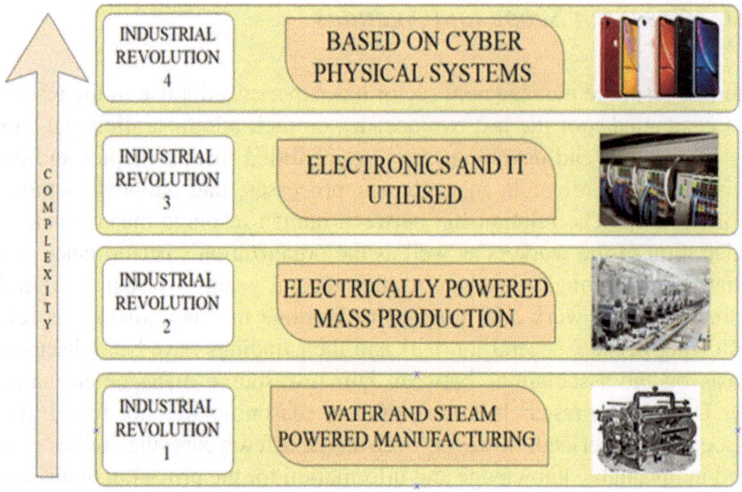

Fig. 16.8 Phases of the transforming industry

the resultants of human resources along with their outcomes and influences on other organizations. The reach of the agenda and the work already done in the research sector has had varied leads in different fields. The reach has been shown in a graph given in Fig. 16.9.

Several fields are yet to be included in the domain. There was also a need felt to review more literature for a better understanding of the challenges in this sector. The presented article turned out to be massively helpful for practitioners working in the industries and also helped in defining agendas for research in the future [28].

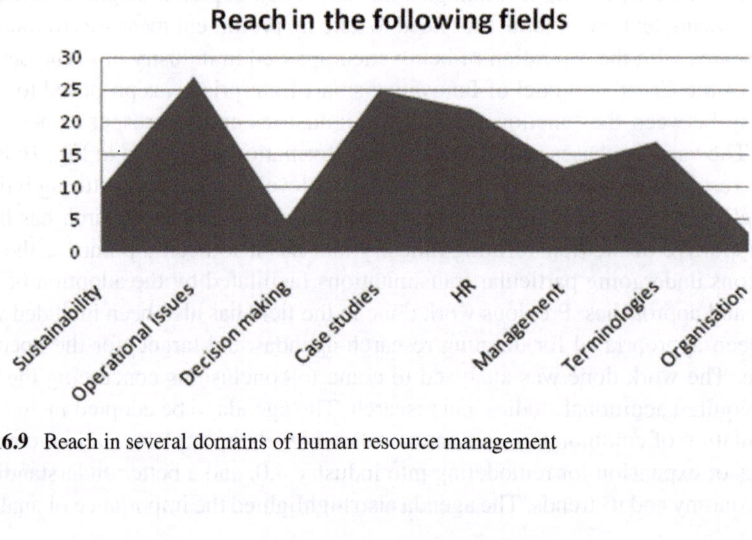

Fig. 16.9 Reach in several domains of human resource management

16.4.2 Implementations of Industry 4.0

The industry has been transforming significantly post the industrial revolution. Whether it was the adoption of the practice mechanizing pieces of machinery, usage of electrical energy, or the process of digitalization, advancing industries have always made use of the best technology available. In the upcoming future, more efficient systems and methodologies where products can decipher the most suitable manufacturing technique happen to be in the vision. The involvement of Industry 4.0 made the systems more IT-based and brought considerable changes in the manufacturing proceedings. The upgradations accompanied by Industry 4.0 also highlighted the dependencies in supply chain systems and proposed to include them in upcoming projects under industry 4.0 [14]. A system that has been shown in Fig. 16.10 could be one of the possible methodologies to be proposed further to facilitate the growth for industry 4.0.

Technologies like big data, data science, and analytics have been described as needs for productivity and innovation in the era of Industry 4.0. Big data has been explained as a vast amount of data that gets generated from different sources and can vary in type. It was described as high velocity and a high variety of data. Subsequently, novel tools and technology were also needed to deal with such a vast amount of data. As the involvement of big data kept increasing in organizations, its analysis and monitoring also became necessary. To successfully deal with data of such a large scale and to also prosper in the field of human resources, there are a set of skills

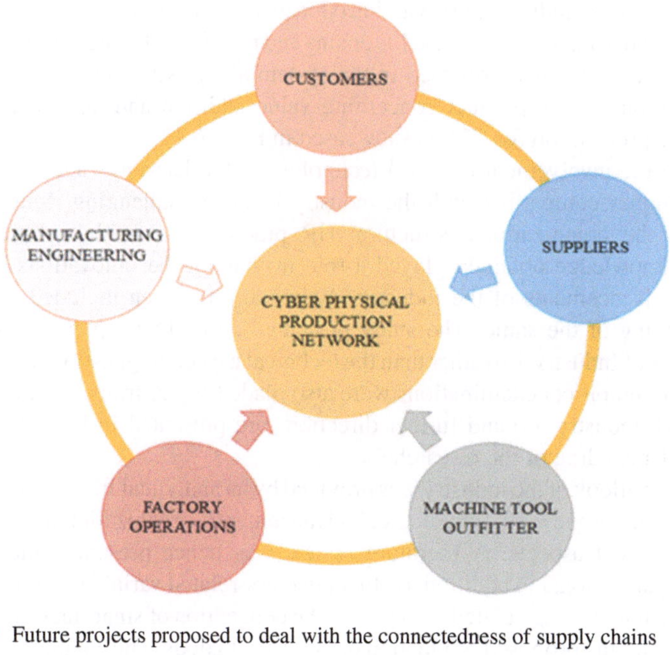

Fig. 16.10 Future projects proposed to deal with the connectedness of supply chains

Fig. 16.11 The requirement of analytical skills

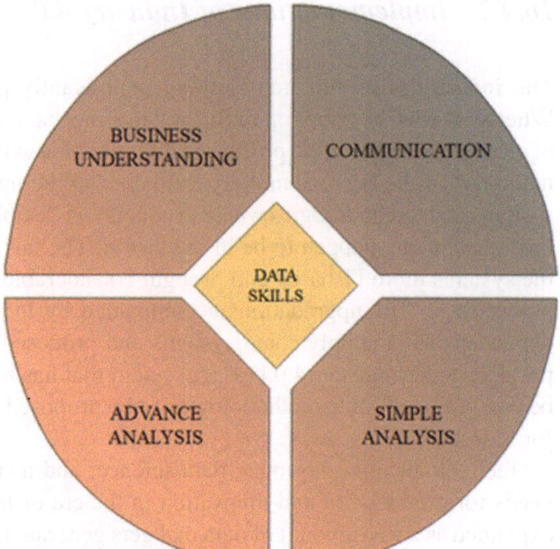

that become essential. Along with the ability to monitor and analyze data, a concrete understanding of business and communication skills also end up playing a major role [15]. The need of different analytical skills is provided in Fig. 16.11.

The Cyber-Physical System (CPS) for the production, smart manufacturing, and the betterment of industry 4.0 was introduced utilizing sensors, microcomputing networks, and a linkage between value chains and machines. Reengineering and other enhancements were also proposed in the system along with suggestions regarding customizations of the product concerning value addition and an effective supply chain. The proposition for CPS is showcased in Fig. 16.12.

Post to the involvement of novel technologies, the learnings and other innovative techniques compatible with the industry became challenging. Multiple fields including the organizational structure, HR practices, leadership strategies, and previous knowledge obtained played a role in making the innovations work well with the advancements of the industry. A diagram has been included for a better understanding of the same. The study presented an outlook upon the managerial approaches of Industry 4.0 rather than the technical aspect. Suggestions regarding the adoption of empirical examinations were also made [29]. A framework for compatibility with industry 4.0 and further directions are presented in Fig. 16.13 for the better understanding of the researchers.

A novel outlook of the industry was provided by an analytical framework to present a comprehensive view of technological advancements, organizational working, and other prominent aspects. A literature review was hence presented that included several organizational in addition to other business-related variables for a technological and manufacturing-related viewpoint. The conception of smart factories was also introduced for the proposed production of intelligent goods. The enablers concerning

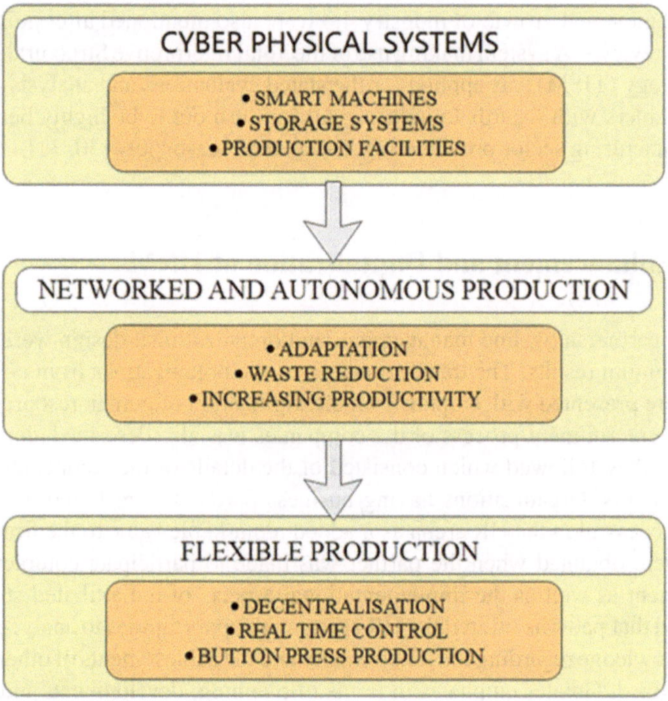

Fig. 16.12 Proposed idea for cyber-physical systems (CPS)

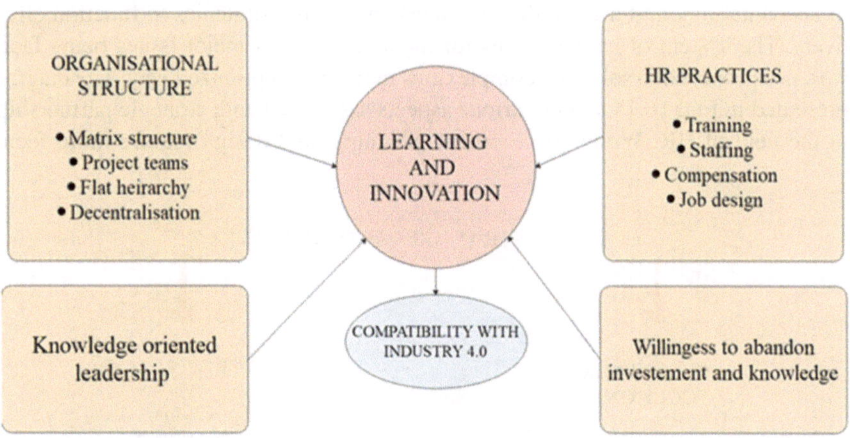

Fig. 16.13 Framework for compatibility and further research

the enrichment and growth of industry 4.0 were also mentioned after going through literature reviews. A system making use of the Total Interpretive Structural Modelling methodology (TISM) was applied for the stated evaluations and analysis. The study of the enablers with regards to industry 4.0 can turn out to be highly beneficial for the manufacturing sector provided it gets executed reasonably [30, 31].

16.5 Enhancement and Digitalization of HRM

Strategic partnerships, line management, and organizational design were discussed to get optimum results. The transformations of the organizations from e-HRM to d-HRM were presented with emphasis on the digitization of human resource management. The recruitment process of the companies was also described along with the procedure they followed which consisted of the details of the companies and other training details. Organizations having business partners were found to be tremendously successful in the HR arena as it added remarkable value to the firm. The best results were obtained when the partner was made to participate completely in the development as well as the implementational aspects of the affiliated strategies. It was found that patterns offered the HR sector with opportunities to analyze as well as obtain knowledge regarding the functionality and the achievements of other strategies being utilized. Greater emphasis in terms of planning, development, and organizational design was sought. Additionally, placing line managers in the administration of HR might work against the benefits of partnering up. It requires a greater deal of accountability and trust in the line managers for the strategies to function effectively. The system of having teams for the addressing of major issues being faced was also proposed to solve the complexities in the most optimized way. The diagram presented in Fig. 16.14 shows various aspects associated with strategic partnerships in the field of HR. While development, planning, and design require more focus,

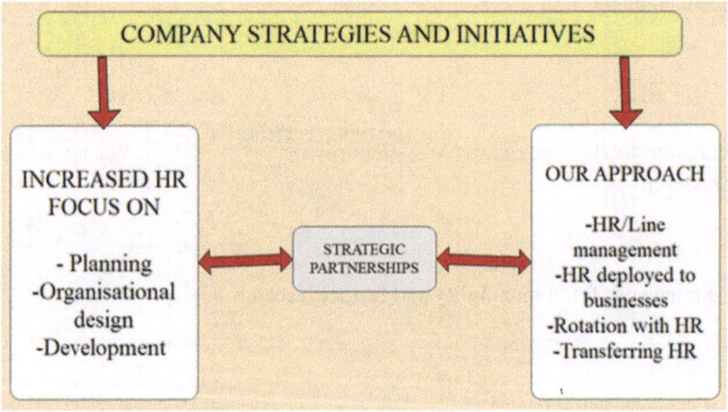

Fig. 16.14 Representation of HR as a strategic partner

the approach while indulging in associations also becomes crucial. Transferring and rotation of jobs in the HR field become significant along with more comprehensive responsibilities and accountability on the line managers [16].

Another study presented the complications with accounts that require HR analytics to ensure the strategies being followed in addition to the performance of the organization. The need for the acknowledgment of the benefits as well as the drawbacks of this field was felt to perform in optimal ways. HR functionalities were also found to be lacking essential management approaches to concur with big data. The practitioners from the field of big data were not found to be well acquainted or comfortable with the domain of HR and vice versa. However, the transforming industry does demand an amalgamation of the two. With effective training, HR professionals could rise to the occasion and strategically benefit the organization as well [32].

The transformation of the human resource industry from the sector of e-HRM to the digital version known as d-HRM was discussed. Digitalization became more of a necessity for HRM when the management sector began entering the smart business field where every aspect was digitalized for efficient outcomes. The association of HRM with the internet of things also proposed several challenges that needed to be dealt with. The expectations from HRM to cater to seekers in business development were also highlighted. It was discovered that with the rise in automation, business reach, and technological advancements, HRM required to emphasize non-routine and confirmation-based tasks. Suggestions regarding upcoming research work were made along with defining the areas that need advancements for the provision of better results [17].

For the effective implementation and adoption of industry 4.0, the employment of IT in the field of human resource management became crucial. The system introduced needed to be transparent with their functionalities and also needed to provide a protected environment which was implemented utilizing blockchain technology. A review describing the adoption of blockchain and its applicability in the human resource domain was presented. The aspects of the provisions of an efficient and cost-effective system post the adoption of blockchain technology was also discussed. A recruitment system based on blockchain technology was introduced called the blockchain-based recruitment management system (BcRMS) which is displayed with the help of Fig. 16.15. The presented system included details from the previous company, institutions of education as well as the training company. The proposed systems provided trusty results and performed better than the existing systems designed for hiring [18].

In [33], a model was introduced to present and explore the association between green innovations and their drivers. Comments were also made regarding the competitive challenges faced by the organization. The responsibilities and roles of data management and HR practices were also included in the article. To gain a competitive advantage and to make the organization follow the path of sustainable development, green products and proceedings were adopted. The adoption of the processes turned out to be dependent on several factors including the market demand, big data routinization, and assimilation, the role of the stakeholders as well as corporate ethics.

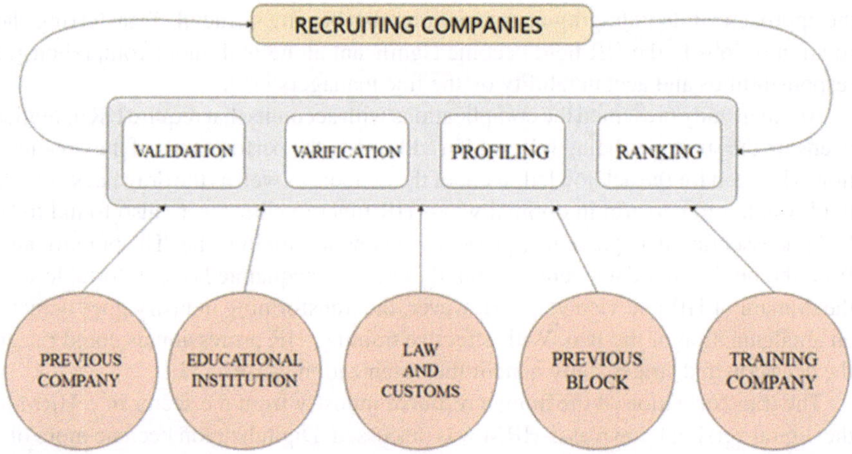

Fig. 16.15 Blockchain technology-based recruitment management system

The framework for the HR practices and training is thus provided in Fig. 16.16 for the awareness of the readers.

The dependence of organizations for the conduction of meetings and other discussions based on the new policies increased massively during the period of COVID-19. The platform faced several challenges ranging from the traffic on the site to the effective handling of stress, anxiety, and tiredness caused by long hours of meetings and discussions. The over-usage of such a platform even resulted in the creation of a new terminology called Zoom fatigue. Other challenges involved the issue of

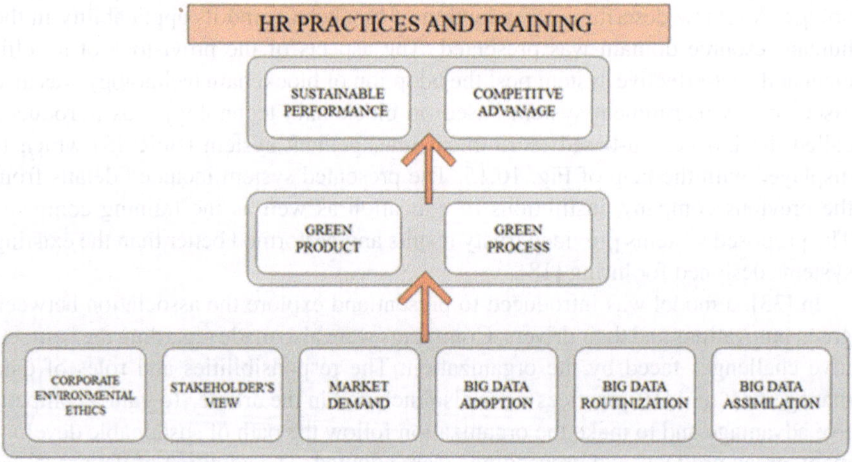

Fig. 16.16 A framework of the proposed model for the HR practices and training

making the conversations synchronized so no detail was lost in online communication. The delays also proposed new issues with such platforms. There also felt a need for the presence of an authoritative individual to carry out such procedures. Therefore, other suggestions to effectively communicate on such platforms were made along with the inclusion of issues faced by the users during meetings. The study discussed the unique features provided on Zoom and how they proved to be helpful for organizational meetings as well as regarding discussions and presentations on new ideas. Challenges faced during initial connectivity were highlighted. The usability of the platform and its functionalities were discussed along with a focus on complementing the qualitative methodologies that already exist. The role of such platforms in upcoming fields of research was also mentioned [19, 20].

16.6 Summary

This chapter introduces the concept of smart HR and its applications to industry 4.0. Human resource management (HRM) is experiencing dramatic changes with the introduction of artificial intelligence and IoT and moves on to the modern coinage called smart HR. The characteristics of HR in industries, along with the need to switch to a smarter domain, are thus quite obvious. The benefits of the innovations adopted and the methodologies proposed have been discussed with respect to market dynamics, as well as working conditions and the overall organizational structure. The industrial changes related to the emerging economy and other innovations brought on by the implementation of novel systems and technologies are also discussed. The study also took on the impact in the field of smart HR of the implementation of new working techniques such as smart work (SW), structural equation modeling (SEM), and smart city projects (SCPs). Several steps taken for the sector's growth and development were also highlighted, along with suggestions that could be taken to further improve work and organizational experience. A lot of research, including supply chain processes and cyber-physical systems, were also included for a better understanding of the concepts. Besides, it also highlighted the identification of domains that require further research. Moreover, strategic partnerships, line management, and organizational design were also discussed to get the best results. The transitions of organizations from e-HRM to d-HRM were also addressed in the emphasis on the digitization of human resource management. The advent of internet of things (IoT) and advancements in communication technology can further add value to the smart HR reforms plugged with artificial intelligence methods.

References

1. Wiederhold, B. K. (2020). *Connecting through technology during the Corona Virus disease 2019 pandemic: Avoiding "Zoom Fatigue"*.

2. Erro-Garcés, A. (2019). Industry 4.0: Defining the research agenda. *Benchmarking: An International Journal.*
3. Henning, K. (2013). *Recommendations for implementing the strategic initiative INDUSTRIE 4.0.*
4. Nosalska, K., Piątek, Z. M., Mazurek, G., & Rządca, R. (2019). Industry 4.0: Coherent definition framework with technological and organizational interdependencies. *Journal of Manufacturing Technology Management.*
5. Davenport, T. H., & Dyché, J. (2013). Big data in big companies. *International Institute for Analytics, 3*, 1–31.
6. Guest, D. E. (2011). Human resource management and performance: Still searching for some answers. *Human Resource Management Journal, 21*(1), 3–13.
7. Archibald, M. M., Ambagtsheer, R. C., Casey, M. G., & Lawless, M. (2019). Using Zoom videoconferencing for qualitative data collection: Perceptions and experiences of researchers and participants. *International Journal of Qualitative Methods, 18*, 1609406919874596.
8. De Mauro, A., Greco, M., Grimaldi, M., & Ritala, P. (2018). Human resources for Big Data professions: A systematic classification of job roles and required skill sets. *Information Processing & Management, 54*(5), 807–817.
9. Lasi, H., Fettke, P., Kemper, H. G., Feld, T., & Hoffmann, M. (2014). Industry 4.0. *Business & Information Systems Engineering, 6*(4), 239–242.
10. Shamim, S., Cang, S., Yu, H., & Li, Y. (2016). Management approaches for Industry 4.0: A human resource management perspective. In *2016 IEEE congress on evolutionary computation (CEC)* (pp. 5309–5316).
11. Habraken, M., & Bondarouk, T. (2017). Smart industry research in the field of HRM: Resetting job design as an example of upcoming challenges. In *Electronic HRM in the smart era*. Emerald Publishing Limited.
12. Kurniawan, R., Budiastuti, D., Hamsal, M., & Kosasih, W. (2020). The impact of balanced agile project management on firm performance: The mediating role of market orientation and strategic agility. *Review of International Business and Strategy.*
13. Shah, N., Irani, Z., & Sharif, A. M. (2017). Big data in an HR context: Exploring organizational change readiness, employee attitudes and behaviors. *Journal of Business Research, 70*, 366–378.
14. El-Kassar, A. N., & Singh, S. K. (2019). Green innovation and organizational performance: The influence of big data and the moderating role of management commitment and HR practices. *Technological Forecasting and Social Change, 144*, 483–498.
15. Liboni, L. B., Cezarino, L. O., Jabbour, C. J. C., Oliveira, B. G., & Stefanelli, N. O. (2019). Smart industry and the pathways to HRM 4.0: Implications for SCM. *Supply Chain Management: An International Journal.*
16. Verma, A., Bansal, M., & Verma, J. (2020). Industry 4.0: Reshaping the future of HR. *Strategic Direction.*
17. Sivathanu, B., & Pillai, R. (2018). Smart HR 4.0—How industry 4.0 is disrupting HR. *Human Resource Management International Digest.*
18. Ferraris, A., Santoro, G., Bresciani, S., & Carayannis, E. G. (2018). HR practices for explorative and exploitative alliances in smart cities. *Management Decision.*
19. Jain, V., & Ajmera, P. (2020). Modelling the enablers of industry 4.0 in the Indian manufacturing industry. *International Journal of Productivity and Performance Management.*
20. Priyadarshini, S. (2020). HR professionals: Key drivers for implementing sustainable business practices. *Strategic HR Review.*
21. Kottursamy, K., Raja, G., Padmanabhan, J., & Srinivasan, V. (2017). An improved database synchronization mechanism for mobile data using software-defined networking control. *Computers & Electrical Engineering, 57*, 93–103.
22. Çengiz, K., Sharma, R., Kottursamy, K., Singh, K. K., Topac, T., & Ozyurt, B. (2021). Recent emerging technologies for intelligent learning and analytics in big data. In *Multimedia technologies in the internet of things environment* (pp. 69–81). Springer.

23. Queiroz, M. M., Pereira, S. C. F., Telles, R., & Machado, M. C. (2019). Industry 4.0 and digital supply chain capabilities. *Benchmarking: An International Journal.*
24. Lawler, E. E., & Mohrman, S. A. (2003). HR as a strategic partner: What does it take to make it happen? *Human Resource Planning, 26*(3), 15–29.
25. Veza, I., Mladineo, M., & Peko, I. (2015). Analysis of the current state of Croatian manufacturing industry with regard to industry 4.0. In *15th International scientific conference on production engineering-CIM'2015: Computer integrated manufacturing and high speed machining.*
26. Isari, D., Bissola, R., & Imperatori, B. (2019). HR devolution in the digital era: What should we expect? In *HRM 4.0 for human-centered organizations.* Emerald Publishing Limited.
27. Raguseo, E., Gastaldi, L., & Neirotti, P. (2016). Smart work: Supporting employees' flexibility through ICT, HR practices and office layout. In *Evidence-based HRM: A Global Forum for Empirical Scholarship* (Vol. 4, No. 3, pp. 240–256). Emerald Group Publishing.
28. Fairhurst, P. (2014). Big data and HR analytics. *IES Perspectives on HR, 2014,* 7–13.
29. Witkowski, K. (2017). Internet of things, big data, industry 4.0—Innovative solutions in logistics and supply chains management. *Procedia Engineering, 182,* 763–769.
30. Angrave, D., Charlwood, A., Kirkpatrick, I., Lawrence, M., & Stuart, M. (2016). HR and analytics: Why HR is set to fail the big data challenge. *Human Resource Management Journal, 26*(1), 1–11.
31. Dolan-Canning, R. (2019). Sensor-based and cognitive assistance systems in industry 4.0: Big data analytics, smart production, and sustainable value creation. *Economics, Management, and Financial Markets, 14*(3), 16–22.
32. Onik, M. H., Miraz, M. H., & Kim, C. S. (2018). *A recruitment and human resource management technique using blockchain technology for industry 4.0.*
33. Rana, G., & Sharma, R. (2019). *Emerging human resource management practices in industry 4.0.* Strategic HR Review.

Chapter 17
Ensuring Security of Digital Voting Through Blockchain Technology

Ankita Bansal, Abha Jain, and Pardeep Kumar

Abstract Recently, blockchains have found a place in all sorts of industries, ranging from healthcare to the government sector, finance, real estate and utilities. Blockchain has proved to be advantageous to the applications which previously required a central authority and a trusted intermediary, as now blockchain allows them to operate in a decentralised manner and helps them to achieve the same functionality much more efficiently. Transactions in blockchains are trustless, that is, the involved parties without knowing each other and without establishing any trust relations, sit in any corner of the world and such transactions thus occur without any trusted middlemen. For authorisation of these transactions, strong cryptographic principles, which form an essential part of the blockchain infrastructure, are used. Smart contracts basically add a layer of logic and computation to the already existing trust infrastructure of the blockchain, and thus allows for appropriate, heavily automated and distributed workflows, to function efficiently. In this chapter, we outline the art of using smart contracts and blockchain technology to implement a decentralized, secure, transparent yet anonymous digital voting system.

17.1 Introduction

The traditional way of transacting money or an object of value has always been through an intermediary such as banks and governments because this trade requires trust and overlying rules to ensure that both parties involved in the trade don't violate

A. Bansal (✉)
Department of Information Technology, Netaji Subhas University of Technology, New Delhi, Delhi, India

A. Jain
Department of Computer Science, Delhi University, New Delhi, India

P. Kumar
Department of Software Engineering, Quaid-e-Awam University of Engineering, Science and Technology (QUEST), Nawabshah, Pakistan
e-mail: pardeep.kumar@quest.edu.pk

P. Kumar et al. (eds.), *A Fusion of Artificial Intelligence and Internet of Things for Emerging Cyber Systems*, Intelligent Systems Reference Library 210,
https://doi.org/10.1007/978-3-030-76653-5_17

its agreement [1]. Therefore, these banks act as intermediaries to help maintain that trust into processes such as transactions and record keeping [2]. A revolution was brought upon in 2008 when a white paper was released under the name, Satoshi Nakamoto. The white paper explained a peer to peer electronic cash system called Bitcoin that enabled online transactions to happen directly, without an intermediary [3]. The release of bitcoin was truly revolutionary but soon it was evident that the innovation was not the currency, but the technology that sits on top of it i.e. the Blockchain.

When we talk about blockchains, the first thing that seems familiar is Bitcoin [4, 5]. However, there is a need to understand that Blockchain and Bitcoin are not the same. While Bitcoin is a virtual currency that runs on the Blockchain principles, Blockchain is the underlying technology that helps Bitcoin create a decentralized money transfer mechanism which is widely popular today. Blockchain however is not limited to just Bitcoin but is a technology that can change the way transactions are done and has potential to change the way business is conducted. As Sally Davies perfectly quotes it, Blockchain is to Bitcoin, what the internet is to email. A big electronic system on top of which you can build applications.

Blockchain can be defined as a time stamped series of rigid records of figures, which is not possessed by only one person like in a centralised system [6]. Here peers can transact directly, irrespective of the fact whether they know each other or not, and where they are located. Blockchain is basically a "trustless system" where trust is established among unknown peers using the consensus mechanism called proof-of-work, which validates transactions that go into the ledger, without any third party, by validating contracts as legal using only the digital signatures. In Proof of work, transactions are validated by solving complex problems, using which the winner is chosen, which hashes blocks together that go into the ledger [7]. The hashing algorithm used to hash blocks makes our blockchain immutable. Basically an algorithm is chosen which obeys the avalanche effect, in which even a small change in the input is reflected vastly in output. This property of the hash function makes out blockchain immutable. Blocks are then connected to each other using strong cryptographic principles to form the chain. Here every block stores the hash value of the preceding block, therefore it is nearly impossible to meddle the data inside the blockchain once stored. This makes the blockchain highly secure, reliable and transparent [8]. Later, Ethereum's inventor proposed a single blockchain, which can be reprogrammed to perform highly complex computational tasks using logic, which led to the invention of ethereum blockchain, also known as smart contract [9]. A smart contract basically adds a layer of computation and logic, to the trust infrastructure supported by blockchain, by allowing for execution of code [10]. The virtual machines on which smart contracts execute are called the ethereum virtual machines—EVM, which ensures that every block in the ethereum network is able to implement the code, disregarding the operating system or the underlying hardware. A smart contract which is written in a HLL is first converted to an EVM bytecode, and then deployed on the EVM [11].

17.1.1 Blockchain Environment

Blockchain can be broadly classified into two types [12, 13].

Public Blockchains
In a public blockchain, no single entity controls the entire blockchain, hence we can say its decentralised. Here any node in the network can take part in the consensus protocol to determine which transactions will get added to the blockchain.

Private Blockchains
In private blockchains, there is a single centralised entity which validates the blocks and transactions will get added to our blockchain network. The whole blockchain system here is maintained by an organisation or individual.

17.2 Uses of Blockchain in Different Areas

Blockchain has tremendous applications in various sectors such as Banking [14], Internet of Things [15, 16], Transport, Voting, Government, Healthcare [17, 18], Retail [19], Real Estate to name a few. This technology has the ability to make the organizations transparent, democratic, decentralized, efficient, and secure. Figure 17.1 highlights the industries where blockchain is disrupting or will soon disrupt. Blockchain has also solved numerous functional and technical problems as depicted in Fig. 17.2.

Following are few domains where Blockchain is being used:

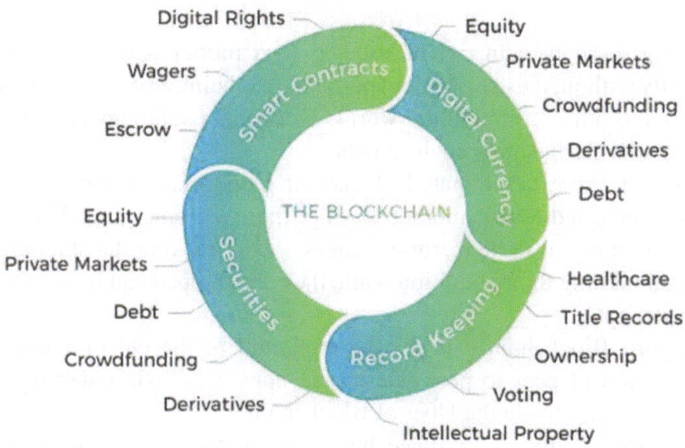

Fig. 17.1 Different domains where blockchain is being used

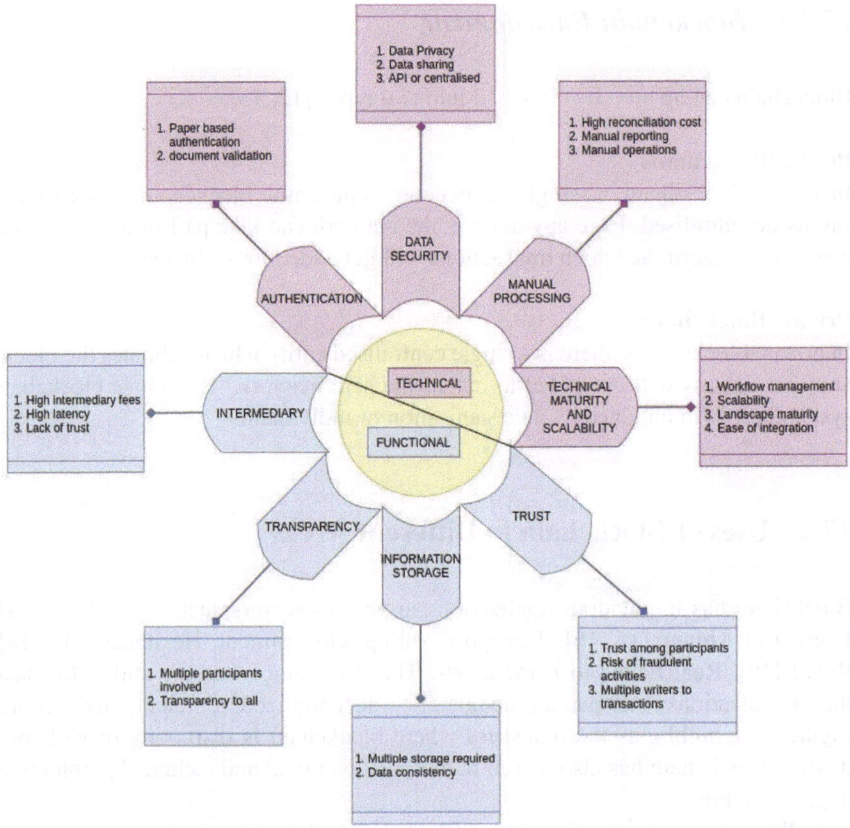

Fig. 17.2 Problems solved using blockchain

1. Banking: Like Bitcoin can be used to send money across the world almost instantly without the use of a third party, Blockchain can be used to give financial services to billions around the world. Blockchain solutions are innovating the money transfer business environment.
2. Internet of Things: Automated Blockchain ledgers can be used to track various things connected to the Internet. Both of these are emerging elds and merging them can turn out to be disruptive ideas. Also, tracking the shipping of goods and maintaining their condition while they are shipped can be another idea that connects both.
3. Transport: Blockchain can also be exploited in the eld of transport by the introduction of peer to peer ridesharing apps. CarTaxi, a startup by Russian developers, has made the Uber of Blockchain.
4. Retail: The ability to eliminate huge intermediaries such as Ebay, Amazon and Flipkart highlights the importance of Blockchain technology. This trustless

mechanism of shopping without the use of middlemen is what lies ahead in the future. Startups like OpenBazaar are working on the same.
5. Voting: The most important area that can be disrupted by Blockchain is voting. The traditional way of voting through ballots has many limitations which can all be eliminated through digital voting, which is our area of study as explained in the later sections.

17.3 Voting Through Blockchain Technology: Motivation

The Electronic Voting has been in use ever since the 1970s with fundamental benefit over paper based voting [20]. Figure 17.3 shows the basic steps involved in the process of E-voting. In 2017, Switzerland based Zug, developed decentralized Ethereum based digital identity system. In 2018, it adopted a blockchain based Voting system for the first time in it's Elections. The countries around the world are in a tight race to adopt the modern technologies like Blockchain, IOT, AI to stand out in the global order. India has a long way to go due to the digital poverty being there in the country. But gradually, India can adopt these technologies to improve transparency, efficiency and productivity of its system. Elections are the foundation of a democratic country. Often there are news of elections being rigged or the results being manipulated. Using

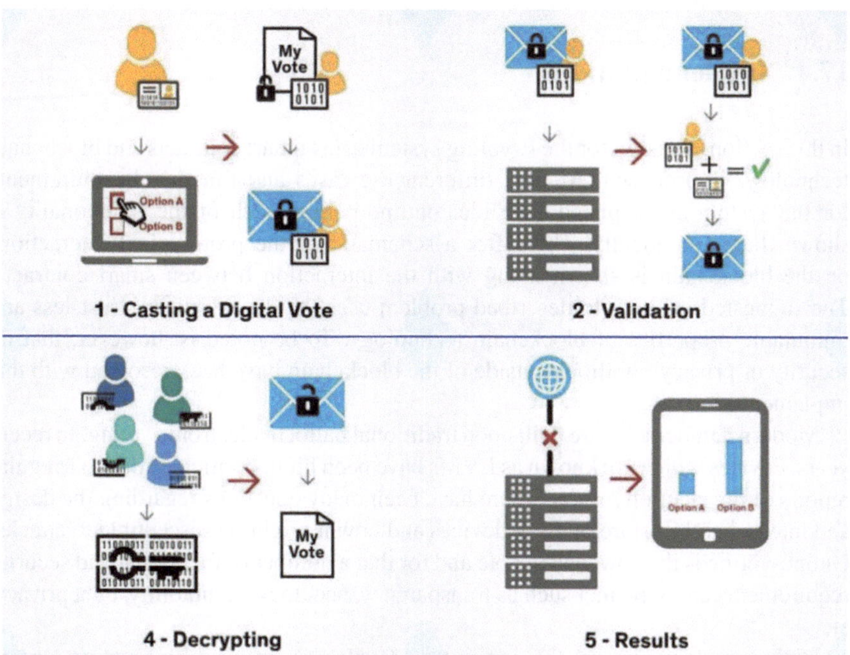

Fig. 17.3 Steps involved in E-voting

Blockchain technology can help find a better method of holding elections. Even though, E-voting was being used in countries like Australia, but there were huge security and privacy concerns. The E-Voting can be made for secure and reliable use of blockchain technology. Blockchain saves a lot of time in the process and thus leaves no scope for manipulation of results due to time lapse. It fastracks the task of counting of votes and making results publicly available thus building public trust in the process. In other words, blockchain provides all the features such as anonymity, prevent modification, double-spending, masquerading and repudiation problems [21, 22]. Nir Kshetri had shed some light on problems occurred in using blockchain in voting application in various attempts such as voters were not given the transparency of the voting process, voters were not offered any way to know whether their choices are accurately recorded, initially assigned key were too easy to hack, voting application had high error rates in authenticating the voters. Estonia was the pioneering country in the implementation of blockchain based E voting system. Such a system takes data about voters from government databases and even if a hacker breaks into the system, he/she cannot vote more than once. Every transaction is linked to the previous transaction, so the vote once casted cannot be changed. All the nodes will be synchronized so in case of any attempt to make changes to the vote casted will be immediately recognised.

This chapter discusses a safe, secure, easy to use and better method of voting using blockchain technology.

17.4 Implementation

In this section, a design for the E-voting system using smart contracts and blockchain technology is proposed. At first, different use-cases and functional requirements for the system are depicted. The idea/outline behind each of the functionality is shown diagrammatically. Thereafter a schematic of the prototypical interactions on the blockchain is shown along with the interaction between smart contracts. The suggested solution to described problem uses the decentralized, trust-less and immutable properties of blockchain technology. To be noted is, however, that no security or privacy liabilities outside of the blockchain have been resolved with this implementation.

Modern democracies are built upon traditional ballot or electronic voting. In recent years, devices which are known as EVMs have been highly criticized due to irregular reports of the election results. There have been many questions regarding the design and internal architecture of these devices and how it might be susceptible to attacks. Online-voting is the new way to vote and for that a number of functional and security requirements are to be met such as transparency, accuracy, auditability, data privacy, etc.

In this implementation, the concept of developing an electronic voting system using blockchain technology is implemented. The blockchain based voting project has two modules to make the whole project integrated and work along. One will be the Election Commission (election authority) who will be responsible for creating elections, adding registered parties and candidates contesting for the election added under the smart contracts. The other end will be the voter's module where each individual can cast a vote for their respective Assembly Constituency. Using an election's REST API hosted on Ethereum's Blockchain, the details are shown at the front-end of the voter for casting the vote. Then, while polling the vote is stored on our blockchain framework of which the Election Commission fetches the vote count. The limitation which we have faced due to not using the traditional way of smart contracts is that the blockchain framework which we have coded cannot run on the main net as it needs to be hosted and a separate web3 provider have to be used for interacting with it and not having a public API of voter ID creates a drawback of not having authentication of a voter.

17.4.1 Use-Cases and Requirements

The implementation outlined in this chapter is limited to two archetypical users: voter land election-authority. Here, the voters are assumed to be individuals eligible to cast votes and election-authority is assumed to be national body responsible for conduct and manage election-process. According to the various user-requirements, the following use-cases have been derived which are shown in the use-case diagram in Fig. 17.4.

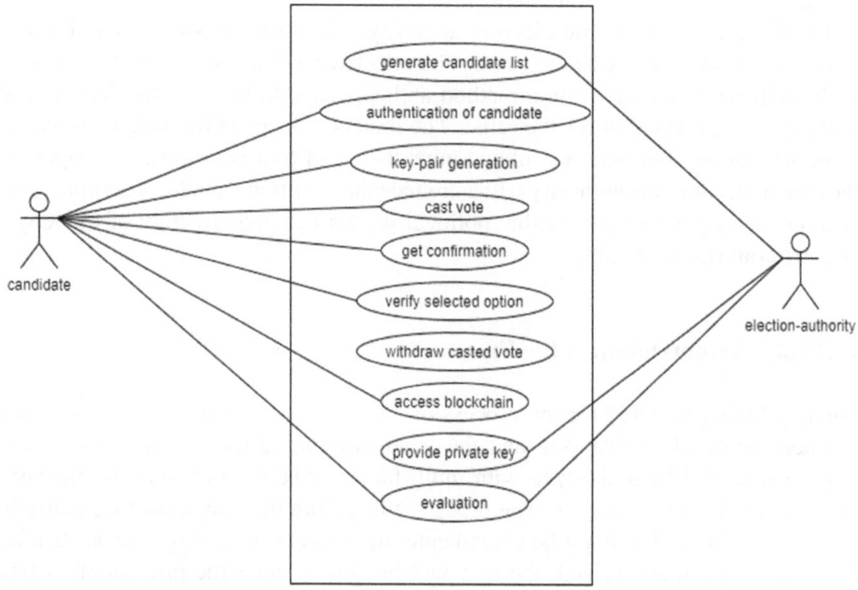

Fig. 17.4 Use case diagram of the proposed system

There are some requirements that apply to the general system and not just to one user specifically. Some of them are described in user-requirements, but for the sake of exhaustiveness and application to users not in the system, they are explicitly written below.

R1. Any unauthorised or ineligible individual must not be able to cast vote.
R2. Only those which have been authorized be able to send the vote transaction.
R3. Election-authority or any other entity must not be able to interfere the blockchain and not be able to trace the casted vote back to individual.
R4. Smart contracts must be exchangeable without needing to re-move the entire system or change addresses to contracts with which humans interact.

17.4.2 *Proposed Design*

In this section, the design, based on the user stories and requirements from the previous section, is described. First of all, the general outline of the methodology is explained. Thereafter, the detailed design for each stage in the process is described.

17.4.2.1 Design Overview

Figure 17.5 provides the general outline of the proposed system. Before election, the election authority will determine the set of candidates that are eligible for voting and provide the list to the general public.

On the election date, the election authority will open the voting portal where selected candidates can cast their vote. The selected candidate will authorize himself with multi-factor authentication method and cast its vote to candidate leaders and will get the confirmation for the same. The miners will select the vote transactions from the waiting transaction pool and mine the set of transactions to the ledger. At the end of election, the authority will announce the results and each concerning party such as voter, government, media, political parties can acquire copy of the ledger and can confirm the results.

17.4.2.2 Authorization of the User

Figure 17.6 is presented for the process of the authorization process after the user has been selected to cast vote and the voter has started the voting process. The eligible user will be authorized with multi-factor authentication process managed by national election-authority. The process will ensure that any malicious entity or ineligible individual will not be able to enter the process of casting vote. Only after the user has been authorized, the user will be able to enter the process of casting vote.

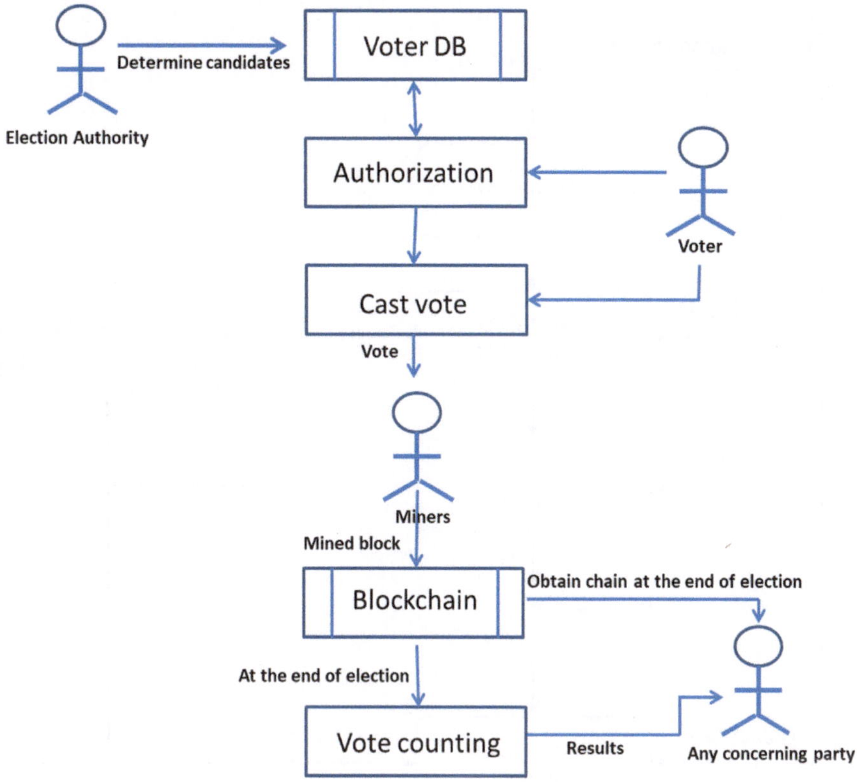

Fig. 17.5 General outline of the method

17.4.2.3 Hiding User Identity

Figure 17.7 is presented to show the process of hiding the user Identity that will be initiated after the authorization and before casting the vote transaction into the transaction pool that will be added into blockchain network.

The key will be generated locally in user's device within the application isolated from other applications. These keys will help in Digital-signature, asymmetric encryption and symmetric encryption methods. With the help of these methods the user will be able to cast vote anonymously and able to verify its own vote and votes casted by other voters. The generation of keys in isolated application will also ensure the secure storage of the key with the device.

17.4.2.4 Casting Vote and Generation of Vote-Transaction Block

Figure 17.8 depicts the process of casting vote and generation of transaction block that will be mined by miners and added to blockchain network. The candidate selected by

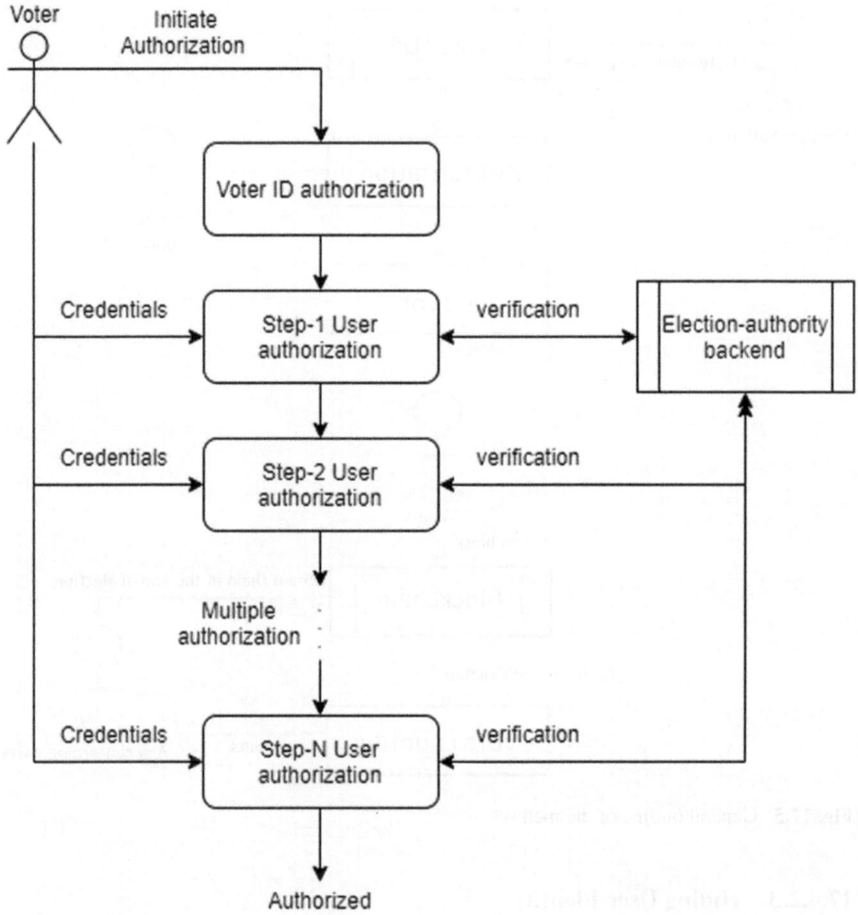

Fig. 17.6 Multifactor authorization process

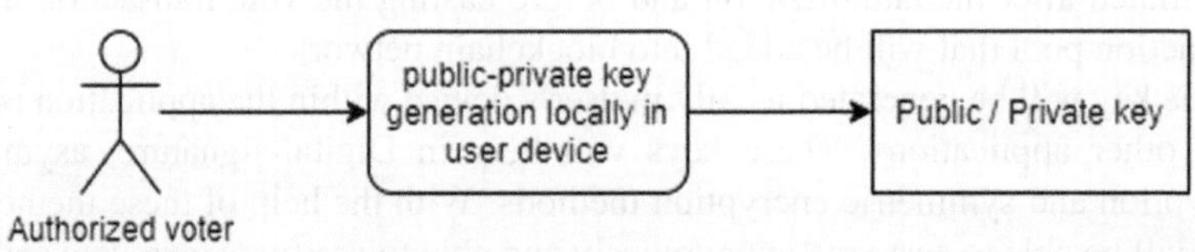

Fig. 17.7 Key generation process

the voter will be encrypted using asymmetric encryption algorithm with public key provided by the election authority, that will hide the selected option from others in the network and will also help in real-time result tracking at the day of voting process and the encrypted message can only be seen by election-authority. This encrypted message is termed as toC in our process. The candidate selected by voter will also be

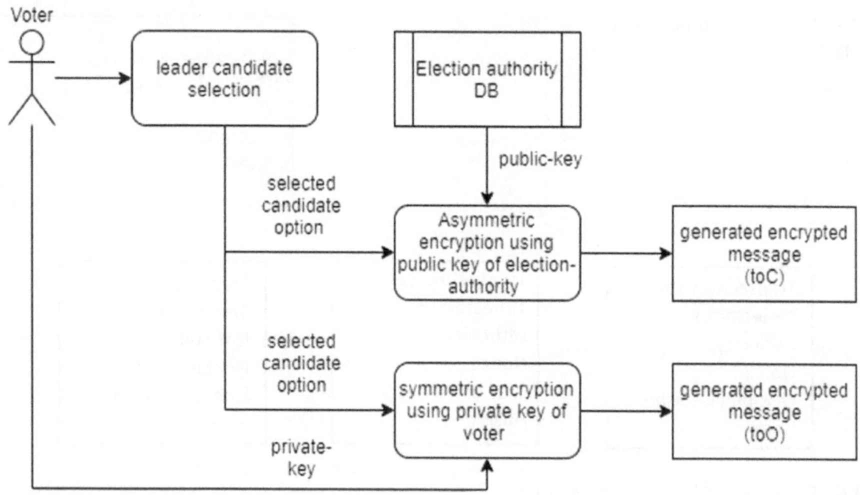

Fig. 17.8 Vote encryption process

encrypted using symmetric encryption algorithm with private key of the user itself, that will help will in confirmation of the added vote to the network. The voter can verify the vote has not been altered by any means and the same vote has been added to the network. This encrypted message is termed as toO in our process.

After the encryption of the selected option the vote-transaction block will be generated. The process of generation and the block is shown in Fig. 17.9.

The toO, toC the timestamp of the time at which voting was done along with the voter's generated public key converted into hash-digest using hash generation schemes and the generated hash is signed with private key of the voter using digital signature schemes. The signed hash in vote-transaction block will provide the proof of origin and any change in the data will result in different hash and the block can be mark invalid. The voter's public key/address, toC toO, timestamp and signed hash are combined to make a vote-transaction block, Fig. 17.10 is shown to describe the structure of vote-transaction block.

Each transaction consists of 5 attributes:

- Voter-public-address is candidate's generated public address after the authorization.
- toC: selected option signed with public key of election authority.
- toO: selected option signed with private key of the candidate.
- timestamp: time of generation of the transaction block in epoch seconds.

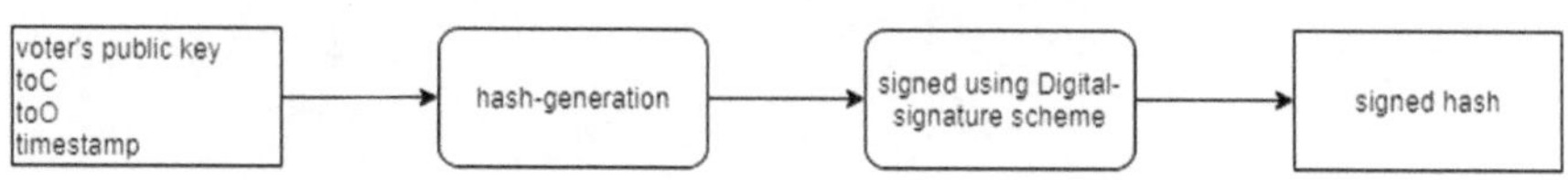

Fig. 17.9 Digital signing process of voter's data

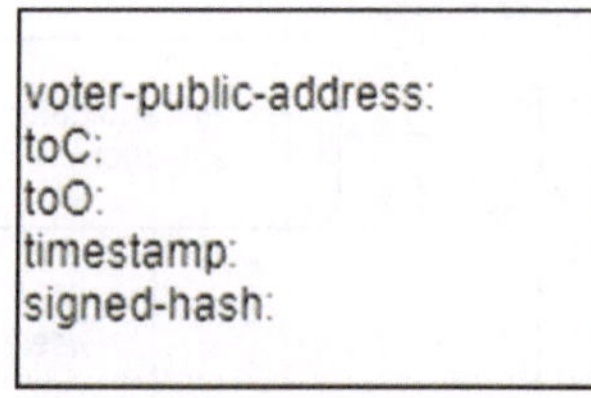

Fig. 17.10 Vote-transaction block

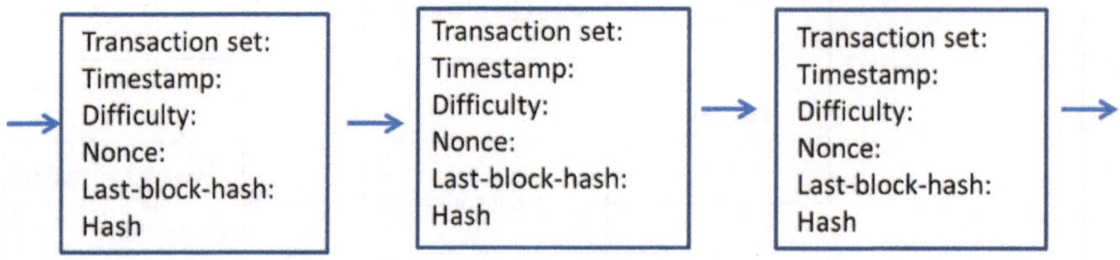

Fig. 17.11 Vote blockchain structure

- Signed-hash: hash digest of the above attributes signed with private key of the candidate.

Such transaction block for each user will be added to the transaction pool to be mined by miners and added to the blockchain network. The structure of such blockchain is shown in Fig. 17.11.

17.4.2.5 Verify Casted Vote

Figure 17.12 shows the process of verifying the casted vote by a voter. The voter will provide his generated public key and transaction block address to search for his vote-transaction block in blockchain network after the block has been found the voter

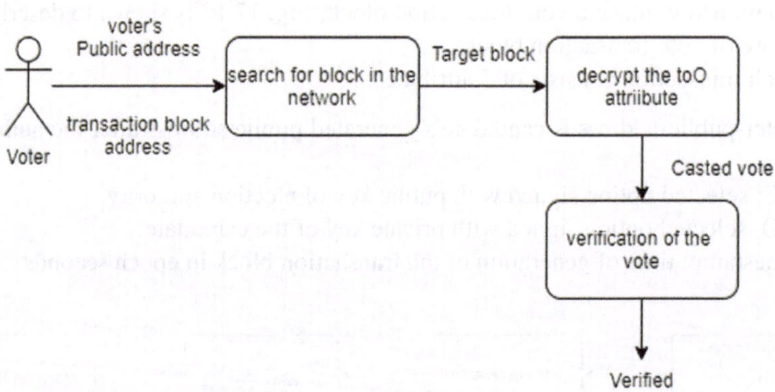

Fig. 17.12 Vote verification process

can decrypt the toO attribute of the transaction block using symmetric decryption scheme with his private key and voter can verify the vote is indeed the same. The voter can verify its voter any-time after the vote-transaction block has been added to the network and has been notified the voter.

17.4.2.6 Verify Result

After the election process, at the result declaration day, the election-authority can evaluate the blockchain network and declare result. The election-authority will also release its private key to general public so that general public can also access the blockchain network and verify the result. Figure 17.13 shows the process of verify result.

After the election-authority have released its private key any interested party can access the blockchain network and decrypt the toC attribute of every transaction block using asymmetric decryption scheme and can verify the declared result.

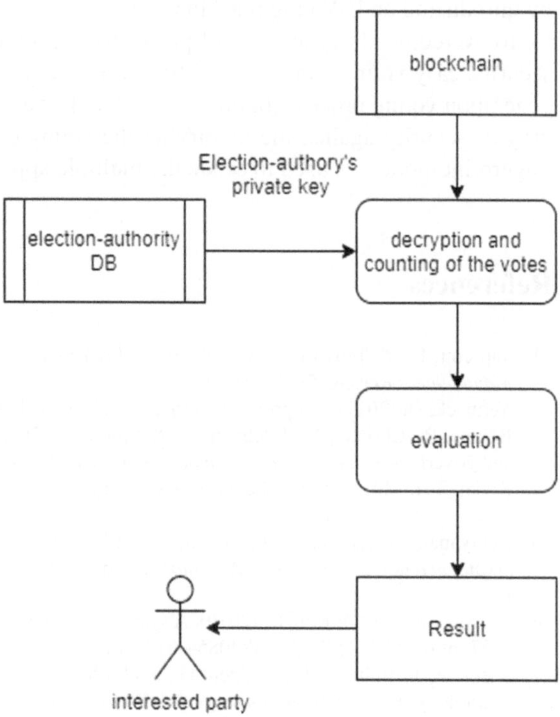

Fig. 17.13 Process of verifying result

17.5 Conclusion

This project has been developed to do a blockchain utilizing using smart contracts to enable secure and cost based electronic voting systems that efficiently election while guaranteeing voters privacy. of the system. It outlines the systems architecture, the design, and a security analysis In the next build of this application, it has been proposed to create separate client designs for various roles such as one for election commission and one for candidates registered to a certain party with the existing voting client design. Also, the current versions lack authentication as we don't have access to current Aadhar or Voter SDK to integrate in our application. Also, it is planned that in the next build notification prompt will be given on the day of voting to all the voters to cast their vote so that the voter turnout is maximum for that election. To maintain the integrity nowadays election commission is formed. In the starting, the election is done by voting to a candidate by the paper and then many vote rigging has been done in this method. But the same person has voted for more than one time and day by day this has increased dramatically in the election process. To avoid this, the election commission has made a great change in the voting process by introducing an E-Voting machine.

To overcome all these types of problems, we are proposing a system with more security, easy voting and vote counting which set free time and money for the next generation voting process. In this system block chain technology is implemented to provide security against the modifying the voting count through hacking, and the fingerprint module is used to evade the multiple appearances of the voters.

References

1. Tapscott, D., & Tapscott, A. (2017). How Blockchain will change organizations. *MIT Sloan Management Review, 58*(2), 10–13.
2. Yermack, D. (2017). Corporate governance and Blockchains. *Review Finance, 21*(1), 7–31.
3. Böhme, R., Christin, N., Edelman, B., & Moore, T. (2015). Bitcoin: Economics, technology, and governance. *Journal of Economic Perspective, 29*(2), 213–238.
4. Kiviat, T. I. (2015). Beyond bitcoin: Issues in regulating blockchain transactions. *Duke Law Journal, 65*(3), 569–608.
5. Narayanan, A., Bonneau, J., Felten, E., Miller, A., & Goldfeder, S. (2016). *Bitcoin and cryptocurrency technologies: A comprehensive introduction* (336 p.). Princeton University Press.
6. Xu, M., Chen, X., & Kou, G. (2019). A systematic review of blockchain. *Financial Innovation, 5*, 27. https://doi.org/10.1186/s40854-019-0147-z
7. Sedlmeir, J., Buhl, H. U., Fridgen, G., et al. (2020). The energy consumption of blockchain technology: Beyond myth. *Business & Information Systems Engineering, 62*, 599–608. https://doi.org/10.1007/s12599-020-00656-x
8. Khatoon, A., Verma, P., Southernwood, J., Massey, B., & Corcoran, P. (2019). Blockchain in energy efficiency: Potential applications and benefits. *Energies*. https://doi.org/10.3390/en12173317.
9. Salanfe. (2018). *Setup your own private Proof-of-Authority Ethereum network with Geth*. Hacker Noon. Available at: https://tinyurl.com/y7g362kd.

10. Christidis, K., & Devetsikiotis, M. (2016). Blockchains and smart contracts for the internet of things. *IEEE Access, 4*, 2292–2303.
11. Andoni, M., Robu, V., Flynn, D., Abram, S., Geach, D., Jenkins, D., McCallum, P., & Peacock, A. (2019). Blockchain technology in the energy sector: A systematic review of challenges and opportunities. *Renewable and Sustainable Energy Reviews, 100*, 143–174. ISSN 1364-0321. https://doi.org/10.1016/j.rser.2018.10.014.
12. Tapscott, D., & Tapscott, A. (2016). *Blockchain revolution: How the technology behind bitcoin is changing money, business, and the world* (368 p.). Portfolio.
13. Kulkarni, A. (2018). *How to choose between public and permissioned blockchain for your project*. Chronicled.
14. Hyvarinen, H., Risius, M., & Friis, G. (2017). A Blockchain-based approach towards overcoming financial fraud in public sector services. *Business & Information Systems Engineering, 59*(6), 441–456.
15. Reyna, A., Martín, C., Chen, J., Soler, E., & Díaz, M. (2018). On blockchain and its integration with IoT. *Challenges and opportunities, Future Generation Computer Systems, 88*, 173–190.
16. Silva, B. N., Khan, M., & Han, K. (2017) Internet of Things: A comprehensive review of enabling technologies, architecture, and challenges. *IETE Technical Review*, pp. 1–16.
17. Engelhardt, M. A. (2017). Hitching healthcare to the chain: An introduction to blockchain technology in the healthcare sector. *Technology Innovation Management Review, 7*(10), 22–34.
18. Radanovic, I., & Likic, R. (2018). Opportunities for use of blockchain technology in medicine. *Applied Health Economics and Health Policy, 16*(5), 583–590.
19. Wang, Y. L., Han, J. H., & Beynon-Davies, P. (2019). Understanding blockchain technology for future supply chains: A systematic literature review and research agenda. *International Journal of Supply Chain Management, 24*(1), 62–84.
20. Hjalmarsson, F., Hreioarsson, G., Hamdaqa, M., & Hjálmtýsson, G. (2018). Blockchain-based E-voting system, pp. 983–986. https://doi.org/10.1109/CLOUD.2018.00151.
21. McCorry, P., Shahandashti, S. F., & Hao, F. (2017). A smart contract for boardroom voting with maximum voter privacy. Available at: https://eprint.iacr.org/2017/110.pdf.
22. Barnes, A., Brake, C., & Perry, T. (2016). Digital voting with the use of blockchain technology. Available at: https://www.economist.com/sites/default/files/plymouth.pdf.

Chapter 18
A Study on Interaction Effect of Demographic Variables on Customer Satisfaction Towards Organized Retailing

K. Soujanya and Daniel Pilli

Abstract Revolutionary changes are taking place in Retail trade in India. There is a growing demand for Organized retailing. An increasing number of the aspiring middle class, favourable demographics, increasing urbanization, increasing number of nuclear families, rising affluence amid consumers and growing preferences for branded products have fuelled this demand. The famous adage that "the customer is always right" is very true in today's business scenario. This implies that customer satisfaction is the very key to the success of any business organization. The demographic variables of the customers exert a significant impact on customer satisfaction. The present paper attempts to examine the interaction effect of various demographic variables on customer satisfaction towards various attributes of organized retailing.

Keywords Organised retailing · Customer satisfaction · Demographic Variables · Increasing urbanization · Favourable demographics

18.1 Introduction

Indian retail industry was mostly unorganized, nascent, and highly fragmented. It may be true when it has been compared with the developed nations. Though it is in the early stages of development, many products are sold in this industry. The retailer is succeeding to reach every corner in the country with their products. The retail sector still has immense market potential to target. In the coming days, retailing will develop as one of India's most significant and vital industries and contributes to its economic development. The retail sector also generates lots of employment opportunities for unemployed people. As organized retailing is evolving as a significant sector, the unorganized retail industry is making attempts to survive in the market in Fig. 18.1. Organized and unorganized retailing in India are having their

K. Soujanya (✉)
Department of K.L. Business School, K.L. University, Vaddeswaram, India

D. Pilli
Department of Management Studies, Narasaraopet Engineering College, Narasaraopet, India

P. Kumar et al. (eds.), *A Fusion of Artificial Intelligence and Internet of Things for Emerging Cyber Systems*, Intelligent Systems Reference Library 210,
https://doi.org/10.1007/978-3-030-76653-5_18

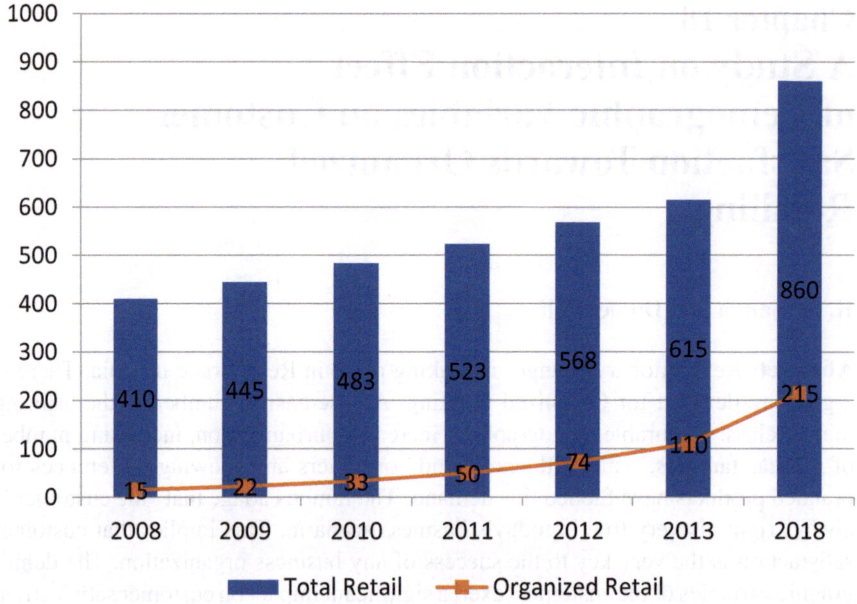

Fig. 18.1 Organized retail share to reach 25% share by 2018. *Source* Technopak analysis

own merits and demerits, but both run next to each other in a different way to satisfy the customers.

18.1.1 Demographic Variables and Customer Satisfaction

The various demographic variables, such as Age, Gender, Income, marital status, Occupation, qualification, etc., play a significant role in determining customer satisfaction. Customers of different age groups differ in their purpose of visiting the store. The older generations visit the store exclusively to shop only. At the same time, the younger generations view the retail stores as the venues where they can spend their leisure time and be entertained. Men and women differ in their perspectives, motives, rationale/purpose of visiting a store. The needs and wants the likes and dislikes, the buying patterns, and the customers' explanations vary based on their gender [1]. Marketing policy plays an integral part in an organization's growth. E-Marketing blend has developed several ways of reaching and involve consumers, including Email, social media ads, search engine optimization, etc. Demographic variables such as Gender, Age, Income, marital status, and education are also used and analyzed to classify customers for better marketing techniques. Demographics thus considered one of the leading market segmentation bases.

The income position has a significant influence on consumer buying behaviour. If the Income is high, then the person will buy more and vice-versa. Different income groups prefer other retail formats as their income levels affect what they can afford. High-income groups may choose hypermarkets, and low- and middle-income groups may like small Kirana stores or supermarkets. Consumers generally prefer less expensive goods when their Income is constricted. As income increases, they tend to purchase more expensive and high appealing goods. Marital status is identified by categories such as single, unmarried, separated, divorced, or widowed [2]. Marital status represents a mental state, which influences the perceived benefits and costs of their product preferences and purchase decisions. Occupation is one of the significant factors determining the customers' satisfaction towards various products they purchase, the store from where they are buying, and how they evaluate the store's marketing and promotional policies. The customers' expectations, understanding, and knowledge of the products, store policies, and promotional strategies depend on their education. It, in turn, influences their satisfaction [3].

18.2 Review of Literature

Surajit Ghost Dastidar and Biplab Datta [4] observed that the males were more risk-taking/innovative than females and younger consumers were more prone to indulge in interpersonal communication about purchases than the aged ones. Thiruvenkadam and Panchanatham [5] examined the effect of various demographics on retail store selection decisions. They concluded that demographics were having a significant impact on the decisions related to retail store selection. Ubeja and Bedia [6] Results revealed that the customers were not very sale promotion conscious, but gender-wise they were quite aware regarding some promotional offers. It was found that; female, those who were dependent or independent were more conscious about sales promotion related to on the spot offers in shopping, which contributed to their satisfaction. Gopala Krishnan and Varadaraj [7] analyzed the aspects of consumer satisfaction and stored loyalty empirically and found a positive relationship between demographics such as gender, marital status, and family and customer satisfaction and found that there was a close relationship between gender and their happiness. Basant Kumar [8] The respondents were young and educated. They were more fascinated with fashion and new developments in technology; we're interested in various products, affordable prices, options in the mode of payment, self-service, and entertainment.

18.2.1 Research Gap

From the literature review, it was found that very few studies were done on examining the impact of demographic variables on customers' satisfaction towards various organized retailing attributes. Very few studies analyzed the differences in customer satisfaction towards organized retailing of varied demographic customer groups. An attempt was made to explore the impact of demographics on customer satisfaction towards organized retailing [7].

18.2.2 Objectives of the Study

1. To examine the impact of demographic variables on the satisfaction of customers towards various attributes of organized retailing.
2. To analyze the variance in the customers' satisfaction levels and identify which demographic variables are getting impacted more.
3. A real and accurate online review of detection is using advanced techniques.

18.3 Research Methodology

The customers who regularly and occasionally make purchases from Spencer's retail outlet located in Guntur District were the current research population. The researcher has selected spencer's retail outlet in the Guntur district for a present research study. The mall culture in Guntur is in the evolutionary stage. Spencer's, which is one of the oldest retail formats in India, has introduced the joy of hyper store shopping experience to the customers of Guntur since 2010 So, the researcher had attempted to study the customer satisfaction towards organized retailing and how satisfaction gets influenced by various demographic variables [8]. A total of 915 questionnaires were distributed, out of which 664 questionnaires were returned filled up in all respects. The present study sample was drawn using the Convenience Sampling method. The researcher finds it easy and convenient to take exit interviews from the customers leaving the store. Primary data was collected with the help of the observation method and survey method. The secondary data was obtained from Books, Journals, Periodicals, Abstracts, Directories, Research reports, Conference Papers, Web Sites, Newspapers, and Magazines. M.A.N.O.V.A. (Multiple Analysis of Variance) was calculated to find whether there is any significant difference in customer satisfaction levels experienced. the respondents belonging to different demographic variables [9]. Post hoc tests, namely Scheffe and Tukey HSD, are used to examine. The demographics which are getting affected more in comparison with the other demographics in particular in Table 18.1.

Table 18.1 Comparison of satisfaction of customers concerning various attributes of organized retail marketing across different demographics factors

		Product attribute satisfaction	Store attribute satisfaction	Sales promotion satisfaction	The behaviour of sales personnel satisfaction
A.G.E	Lambda	0.981	0.988	0.998	0.987
	F value	4.136	2.648	0.491	2.198
	p-value	0.006	0.048	0.612	0.068
Gender	Lambda	0.998	0.996	0.994	0.995
	F value	0.52	0.799	1.813	0.812
	p-value	0.669	0.495	0.164	0.518
Marital status	Lambda	0.999	0.992	0.996	0.995
	F value	0.144	1.691	1.216	0.764
	p-value	0.933	0.168	0.297	0.549
Educational Qualification	Lambda	0.979	0.954	0.99	0.995
	F value	4.7	10.498	3.165	0.852
	p-value	0.003	0.0001	0.053	0.492
Occupation	Lambda	0.994	0.999	0.996	0.993
	F value	1.286	0.280	1.302	1.194
	p-value	0.278	0.840	0.273	0.011
Monthly Income	Lambda	0.996	0.973	0.997	0.996
	F value	0.964	6.064	0.871	0.615
	p-value	0.409	0.002	0.419	0.001

18.3.1 Analysis and Interpretation of Results

18.3.2 Comparison of Customer Satisfaction Concerning Different Attributes of Organized Retail Marketing Based on Age

The bundle of benefits that a customer considers in a product, the store's features that are looked at, the sales promotions, and the preferred sales personnel's etiquettes will differ based on the customers' age. Younger people are more interested in the customer-first marketing business approach. That is why retail organizations are putting customers' needs and wants above their own business goals [10]. The calculated *p* values are 0.006, 0.048, 0.612 and 0.068. The *p*-value corresponding to satisfaction towards Product attributes and store attributes is less than 0.05. The *p* values corresponding to satisfaction towards sales promotion attributes and behaviour of sales personnel are greater than 0.05. It shows a significant variance in satisfaction related to product attributes and store attributes based on age. There is no significant variance in satisfaction related to sales promotion attributes and behaviour of sales personnel based on age [11–14]. As a significant variance was found in satisfaction levels towards product attributes and store attributes based on Age, Post Hoc test

Table 18.2 Test of homogeneity of variances

	Levene statistic	df1	df2	Sig.
Product attributes	1.089	3	660	0.353
Store attributes	1.115	3	660	0.342

was done to examine which specific age group is getting influenced more with these attributes [15–17]. The following are the results of the post hoc test in Table 18.2.

The homogeneity of variances of product attributes and store attributes are 0.353 and 0.342, respectively, which are more significant than 0.05. Therefore, the post-hoc tests named Scheffe and Tukey HSD are performed [18–21].

18.3.3 Homogeneous Subsets

The mean scores of the age groups "Below 25, 25–35, and 35–45" are 18.43, 18.19, 18.09, respectively. In Table 18.3, It is evident that the respondents belonging to these age groups are getting more influenced by the product attributes. Those of the respondents in the age group of 'Above 45" as the mean score of this group are 16.71, which is less than the other age groups [22].

Table 18.4 shows that the respondents with age groups "Below 25 and 25–35" have 47.92 and 47.58 respectively, which are higher than those of ages 35–45 and Above 45. Table 18.4 indicates that the respondents of these age groups are more likely to get influenced by the store attributes comparatively than the other age groups [23].

Table 18.3 Comparison of satisfaction towards product attributes based on age

	Age	N	Subset for alpha = 0.05	
			1	2
Tukey HSD[a,b]	Above 45	35	16.71	
	35–45	58		18.09
	25–35	296		18.19
	Below 25	275		18.43
	Sig.		1.000	0.859
Scheffe[a,b]	Above 45	35	16.71	
	35–45	58		18.09
	25–35	296		18.19
	Below 25	275		18.43
	Sig.		1.000	0.891

a—is the larger of the two means being compared
b—is the smaller of the two means being compared

Table 18.4 Comparison of satisfaction towards store attributes based on age

	Age	N	Subset for alpha = 0.05		
			1	2	3
Tukey HSD[a,b]	Above 45	35	45.34		
	35–45	58	45.67		
	25–35	296		47.58	
	Below 25	275		47.92	
	Sig.		0.969	0.965	
Scheffe[a,b]	Above 45	35	45.34		
	35–45	58	45.67	45.67	
	25–35	296		47.58	47.58
	Below 25	275			47.92
	Sig.		0.976	0.076	0.974

a—is the larger of the two means being compared
b—is the smaller of the two means being compared

18.3.4 Comparison of Customer Satisfaction Concerning Different Attributes of Organized Retail Marketing Based on Gender

The needs and wants, the likes and dislikes, the buying patterns and motives of the customers varies based on their gender. Whether men and women differ in their satisfaction related to the product, store, sales promotion, and sales personnel behaviour was examined. The results showed that the respective calculated p values, 0.669, 0.495, 0.164, and 0.518 are more significant than 0.05. It shows no considerable variance among the satisfaction levels related to organized retailing attributes concerning gender.

18.3.5 Comparison of Customer Satisfaction Concerning Different Attributes of Organized Retail Marketing Based on Marital Status

Marital status is identified by categories such as single, unmarried, separated, divorced, or widowed. Marital status represents a mental state, which influences the perceived benefits and costs of their product preferences and purchase decisions. Marital status influences the value for money, customer satisfaction, and repurchase intent of the customers. The researcher has attempted to determine whether the customers' satisfaction will get influenced by their marital status. The calculated p

Table 18.5 Test of homogeneity of variances

	Levene statistic	df1	df2	Sig.
Product attributes	1.933	3	660	0.123
Store attributes	3.785	3	660	0.060

values are 0.933, 0.168, 0.297 and 0.549 are greater than 0.05. It shows no significant variance among the satisfaction related to organized retail marketing attributes concerning marital status [24].

18.3.6 Comparison of Customer Satisfaction Concerning Different Attributes of Organized Retail Marketing Based on Education Qualifications

The customers' expectations, understanding, and knowledge of the products, store policies, and promotional strategies depend on their education. It, in turn, influences their satisfaction. The calculated p values are 0.003, 0.0001, 0.053 and 0.492. The p values corresponding to satisfaction related to product attributes, store attributes less than 0.05, show significant variance in customer satisfaction among these attributes concerning Educational qualification. The p-value corresponding to attributes of sales promotion and behaviour of sales personnel was greater than 0.05. It shows no significant variance in satisfaction level related to sales promotion attributes and behaviour of sales personnel concerning education qualification. As a significant variance was found in satisfaction levels towards product attributes and store attributes based on the respondents' education qualification, a Post Hoc test was done to examine which specific qualification is influenced more by these attributes.

The following are the results of the post hoc test in Table 18.5.

The homogeneity of product attributes and store attributes' variances is 0.123 and 0.060, more significant than 0.05. Therefore, the post-hoc tests named Scheffe and Tukey HSD are performed.

18.3.7 Homogeneous Subsets

From Table 18.6, it was clear that the respondents with Post Graduation and Schooling qualifications are getting more influenced by the product attributes as their mean scores are 18.56, 18.45 respectively, which are higher when compared with the sample respondents whose stuff is Graduation and Diploma in Table 18.6.

The results showed that the respondents with the educational qualifications Schooling and Graduation are getting impacted more with the store attributes as their mean scores are 48.03, 47.60 respectively, which are higher than the mean

Table 18.6 Comparison of satisfaction towards product attributes based on education qualification

	Qualification	N	Subset for alpha = 0.05	
			1	2
Tukey HSD[b]	Diploma	45	17.40	
	Graduate	365	18.08	18.08
	School level	132		18.45
	Post graduate	122		18.56
	Sig.		0.270	0.597
Scheffe[a,b]	Diploma	45	17.40	
	Graduate	365	18.08	18.08
	School level	132	18.45	18.45
	Post graduate	122		18.56
	Sig.		0.052	0.670

a—is the larger of the two means being compared
b—is the smaller of the two means being compared

scores of respondents with Post Graduation and Diploma Qualifications shown in Table 18.7.

Table 18.7 Comparison of satisfaction towards store attributes based on education qualification

	Qualification	N	Subset for alpha = 0.05	
			1	2
Tukey HSD[a,b]	Post graduate	122	46.30	
	Diploma	45	47.42	47.42
	Graduate	365	47.60	47.60
	School level	132		48.03
	Sig.		0.181	0.779
Scheffe[a,b]	Post graduate	122	46.30	
	Diploma	45	47.42	
	Graduate	365	47.60	
	School level	132	48.03	
	Sig.		0.065	

a—is the larger of the two means being compared
b—is the smaller of the two means being compared

Table 18.8 Test of homogeneity of variances

	Levene statistic	df1	df2	Sig.
The behaviour of sales personnel	2.088	5	658	0.065

18.3.8 Comparison of Customer Satisfaction Concerning Different Attributes of Organized Retail Marketing Based on Occupation

Occupation is one of the significant factors determining the customers' satisfaction towards various products they purchase, the store from where they are buying, and how they evaluate the store's promotional policies. The calculated *p* values are 0.278, 0.840, 0.273, which are more significant than 0.05. It shows no considerable variance in the customers' level of satisfaction towards different attributes of product, store, and sales promotion concerning Occupation. The calculated *p*-value of Behaviour of sales personnel is 0.011, which was less than 0.05, indicates that there is variance in satisfaction levels of customers based on sales personnel's behaviour concerning Occupation. The following are the results of the post hoc test.

The homogeneity of variances of the behaviour of sales personnel is 0.065 which is more significant than 0.05. Therefore, the post-hoc tests named Scheffe and Tukey HSD are performed, as shown in Table 18.8.

The results can be understood as the students' satisfaction is getting more influenced by sales personnel's attributes compared with the other occupations as the mean score of this group is higher than the other fields as shown in Table 18.9.

18.3.9 Comparison of Customer Satisfaction Concerning Different Attributes of Organized Retail Marketing Based on Income

The customers' Income plays a significant role in determining customer satisfaction as the respondents' Income is important in making purchase decisions. The calculated *p* values are 0.409, 0.002, 0.419 and 0.001. The *p* values corresponding to store attributes and attributes of sales personnel are less than 0.05, which shows a significant difference in customer satisfaction relating to store attributes and behaviour of sales personnel based on Income. There was a significant variance in the satisfaction levels of store attributes and attributes of sales personnel based on income [25]. To know which income group is getting affected more by these attributes Post hoc test was done. The following are the results of the test.

The homogeneity of store attributes and attributes of sales personnel's behaviour is 0.129, and 0.137 which are more significant than 0.05. Therefore, the post-hoc tests named Scheffe and Tukey HSD are performed, as shown in Table 18.10.

Table 18.9 Comparison of satisfaction towards the behaviour of sales personnel based on occupation

	Occupation	N	Subset for alpha = 0.05	
			1	2
Tukey HSD[a,b]	Farming	19	17.32	
	Business	100		19.45
	Professional	283		19.27
	Housewife	85		19.44
	Unemployed	38		19.07
	Student	139		19.83
	Sig.		1.000	0.737
Scheffe[a,b]	Farming	19	17.32	
	Business	100	19.07	19.45
	Professional	283		19.27
	Housewife	85		19.44
	Unemployed	38		19.07
	Student	139		19.83
	Sig.		0.069	0.861

a—is the larger of the two means being compared
b—is the smaller of the two means being compared

Table 18.10 Test of homogeneity of variances

	Levene statistic	df1	df2	Sig.
Store attributes	1.792	4	659	0.129
The behaviour of sales personnel	1.752	4	659	0.137

From the results, the respondents belonging to the income groups of Rs 5000–10,000, Rs 10,000–15,000 are getting more affected by store attributes compared with the respondents belonging to the other income groups shown in Table 18.11.

The satisfaction of the respondents belonging to the income group of 15,000–20,000 is most likely to get influenced by the attributes of sales personnel when compared with the respondents of other income groups as this group has a mean score that is higher than the remaining income groups as shown in Table 18.12.

18.4 Conclusion

From the findings, it can be concluded that demographic variables play a vital role in influencing the customers' satisfaction. Therefore, it can be interpreted that the

Table 18.11 Comparison of satisfaction towards store attributes based on income

	Income	N	Subset for alpha = 0.05
			1
Tukey HSD[a,b]	Below Rs. 5000	23	46.26
	Above 20,000	152	46.60
	15,000–20,000	122	46.80
	10,000–15,000	137	48.04
	5000–10,000	230	48.08
	Sig.		0.108
Scheffe[a,b]	Below Rs. 5000	23	46.26
	Above 20,000	152	46.60
	15,000–20,000	122	46.80
	10,000–15,000	137	48.04
	5000–10,000	230	48.08
	Sig.		0.206

a—is the larger of the two means being compared
b—is the smaller of the two means being compared

Table 18.12 Comparison of satisfaction towards behaviour of sales personnel based on income

	Income	N	Subset for alpha = 0.05
			1
Tukey HSD[a,b]	5000–10,000	230	18.85
	Above 20,000	152	19.10
	Below Rs. 5000	23	19.26
	10,000–15,000	137	19.71
	15,000–20,000	122	20.11
	Sig.		0.060
Scheffe[a,b]	5000–10,000	230	18.85
	Above 20,000	152	19.10
	Below Rs. 5000	23	19.26
	10,000–15,000	137	19.71
	15,000–20,000	122	20.11
	Sig.		0.131

a—is the larger of the two means being compared
b—is the smaller of the two means being compared

marketer needs to pay attention to design the marketing and promotional strategies differently for different customers to maximize the fulfilment of the customers. The findings give a partial indicator of future health needs for multiple white-collar workers. It is valuable knowledge to designate health plans depending on the intensity of employee coping resources. Demographic research is demographic research focused on variables including age, ethnicity, and sex. Demographic data relates to statistically articulated socio-economic statistics, including jobs, schooling, wages, marriage rates, birth/death rates, and more variables. Demographic data helps companies consider how to sell and strategically prepare for potential customer demand patterns. The *p*-value of product attributes and sales promotion Behavior is more significant than 0.005; this shows no considerable variance in customer satisfaction related to these attributes concerning Income. The significant disagreement was found in the respondents' satisfaction levels belonging to different Occupations towards details of the sales personnel's behaviour. Post Hoc test was done to examine which specific Occupation is getting influenced more by these attributes.

References

1. Dastidar, S. G., & Datta, B. (2009). Demographic differences in consumer exploratory tendencies: An empirical examination. I.I.M.B. *Management Review, 21*(4), 297–312.
2. Thiruvenkadam, T., & Panchanatham, D. (2011). Influence of demography on store patronage behaviour of chennai shoppers. Chief Patron Chief Patron.
3. Ubeja, S. K., & Bedia, D. D. (2012). A study of customer satisfaction from organized retailing (with reference to Indore city). *Prestige International Journal of Management and Research, 5*(1), 1–10.
4. Gopalakrishnan, L., & Varadaraj, D. S. (2012). An empirical analysis on consumer satisfaction and store loyalty in organized food and grocery retailing with special reference to Coimbatore city. *Namex International Journal of Management Research, 30*, 29–37.
5. Kumar D. P., Rajyalakshmi, K., & Asadi S. S.(2017). Analysis of mobile technology switching behavior of consumer using chi-square technique: A model study from Hyderabad. *International Journal of Civil Engineering and Technology, 8*(9), 99–109.
6. Murthy, S., Babu, P. R., & Reddy, P. S. (2017). The impact of organized retail stores on unorganized retail outlets (with reference to FMCG's in Undivided-Andhra Pradesh). *International Journal of Economic Research, 14*(10), 109–117.
7. Kishore Babu, B., Tulsi Ram Naveen, A., & Sampath, (2018). Perception of the engineering students towards 7ps of organized retail stores and unorganized retail stores, Vijayawada-a comparative study. *International Journal of mechanical Engineering and Technology, 9*(1), 841–851.
8. Kishore Babu, B., Rajeswari, N., & Mounika, N. (2018). An empirical study on consumer green buying behaviour, Vijayawada, Andhra Pradesh. *International Journal of Civil Engineering and Technology, 9*(3), 648–655.
9. Patnaik, R., Babu, B., & Narasimha Roan, M. (2019). E-shopping buying behavior of engineering students' (with special reference to Vijayawada). *International Journal of Innovative Technology and Exploring Engineering, 8*(12), 2885–2889.
10. Kishore Babu, B., & Pavani, P. (2019). Engineering students perception towards social media advertising for social causes, Vijayawada: An empirical study. *International Journal of Recent Technology and Engineering, 7*(6), 1901–1907.

11. Patnaik, R. P. K., Babu, B. K., & Narasimha Rao, M. V. A. L. (2019). E-shopping buying behavior of engineering studentsâ€™ (with special reference to Vijayawada). *International Journal of Innovative Technology and Exploring Engineering, 8*(12), 2885–2889.
12. Mansour, N., & Houri, M. (2005). Testing web applications. *Information and Software Technology*, 31–42. https://doi.org/10.1016/j.infsof.2005.02.007.
13. Al-Yasiri, D. S. Q., & Obaid, A. J. (2018). A new approach for object detection, recognition and retrieving in painting images. *Journal of Advance Research in Dynamic and Control System, 10*(2), 2345–2359.
14. Al-asadi, T. A., & Obaid, A. J. (2016). Object detection and recognition by using enhanced speeded up robust feature. *International Journal of Computer Science and Network Security (IJCSNS), 16*(4), 66–71.
15. Rao, A. N., Vijayapriya, P., Kowsalya, M., & Rajest, S. S. (2020). Computer tools for energy systems. In *International Conference on Communication, Computing and Electronics Systems* (pp. 475–484). Singapore: Springer.
16. Gupta, J., Singla, M. K., Nijhawan, P., Ganguli, S., & Rajest, S. S. (2020). An IoT-based controller realization for P.V. System monitoring and control. In: Haldorai, A., Ramu, A., & Khan, S. (Eds.), *Business Intelligence for Enterprise Internet of Things*. E.A.I./Springer Innovations in Communication and Computing. Cham: Springer.
17. Anson, O., Paran, E., Neumann, L., & Chernickovsky, D. (1993). Gender differences in health perceptions and their predictors. *Social Science and Medicine, 36*(4), 419–427. https://doi.org/10.1016/0277-9536(93)90404-R.
18. Datta, D., Mishra, S., & Rajest, S. S. (2020). Quantification of tolerance limits of engineering system using uncertainty modeling for sustainable energy. *International Journal of Intelligent Networks, 1*, 1–8. https://doi.org/10.1016/j.ijin.2020.05.006.
19. Al-asadi, T. A., Obaid, A. J., Hidayat, R., & Ramli, A. A. (2017). A survey on web mining techniques and applications. *International Journal on Advanced Science Engineering and Information Technology, 7*(4), 1178–1184.
20. Al-asadi, T. A., & Obaid, A. J. (2016). Object based image retrieval using enhanced SURF. *Asian Journal of Information Technology, 15*(16), 2756–2762.
21. Meshram, C., Ibrahim, R. W., Obaid, A. J., Meshram, S. G., Meshram, A., & Abd El-Latif, A. M. (2020). Fractional chaotic maps based short signature scheme under human-centered IoT environments. *Journal of Advanced Research.*
22. Santoso, L. W., Singh, B., Suman Rajest, S., Regin, R., & Kadhim, K. H. (2021). A genetic programming approach to binary classification problem. *E.A.I. Endorsed Transactions on Energy, 8*(31), 1–8. https://doi.org/10.4108/eai.13-7-2018.16552.
23. Satish Kumar, D., Bindu, P., Rao, D. S., Anusha, S., & Srinivas, J. (2018). A comparative study of factors influencing purchasing decisions at D-mart and metro in Vijayawada city. *International Journal of Engineering and Technology(U.A.E.),7*(4.10 Special Issue 10), 221–224.
24. Satish Kumar, D., Anusha, S., Rao, D. S., & Niranjan, H. (2018). A Study on consumer behavior at corporate retail stores in Vijayawada city. *Journal of Physics: Conference Series, 1139*(1).
25. Vinaya Laxmi, K., Vijaya Kumar, P., & Narayana, M. S. (2019). An empirical evaluation on customer satisfaction in retail banking services of SBI bank in Hyderabad district. *International Journal of Innovative Technology and Exploring Engineering, 8*(10), 925–931.

Chapter 19
Challenges in the Adaptation of IoT Technology

Neha, Pooja Gupta, and M. A. Alam

Abstract At present, The Internet of Things (IoT) and its relevant technologies are moving towards the subsequent step of evolution where almost every object in our vicinity would be integrated with the internet allowing them to communicate and share data. This chapter begins with a brief overview of the core concept of IoT followed by various IoT based smart-environment, applications, and services areas of IoT in the smart environment. Despite the countless possibilities that IoT along with other emerging technologies has to offer, various hindrances will be encountered with such large number of devices connected to each other transmitting sensitive data. Several such challenges faced in the adoption of IoT in areas such as smart city development, smart education, healthcare and medicine sector, manufacturing, smart grid, agronomy and telecommunication are investigated next in this chapter. Blockchain technology is gaining popularity for its decentralization abilities and absence of intermediaries and hence can prove to be a panacea for the concerns related to the adaption of IoT. This chapter introduces the use of blockchain to address these challenges arising due to the centralized structure of IoT and the benefits of integrating blockchain with IoT environment. Keeping in mind the recent developments related to Covid-19 and the vulnerabilities caused henceforth, in the last segment of this chapter, we have briefly reviewed various domains where use of IoT can prove beneficial.

Keywords COVID-19 & IoT · Smart city · Smart education · Smart manufacturing · Smart grid · Smart agriculture · Smart healthcare · Blockchain in IoT · Challenges in smart environment

19.1 Overview of IoT

The digital revolution is the new industrial revolution. At present, The Internet of Things (IoT) and its relevant technologies are moving towards the subsequent step

Neha · P. Gupta · M. A. Alam (✉)
Department of CSE, Jamia Hamdard, New Delhi, India
e-mail: aalam@jamiahamdard.ac.in

P. Kumar et al. (eds.), *A Fusion of Artificial Intelligence and Internet of Things for Emerging Cyber Systems*, Intelligent Systems Reference Library 210,
https://doi.org/10.1007/978-3-030-76653-5_19

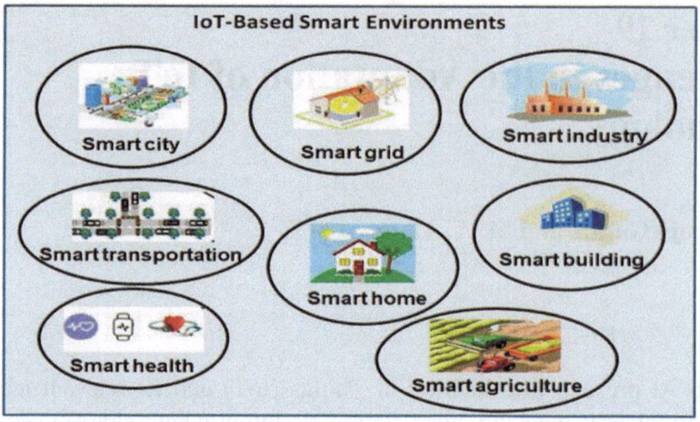

Fig. 19.1 IoT based smart environments

of evolution. These emerging technologies have been grasping a lot of attention of scholars .The internet of things (IoT) is a latest standard to transform the conventional living standards into digital. The Internet of Things IoT is a collaborative structure of various electronic devices that are capable of effective communication. In the coming era, IoT enabled system will furnished and upgrade the existing services in various sectors like agriculture, medical, education, industrialization etc. This will allow the system to be more self-regulated. It have the ability to integrate within the system and provide a better analysis opportunities. IoT is an expanding field for sensing, networking etc. [1]. With the employment of IoT, we can design real smart environment around us. So it is a rising edge of technical and socio-economic importance. Also it has become an important issue in designing and monitoring the system. IoT has become a huge part of our life that remodelled it into more versatile and smart ways. However, at the same time, the Internet of Things faces significant challenges that may stand in the way of realizing its potential advantages.

Figure 19.1 shows some IoT based smart envious. Major components in IoT are as follows:

1. Power saving system: It employ powder efficient system with better throughput.
2. Cloud computing: During the course of communication a large amount of information is generated. It required a reliable structure for data handling, advance processors and Cache. Cloud computing helps in the data management.
3. Network Connection: Internet Connectivity is required for the communication of physical devices. IoT depends on sensors and operates for real time. Network flexibility and integration is also another aspect in IoT infrastructure development.

19.2 Major Challenges in IoT

Despite the scope and capabilities of IoT, there are some apprehensions in the industrial sector due to the challenges and hurdles in adapting IoT [1–3]. A few of such concerns are mentioned below (Fig. 19.2).

19.2.1 Privacy

With the growth in the quantity of Internet-connected devices, the data traffic generated by them will increase significantly. Since these devices can process large volumes of data, the possibility of security breaches will also increase. For example, uninformed data collection, increased surveillance etc. Hence, to keep up with advancement in technologies and to deal with the concerns thus arising, new legislations for data privacy are required.

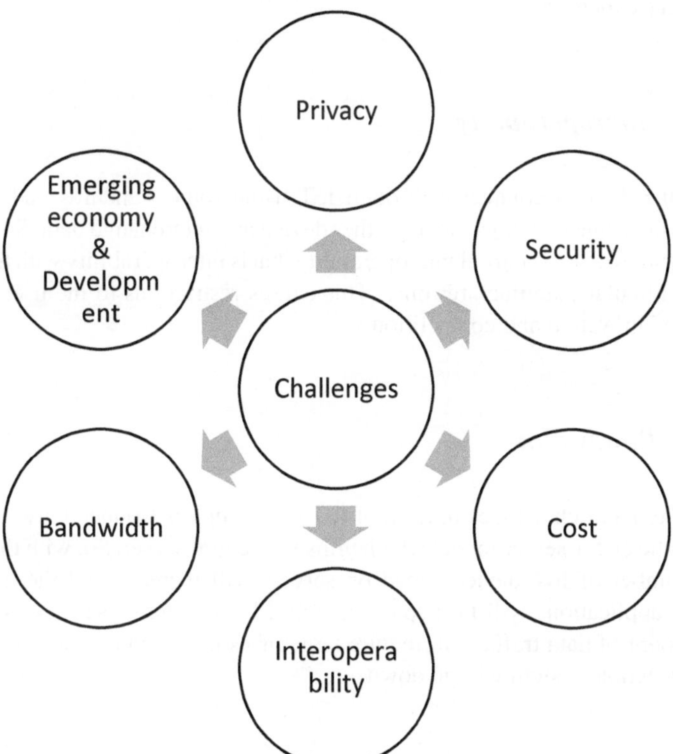

Fig. 19.2 Various challenges in IoT

19.2.2 Security

Although security issues have always been there in the field of information technology, IoT implementations pose different set of challenges related to security.

The interconnected nature of IoT devices means that every poorly secured device that is connected online potentially affects the security and resilience of the Internet globally.

Data thefts and cyber-attacks can be very easily done on IoT devices on which proper security arrangements are not there. Even one security vulnerability can led to unauthorized access by cybercriminals and it can compromise the whole ecosystem. Hence handling these security challenges are of utmost importance for IoT devices.

19.2.3 Cost Factor

IoT is based on a central cloud based system and the implementation of this system can be very expensive.

19.2.4 Interoperability

Implies the devices connected through IoT using same standards and protocols. They are thus able to connect to any other device and interchange data. Some device manufacturers provide partial interoperability that is interoperability with the devices of that particular manufacturer only. This causes restrictions to the user choice as well as to innovation and competition.

19.2.5 Bandwidth

One of the major challenges in front of IoT technology is Connectivity. At present, IoT uses the client-server model which utilises a central server. So, with the increase in the number of IoT devices, load on servers will increase and the bandwidth-intensive applications will face space crunch. Maintaining these servers to handle huge amount of data traffic will require large investments and if the server crashes somehow, whole system will go down.

19.2.6 Legal, Regulatory, and Rights

The use of devices connected with IoT raises many legal and regulatory concerns such as the cross border transfer of data and differing laws and standards in different jurisdictions, conflicts between laws and rights of people. As the technology changes rapidly, laws and standards become outdated. This leaves people and businesses in confusion since the laws have a slower place of evolution than the technology. With a considerably large number of devices connected through IoT, absence or out datedness of laws can be the reason of disastrous scenarios.

19.2.7 Emerging Economy and Development Issues

The Internet of Things can be of great use in areas such as smart healthcare, smart agriculture and water management etc. Developing and less developed regions can utilise these advantages offered by IoT to usher in the areas of societal development. However, these regions need to address the issues related to infrastructure, investment, skill sets and policies etc.

19.3 Applications and Services of IoT in Smart Environments

The Integration of IoT with smart environments has achieved various goals. This section will focus on key areas suggested by smart environments and illustrated the working structure of IoT.

19.3.1 Real Time Data Circulation

Depending on the decisions taken, the smart environment can rapidly adapt itself and improve operational efficiency resulting in higher user satisfaction.In IoT empowered environment data tracked, collected, analyzed and processed with relevant design of network.

19.3.2 Economical Infrastructure

The major function of reliable IoT infrastructure is proper decision making, data collection & predictions. It is achieved with the help of cloud servers. Economical, flexible, scalable and secure cloud-based features create a smart environment.

19.3.3 Performance

Due to the significant developments in IoT devices, amount of information within the network has also increased. To control such massive data with optimal throughout is a tedious job. Integration with the cloud can produce a supporting smart design to deal with this issue.

19.3.4 Opportunities to New Industrial Model

IoT could provide a digital competitive revolutionary environment for existing business models. IoT can validate current industries to design modern revenue models. The adoption of IoT in current business prototype will streamlined the profits, improve customer services, expand the market and create a better understanding of the system.

19.4 Challenges of IoT with Respect to Various Smart Environments

This section will explain the difficulties faced by IoT in various smart environments:

19.4.1 Agronomy

The uses of IoT in agronomy includes lot of model and simulations. Sensors has great importance in farming. Current developments in sensor technology have contributed greatly to the technological advances of conventional cultivation to precision and micro precision agronomy [4]. Wireless Sensor Networks have been used in monitoring soil and environmental attributes. Data flow through various kind of sensors like climate, radiation sensor etc. are used for monitoring, mining, logic and control purposes. Moreover, there is a captivating demand for high standard and safe farming products. This drift the requirement for more interoperable, appropriated, robust and precise logistics traceability systems. The applications of IoT in agribusiness could be marked in various way like Monitoring & Control, Open field agriculture, tracking of food supply chain, in Green house cultivation etc. The IoT family come up with all the appropriate and suitable tools and features for building and maintaining such infrastructure and services [5].

The implementation of IoT in agribusiness faces some difficulties that are referenced as.

19.4.1.1 Security Challenges

The essentials for Security could be summed up as Authentication, privacy and access control. The physical security of hardware is one of the concerning points as majority of the equipment are positioned in open area. Also, they operate without periodic inspections. It seems that it cannot be managed by the same security policies in all places. There is usually unauthorized data access that leads to information leakage. This requires data acquisition security criteria [6, 7]. Some other attacks can also be seen i.e. interference, eavesdropping and replay attacks.

19.4.1.2 Networking Issues

As environmental characteristics change from one place to another. This creates some difficulties in networking. Wireless communication is commonly used in agricultural science in IoT implementation because it is efficient in nature. Variable environmental characteristics such as humidity, rainfall, temperature, etc. influence wireless. communication performance. Wireless mode passes through multi-path propagation, low link quality, and many types of noise and disturbances due to many environmental factors [8]. According to the requirements and challenges of the rural environment, data has to be transferred using robust and reliable technologies. Handling and routing of data in the network is also a major concern.

19.4.1.3 Organizational Issue

When logistics is accounted for the food and farming area, this infrastructure encourage information trade and transportation of merchandise, enhancing the streamlined production and inventory network all around the world. IoT provides more precise and live progression of materials and products that brings gradual changes in agribusiness [9]. When IoT integrates with cloud computing, it improves service quality. A large amount of information is distributed by the system, which significantly increases the cost and speed to move to the cloud. Obviously, there must be an adaptive balance between storage and processing blocks. In addition, some part of the work is processed in the cloud, which is also an important concern.

19.4.1.4 Infrastructural Issues

Concerning IoT infrastructure in farming, a few difficulties emerge. Most of the segments have to face unharmonious environmental conditions like rain, humidity, atmospheric pressure, air turbulence etc. Such unfavorable surroundings disrupt the operation of the electronic devices and may prompt to improper outcomes and damage to the circuit. Also the associated equipment should remain dynamic and produce reliable output throughout the process. They should operate with optimal power and resources for significant amount of time. Hence proper programming and optimal operational power is required, since the regular battery substitution or reset of the stations isn't simple in the large field. Power harvesting can relatively fix such an issue. Besides, the huge number of interconnected devices delivers tremendous amount of information, which may be beyond the scope of processing limits of small scale IoT infrastructures [10].

19.4.2 Smart Healthcare

Health is an essential part of life. With the increasing graph of the population, the load on the current medical system is shoot-up. And the current system has proved inefficient to meet basic health needs. Meanwhile the ongoing situation due to the novel coronavirus has clearly indicated the need for dynamic, proactive, robust, adaptable, scalable, sustainable and agile medical services. The Internet of Things (IoT) has significantly changed health services in a relatively short period of time [11, 12]. The Internet of Things (IoT) has been widely recognised as a potential solution to reduce pressure on the health care system [13, 14].The ease of using IoT in medical can be listed as (Fig. 19.3).

19.4.2.1 Real Time Reporting and Monitoring

With the help of various connected devices in IoT infrastructure, real time data monitoring is possible. It is helpful in the case of many medical emergencies like heart failure, diabetes etc. With the help of Smart health wearables constant observations about various health vitals like Blood pressure, oxygen saturation levels, heart rate, electrocardiograms (ECGs), body temperature, blood glucose and weight etc. are achieved. As per the access permission, the connected IoT devices are able to collect and transfer health data with licensed authorities like physician, health consultant, registered medical clinics or insurance agencies. Furthermore, the data collected is available to the licensed person/organization anytime, anywhere.

Fig. 19.3 Overview of IoT in infrastructure in health care

19.4.2.2 Medical Assistance at Remote Location

This is a pathetic situation for a patient to avail proper clinical help at remote locations. It may be because of unbalanced medical facilities at that location. In such cases the IoT enabled medical infrastructure can dispense relevant medical support to such patients. Medical practitioners can deliver prescriptions through healthcare delivery chains to patients through IoT devices.

19.4.2.3 Continuous Follow-Ups and Alerts

Live tracking and alerts could be a boon in case of critical health conditions. Real-time notifications for proper monitoring, analysis and diagnosis to protect the health of a patient is feasible with IoT devices. This allows for hands-on treatment, precise and appropriate medication by specialists. Subsequently promoting the thorough patient care.

19.4.2.4 Budget Friendly and Proper Connectivity

Healthcare mobility solution assist in automating the healthcare workflow. IoT in medical care empowers interoperability, device to device communication, exchange of data and also routing of the medical data. This prompts a practical and affordable distribution of medical services. It also bring down the avoidable visits to the clinics/hospitals and aids to efficient resource utilisations.

19.4.2.5 Data Selection and Analysis

Data collected via IoT devices can be examined through mobility solutions powered by IoT. This will curb the further classification of data. It also enhances critical healthcare analytics and data-driven experiences, which decreases the mistakes and accelerates decision making.

The coin has two sides. IoT application in medical field have faced several challenges which are as follows [14, 15].

19.4.2.6 Outdated Medical Infrastructure

Despite of the fact that retrofitting can revive aging infrastructure, IoT is difficult to exploit if the infrastructure of a facility is obsolete. Outdated infrastructure is a known problem in healthcare. At the point when emergency clinics are in critical condition and requires renovation, they also experience various challenges to hire well trained staffs and up gradation. As there is significant demand for potential candidates and they might not have any desire to cope up with outdated medical infrastructures.

19.4.2.7 Data Security and Privacy

Data security and privacy is always being a point of concern for any system.IoT-enabled mobile devices capture data in real-time.But most of them don't have compliance with protocols and standards. Therefore there is a high possibility of unauthorised access of the information. This makes data more vulnerable to cybercriminals. There is a possibility to compromise with personal health information due to hacking.

19.4.2.8 Data Management

The medical sector will produce tremendous amount of information. And because of unregulated of information and communication protocols, it is difficult to collect data for critical insights and analysis. In addition, the health industry has to be very careful to treat patient from IoT devices. IoT devices accumulate data in the large volume. For accurate examinations/scanning, the information must be partially analyzed without overloading. The surge of information produced by IoT devices implemented in medical industry can also lead to unexpected problems and influence decision-making methods if organizations are not ready to deal it appropriately and verify its quality. So data management, its accuracy and precision is an important concern.

19.4.2.9 Device Integration

Integration of many types of devices causes obstacles in the implementation of IoT in the medical ecosystem. This is because of lack of protocols and standards. There is no such synchronous protocol that accompanied data aggregation. This also scale down the scalability of IoT in medicine.

19.4.3 Smart Education

Recent developments and adaptation in Information Communication Technology (ICT) have improved many aspects of the education sector. The recent novel Coronavirus hits the world which has given impetus to the Internet of Things (IoT). Information and communication technology has transformed the conventional learning approach to digital-based learning methods [16, 18]. IoT is one of the paradigms that come up with more smart environments at different stages of society. IoT upgrade the nature of existing learning methods. It has a capability to design a collaborative content producing smart environment for educators and learners. Additionally, It assists educators to formulate smart lessons plans. It could lay out customised study materials and realise students learning Characteristics and performances. This practice has initiated new channels of communication and aids latest dimensions of skills. During this collaboration students can seek guidance and support from their teachers to solve their educational problems. [17] The IoT system empowered learners to be a part of education system from any location at any time. It has promoted personal learning through its choice and frequent lectures. It expands the proficiency of the organisation, classrooms, students and administrators [19]. Students and teachers can enjoy the benefits of IoT devices such as interactive white boards, tablets, smart phones, e-books, 3D printers, virtual and augmented reality headsets, sensors, fitness bands, cameras and videos and more. IoT devices drive a fast, streamlined, systematic and productive learning process. IoT tools provide a comfortable environment for students, teachers and administrators by magnifying their proficiency. IoT devices are exceptionally convenient for managing records of students, teachers and administrators.

Various research exhibits merits of using IoT in education la few of them are as [20, 21]:

i. Efficient and cost effective use of resources.
ii. Improved learning speed.
iii. Ease in record management.
iv. Ease in monitoring & control.
v. Better physical security.
vi. Availability of assignments, grades, lessons in digital format.
vii. Expand the reach of education.
viii. Escalates co-curricular activities.

ix. Stimulate real world learning scopes and practices.
x. Well organised task execution.

Conversely, there are several challenges that relate to the applications of IoT in education are:

i. Development complexities: IoT is as yet in its beginning phases and numerous issues are as yet unsolved, for example, coverage, high sensors price, and battery life among others [7]. In like manner, IoT specialists need to think about these issues to encourage the work of IoT applications.
ii. Need for proper monitoring & management of IoT devices.
iii. Security and privacy concern: Sometimes due to security violations personal information of students or educators leaked.
iv. Implementation and device availability.
v. Necessity of sufficient bandwidth and reliable internet connection: High speed connectivity is required for accessing study materials.
vi. Proper Teacher Training: This is required for better collaboration and effective use of the IoT devices while imparting knowledge.
vii. Ethical issues.
viii. Financial issues: Financial resources is assumed to be a important part in the integration of IoT devices with classroom attributes. Adequate funds will permit the use of appropriate IoT devices. Here information is collected saved and distributed among educators and learners through reliable network. Therefore, adequate budget is required for the smooth adoption of IoT in the education sector.
ix. Changes in existing curriculums: Departments like software engineering, computer sciences are required to build up their educational programs to consolidate IoT courses to qualify their new students with ample capacities to head various IoT ventures. Also, Educational institutions need to improve their procedures by incorporating IoT into their developed programme. By With the objective of professional development for trainees and students institutes have to organise orientation sessions for their staff with respect to the aces of IoT.

19.4.4 Smart Manufacturing

Smart manufacturing is the use of devices connected via IoT and technologies such as big data, artificial intelligence, machine learning etc. in the manufacturing process. It offers umpteen benefits e.g. increased efficiency, improved productivity, real time supply chain management, optimized inventory management etc. thus providing an edge over competitors [22] (Fig. 19.4).

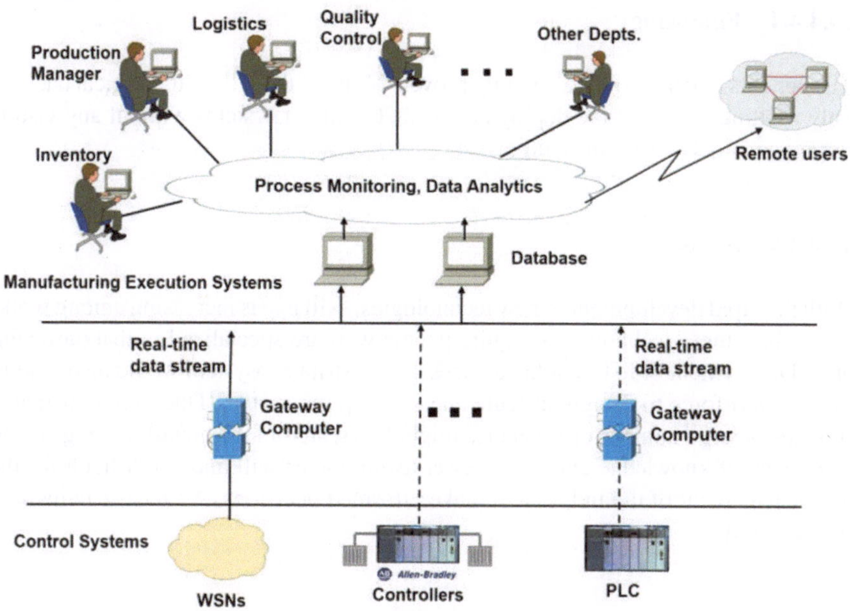

Fig. 19.4 IoT enabled smart manufacturing system

19.4.4.1 Challenges

Despite offering many advantages, like any other technology, this one also suffers from various challenges. A few such challenges are discussed here [23, 24].

19.4.4.2 Integration

Traditionally, machines in the industries are used for a long time with repairs and maintenance whenever needed. To make the manufacturing ecosystem compatible with the newer technologies, these machines need to be replaced. This would cause huge expenses. So, the businesses especially smaller ones does not find it rational to spend on new machines when old ones are working fine.

19.4.4.3 Connectivity

In manufacturing process enabled by IoT, communication happens through wireless medium unlike the existing wired communication. This requires alteration in the network infrastructure causing a burden on the finances of the organisation.

19.4.4.4 Financing

The benefits of using IoT are not fully proven. Many of these have theoretical backing only. Getting funds for the deployment of IoT enabled systems without any visible improvements is a very difficult task.

19.4.4.5 Skills

With the rapid development of new technologies, skill gap is increasing among workforce since these technologies require people who are specialized in that particular area. Deployment of IoT enabled devices in industrial ecosystem would also require skilled workforce to design, execute and keep up the system. Due to the mismatch of skills, a large number of vacancies might be created in the manufacturing sector. Also, lack of knowledge about the newer technologies will make it difficult for the top management of the industry to make informed decisions. All these can hamper the economy.

19.4.4.6 Lost Manufacturing Jobs

The deployment of smart industry requires highly specialized workforce. This would leave a lot of existing workers' skills as obsolete and may render them jobless.

19.4.4.7 Security

Security is the major concern of the IoT enabled devices. In the production process, there are several components and data transfer take place between them. If due to improper security solution, attackers get access to the network, manufacturing process can be compromised, machines can be hacked, and damaged.

19.4.5 Smart Grid

In the near future, IoT could be used extensively in the field of smart grid.Smart grid allows surveillance and control over the entire power supply chain starting from the power plant to the end users with the help of intelligent devices such as smart meters, smart appliances etc. However, being a critical infrastructure, use of IoT for power grid poses many issues and challenges a few of which are mentioned here [25].

19.4.5.1 IoT-Based Smart Grid's Security Issues

Using IoT in the power grid system is prone to various security issues.

19.4.5.2 Spoofing

Spoofing is when an attacker gets unauthorized access by impersonating a legitimate user. An attacker may tamper with the consumption details by spoofing the smart meter of some other user.

19.4.5.3 Eavesdropping

Devices in the smart-grid system exchange information over the communication network. So, the attacker can have illegal access to this data by compromising the network.

19.4.5.4 Privacy Issue

By eavesdropping, an attacker can get access to the data exchanged in the smart-grid system. This data can further be harvested to gain sensitive information about the end user such as their sleeping schedule, their behaviour pattern, type of appliances used or if they are away from home for a certain period of time. This data can also be used for commercial benefits such as targeted advertisements, without the knowledge and consent of the user.

19.4.5.5 Data Tampering

An attacker can not only look into the data transmitted, but can also tamper with this data. Attacker can tamper with the key parameters set in the smart meter such as the dynamic prices for increased electricity consumption, thus paying less than their actual consumption.

19.4.5.6 Authorization and Control Access Issues

The devices in the smart grid system can be controlled remotely. An attacker, if gets unauthorized access, can damage the physical infrastructure thus causing power failure.

19.4.5.7 Denial of Service Attack

In the smart power grid system, even the critical components are connected through IOT. These components if attacked can render them unavailable.

19.4.5.8 IoT-Based Smart Grid's Challenges

Using IoT in the power grid system can pose unique challenges. These challenges can inhibit the adoption of smart-grid based on IoT by the end users.

19.4.5.9 Scalability

The Smart-grid system spans over wide areas and has a considerably large number of devices connected. So, any security technique to be deployed must be scalable to the future increase in devices.

19.4.5.10 Interoperability

If the devices in the smart-grid system does not follow homogenous protocols and standards, establishing communication among them could be problematic.

19.4.5.11 Non-homogenous Resources

Interoperability issues can also arise if the devices in the smart-grid are having heterogeneous resources e.g., memory capability, bandwidth etc. Establishing communication among such devices can be very challenging.

19.4.5.12 Resource Constraint

Many devices in the grid system have inadequacy of resources. So, the deployment of security techniques on these devices needs to take care of this constraint. This can pose difficulties while applying existing security technologies.

19.4.5.13 Legacy Systems

Components and devices of the classical power-grid have negligible security infrastructure support. Replacement or updation of these devices in order to make them smart-grid compatible is a tedious task.

19.4.5.14 Trust Management

Different components in the smart-grid system are managed by different entities e.g., smart appliances are managed by the users, substations and smart meters by the grid operator. In order to maintain a smooth communication between these components, trust between entities is a must. Building trust is a huge challenge for such a huge network.

19.4.6 Smart City

An urban area can be transformed into a smart city under the umbrella of electronic devices and sensors. A smart city is an environment that utilizes IoT connectivity and smart appliances to achieve automation and intelligence capability. Smart city offers a solution to the problems caused by rapid urbanization and increase in population. Its aim is to improve the efficiency of services offered to citizen thus easing their life.

19.4.6.1 Challenges in the Adoption of Smart Cities

Despite numerous advantages of IoT in the implementation of smart cities, this concept also suffers from various challenges a few of which are discussed here [26–28].

19.4.6.2 Governance Issue

Governance is a major concern in developing a smart city. A strong political will and clear vision will be required to make efficient long term plans. Implementation of these plans is the responsibility of local authorities and if effective coordination is not maintained among them, it will make the task very difficult. Trust between the government and people is also a matter of concern which can get into trouble if privacy and security issues are not addressed properly.

19.4.6.3 Economic Issue

Smart city ecosystem requires large-scale infrastructure e.g. sensors, smart appliances, smart-grid etc. Infrastructure development at this scale requires huge financial expenditure. Functioning and maintenance of this system will also require high expenses. Deployment and development of smart city ecosystem requires workforce with highly specialized skillset. Also, workers need to be given regular trainings and workshops to keep up with the advancement in technologies.

19.4.6.4 Technology Issues

Development of smart cities involves research, development and new and efficient technological inventions. Planning and executing authorities lack knowledge needed for developing such technology intensive ecosystem.

Majority of the Citizen lack technological knowledge as well as access to technology. Making arrangements for their training and education is a huge challenge. Data privacy and security is a major concern since a lot of personal data about the citizens will be transmitted over the network which if compromised can prove to be a disaster. This system involves integrating various components which is quite challenging since these components does not use homogenous protocols.

19.5 Blockchain as a Solution

Despite the fact that, IoT has many advantages, some challenges also exist like security & privacy issues, Legal, regulatory issues, Interoperability issues, IoT standards issues, developmental problems etc. In IoT, the process of data flow is vulnerable at many stages. There is a high probability of data mishandling and negative control on data flow process by hackers. Blockchain would be a useful technology to address such security and privacy issues in IoT. This is due to the fact that blockchain is based on decentralised network concept that stamp out the central server dependency of IoT. Blockchain allows information flow through distributed ledger for every transaction with legitimate confirmation [29] (Figs. 19.5 and 19.6).

Most IoT systems, which have been implemented so far, are based on the centralised server concept. In IoT, most of the collected data is kept up in a central server. If the devices need to retrieve data to interact they need to collaborate with central nod and the information stream will occur through it. In IoT systems, sensor devices collect information and transmit it to the central server via the Internet. From a centralized server, analytics were proceeded according to the client necessities and comfort. In comparative, the enormous scope IoT system aspires to perform analysis,

Fig. 19.5 Data flow in IoT

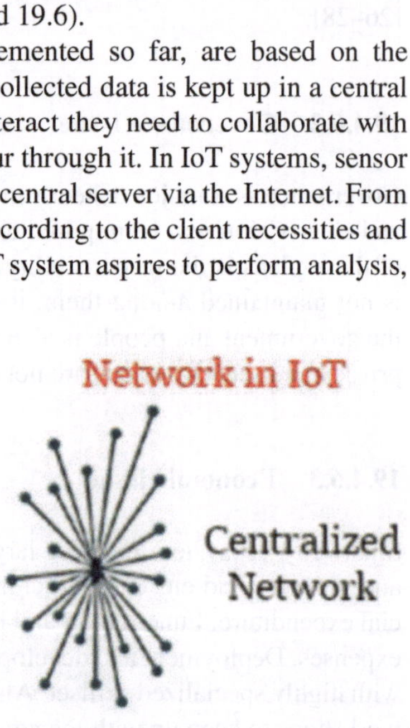

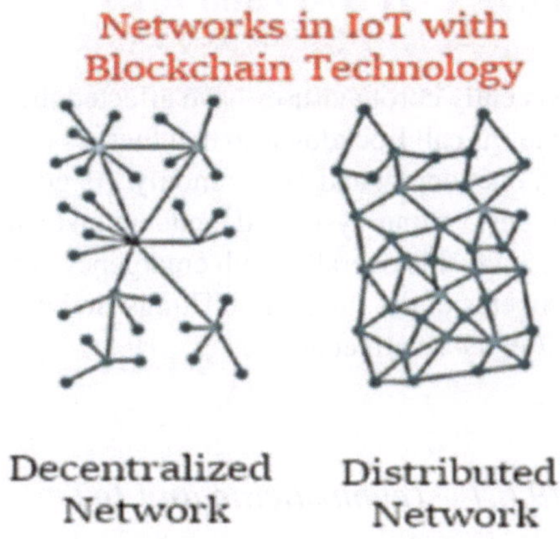

Fig. 19.6 Data flow in IoT with Blockchain

the processing capacity of existing Internet structures may not help adequately. To handle the huge data processed in large-scale IoT systems, the infrastructure of the Internet needs to be enhanced. One most ideal approach to understand this is to have decentralized or appropriated networks.

Blockchain technology is powered by peer-to-peer networking (PPN), distributed file sharing (DFS), and autonomous device coordination (ADC), capabilities that permits the IoT to follow to a large number of associated devices. Also, Blockchain permits the IoT to handle data exchanges between the devices in communication [30, 31]. The distributed network of blockchain makes peer to peer messaging faster and also reduces the dependency on the central server. In IoT with blockchain, distributed ledger is carefully designed which doesn't permit in distortion, wrong validations in information. The adoption of blockchain in IoT will make the information stream will turn out to be more reliable and secure.

Subsequently integration of Blockchain technology with IoT also adds following benefits [31]:

i. Tamper-evident information.
ii. Helps in the development of cost efficient infrastructure.
iii. Stamp out single authority control.
iv. Increase confidentiality.
v. Build system more reliable and Robust.
vi. Keep Track record of old transactions.
vii. Records the notable activities.
viii. Distributed record sharing.
ix. Improve privacy of the information.
x. Empowered for self-directed functioning.

19.6 COVID-19 and IoT

Recently coronaviruses have affected the whole world in incalculable ways, various nations call Lockdown so that there is no loss of life. It has an unprecedented impact on our society and our economy. As governments go hand-in-hand to address these issues, various systems dependent on technology such as IoT have jumped in to help manage this overall health emergency. Therefore, COVID-19 may well be a definite impetus of the Internet of Things (IoT). Some areas where IoT will thrive due to the COVID-19 effect are.

19.6.1 Telemedicine and IoT

The telemedicine sector has consistently collapsed while reaching the market demands. However, in collaboration with medical devices, it has the potential to convert significant proportions of physical health interactions into remote interactions that lead to relief from pressure on the prolonged medical system.

Health care structures in the world are confronting difficulties in providing proper medical/clinical care and lessen the risk of exposure. Contactless investigations/communications are helpful in preventing this epidemic from rapidly outspreading. The emphasis on contactless clinical assistance drives, the medical care communities to pick up the IoT arrangement providers for a powerful approach to combat diseases. With cloud and Artificial Intelligence, IoT offers medical care experts to screen their patients, access the information, and give therapy from a far-off area, this is conceivable by taking advantage of smart devices like a smart thermometer, smart wearable, and various trace apps.

19.6.2 E-commerce and IoT

COVID-19 results in a complete lockdown in several nations. There was turmoil not only in the medical sector but also in transportation and businesses. With the dissolution of the supply chain network due to COVID-19, a major issue of inventory control comes into the picture. This issue could persist through the end of the year. The use of NFC labels could ease this challenge to a great extent.

Then the trace and track system employed by some organizations has been shown to be essential for keeping online business in full action and managing delays in real-time delivery. In the end, IoT has become an approach to contribute a swift and more straightforward support to the consumers. If there is a change in consumption habits, this could be one of the new requests.

19.6.3 Work from Home and IoT

In several sectors, Work home is a growing trend from the past few months due to the pandemic.

It provides us greater adaptability and flexibility, saves home to work travel time. Additionally, it also permits organizations to save money on physical locations. It also enhances the budget in teams working in different locations. IoT comprises 4 segments: Sensors, Networks, Cloud, and Applications (SNCA). Earlier to this health alarming situation, sensors (84%), information handling (77%), and cloud stages (76%) were the premium to organizations dealing with IoT.

19.6.4 Blockchain and IoT

Blockchain technology provides a tamper-proof environment for transactions. For this case a blockchain network where the World Health Organization (WHO), Health Ministry of each nation, all important nodal clinics of every nation, were associated, sharing their real time data constantly, about any new transmittable diseases, at that point the world may have woken up significantly sooner. The WHO and other technical companies are working with blockchain on a program to help pass-on information about the ongoing pandemics under the name MiPasa.The program is a distributed laser technology (DLT) that will aid in exposure of viruses in advance and identification of carriers and hotspots.

19.7 Conclusion

Internet of Things is a step towards creating an interconnected system of devices that are able to communicate without any human interference. It has enormous potential that can be utilized in almost every area e.g. healthcare, agriculture, education, etc. However, there are various concerns and challenges associated with this technology that is responsible for hindering the deployment of IoT based systems. In this chapter, we have given a brief overview of IoT and its applications in various smart environments. We have discussed the challenges in the integration of IoT in various environments such as smart-grid, smart-city, etc. Finally, we have reviewed blockchain technology and how it can be a remedy for the challenges of IoT. The chapter also elucidates some of the domains that flourish during the pandemics with the help of IoT.

References

1. Rose, K., Eldridge, S., & Chapin, L. (2015). The internet of things: An overview. *The Internet Society (ISOC), 80*, 1–50. https://www.internetsociety.org/wp-content/uploads/2017/08/ISOC-IoT-Overview-20151221-en.pdf.
2. .Buntz, B. (2016). Top 10 reasons people aren't embracing the IoT. *Internet of Things Institute*. https://www.IoTworldtoday.com/2016/04/20/top-10-reasons-people-aren-t-embracing-IoT/.
3. Banafa, A. (2017). Three major challenges facing iot. *IEEE Internet of things*. https://www.bbvaopenmind.com/en/technology/digital-world/3-major-challenges-facing-IoT/.
4. Kacira, M., Sase, S., Okushima, L., & Ling, P. P. (2005). Plant response-based sensing for control strategies in sustainable greenhouse production. *Journal of Agricultural Meteorology, 61*(1), 15–22.
5. Talavera, J. M., Tobón, L. E., Gómez, J. A., Culman, M. A., Aranda, J. M., Parra, D. T., Quiroz, L. A., Hoyos, A., & Garreta, L. E. (2017). Review of IoT applications in agro-industrial and environmental fields. *Computers and Electronics in Agriculture, 142*, 283–297.
6. Tzounis, A., Katsoulas, N., Bartzanas, T., & Kittas, C. (2017). Internet of Things in agriculture, recent advances and future challenges. *Biosystems Engineering, 164*, 31–48.
7. Sicari, S., Rizzardi, A., Grieco, L. A., & Coen-Porisini, A. (2015). Security, privacy and trust in Internet of Things: The road ahead. *Computer Networks, 76*, 146–164.
8. Thelen, J., Goense, D., & Langendoen, K. (2005). Radio wave propagation in potato fields. In *1st Workshop on Wireless Network Measurement, Riva del Garda, Italy, April 2005* (pp. np–np).
9. Atzori, L., Iera, A., & Morabito, G. (2010). The internet of things: A survey. *Computer Networks, 54*(15), 2787–2805.
10. Gope, P., & Hwang, T. (2015). BSN-care: A secure IoT-based modern healthcare system using body sensor network. *IEEE Sensors Journal, 16*(5), 1368–1376.
11. Zhu, N., Diethe, T., Camplani, M., Tao, L., Burrows, A., Twomey, N., & Craddock, I. (2015). Bridging e-health and the internet of things: The sphere project. *IEEE Intelligent Systems, 30*(4), 39–46.
12. Chang, S., Chiang, R., Wu, S., & Chang, W. (2016, May–June) A context-aware, interactive M-health system for diabetics. *IT Professional, 18*(3), 14–22. https://doi.org/10.1109/MITP.2016.48.
13. Pasluosta, C. F., Gassner, H., Winkler, J., Klucken, J., & Eskofier, B. M. (2015). An emerging era in the management of Parkinson's disease: wearable technologies and the Internet of Things. *IEEE Journal of Biomedical and Health Informatics, 19*(6), 1873–1881. https://doi.org/10.1109/JBHI.2015.2461555
14. Fan, Y. J., Yin, Y. H., Xu, L. D., Zeng, Y., & Wu, F. (2014). IoT-based smart rehabilitation system. *IEEE Transactions on Industrial Informatics, 10*(2), 1568–1577. https://doi.org/10.1109/TII.2014.2302583
15. Banica, L., Burtescu, E., & Enescu, F. (2017). The impact of internet-of-things in higher education. *Scientific Bulletin-Economic Sciences, 16*(1), 53–59.
16. Al-Emran, M., Malik, S. I., & Al-Kabi, M. N. (2020) A survey of Internet of Things (IoT) in education: Opportunities and challenges. In Hassanien, A., Bhatnagar, R., Khalifa, N., & Taha, M. (Eds.), *Toward social Internet of Things (SIoT): Enabling technologies, architectures and applications. Studies in computational intelligence* (Vol. 846). Cham: Springer. https://doi.org/10.1007/978-3-030-24513-9_12.
17. Malik, S. I., & Al-Emran, M. (2018). Social factors influence on career choices for female computer science students. *International Journal of Emerging Technologies in Learning, 13*(5).
18. Salloum, S. A., Al-Emran, M., & Shaalan, K. (2018, August). The impact of knowledge sharing on information systems: A review. In *International Conference on Knowledge Management in Organizations* (pp. 94–106). Cham: Springer.
19. Pundir, Y., Sharma, N., & Singh, Y. (2016). Internet of things (IoT): Challenges and future directions. *International Journal of Advanced Research in Computer and Communication Engineering, 5*(3), 960–964.

20. Barakat, S. (2016). Education and the internet of everything. *International Business Management, 10*(18), 4301–4303.
21. Safdar, M., Safdar, G. A., Hafeez, A., & Malik, M. A. (2019). Promises and challenges of IoT in education.
22. .The Connection Between Smart Manufacturing and IoT (2018). https://www.manufacturing.net/industry40/article/13123584/the-connection-between-smart-manufacturing-and-IoT.
23. Yang, H., Kumara, S., Bukkapatnam, S. T., & Tsung, F. (2019). The internet of things for smart manufacturing: A review. *IISE Transactions, 51*(11), 1190–1216.
24. James Blackman (2018) Six challenges for smart manufacturing and industrial IoT. https://enterpriseIoTinsights.com/20180813/channels/fundamentals/six-challenges-for-industrial-IoT.
25. Bekara, C. (2014). Security issues and challenges for the IoT-based smart grid. *Procedia Computer Science, 34*, 532–537.
26. Sharma, M., Joshi, S., Kannan, D., Govindan, K., Singh, R., & Purohit, H. C. (2020). Internet of Things (IoT) adoption barriers of smart cities' waste management: An Indian context. *Journal of Cleaner Production, 270*, 122047.
27. Silva, B. N., Khan, M., & Han, K. (2018). Towards sustainable smart cities: A review of trends, architectures, components, and open challenges in smart cities. *Sustainable Cities and Society, 38*, 697–713.
28. Rana, N. P., Luthra, S., Mangla, S. K., Islam, R., Roderick, S., & Dwivedi, Y. K. (2019). Barriers to the development of smart cities in Indian context. *Information Systems Frontiers, 21*(3), 503–525.
29. Dickson, B. (2016). Decentralizing IoT networks through blockchain. *TechCrunch, TechCrunch.*
30. Banafa, A. (2017). IoT and blockchain convergence: Benefits and challenges. *IEEE Internet of Things.*
31. Kumar, N. M., & Mallick, P. K. (2018). The Internet of Things: Insights into the building blocks, component interactions, and architecture layers. In *Procedia Computer Science* (Vol. 132, pp. 109–117). Elsevier B.V. https://doi.org/10.1016/j.procs.2018.05.170

Chapter 20
EEG Signals Based Choice Classification for Neuromarketing Applications

Ahmed Ali, Toufique A. Soomro, Faiza Memon, Muhammad Yousaf Ali Khan, Pardeep Kumar, Muhammad Usman Keerio, and Bhawani Shankar Chowdhry

Abstract Recently, the promising field of neuromarketing, which uses neuroscience in the decision-making has attracted growing interest from various end-user industries and geography. As neuromarketing helps with effective advertising, brand creation, and product design. Conventional marketing techniques (e.g. newspaper publishing and television commercials) are sometimes ineffective in selling products, as these techniques do not strongly inspire customers to purchase certain products. Such traditional marketing techniques seek to assess the customer's tendency towards a product, which may not necessarily imply an attitude towards the brand and its purchase intentions. It is possible that the seller misinterprets the behaviour of the customer, since the expected behaviour does not always reflect the actual purchase behaviour of the customer. The book chapter focuses on linking the gap between neuromarketing techniques, reflecting implicit customer behaviour, and conventional market techniques, which involve explicit consumer behaviour. In this chapter, the authors have thoroughly reviewed neuromarketing techniques for the recognition of choice based on EEG signals. In addition, the authors also identified another area

A. Ali · F. Memon
Electrical Engineering Department Sukkur, IBA University, Sukkur, Pakistan

T. A. Soomro (✉)
Department of Electronic Engineering, Quaid-E-Awam University of Engineering, Science and Technology, Larkana, Pakistan

M. Y. A. Khan
Department of Electrical Engineering, Gomal University, D.I Khan, Pakistan

P. Kumar
Department of Software Engineering, Quaid-e-Awam University of Engineering, Science and Technology (QUEST), Nawabshah, Pakistan
e-mail: pardeep.kumar@quest.edu.pk

M. U. Keerio
Department of Electrical Engineering, Quaid-E-Awam University of Engineering, Science and Technology, Nawabshah, Pakistan

B. S. Chowdhry
Department of Electronic Engineering, Mehran University of Engineering and Technology, Jamshoro, Sindh, Pakistan

P. Kumar et al. (eds.), *A Fusion of Artificial Intelligence and Internet of Things for Emerging Cyber Systems*, Intelligent Systems Reference Library 210,
https://doi.org/10.1007/978-3-030-76653-5_20

under-explored in neuromarketing studies i.e. data mining techniques for classifying and predicting customer choice recognition. This book chapter therefore provides a comprehensive review of the reported machine learning and deep-learning techniques for tracking customer choice recognition based on EEG data involving valence features and power spectral density. We found that the KNN + Fourier transform algorithm has the highest reported accuracy and is more reminiscent of support vector machine techniques and the nearest k-nearest neighbour.

Keywords Customer choices · Electroencephalogram (EEG) · Neuromarketing · Brain-computer interface (BCI)

20.1 Introduction

Consumer neuroscience is a growing sub-field of neuroscience that connects the affective and cognitive domains of customer behavior. Over the last few years, consumer neuroscience has rapidly gained popularity across the advertising and marketing field, and neuroscience approaches are being used to advertise a variety of products and brands.

Consumer neuroscience could help the seller to understand the neuronal activities of the customer for different products and to identify the elements that influence the customer's preferences when purchasing products. It is well documented in the neuromarketing studies that customers do not always understand what happens in their unconscious neuronal activities. In addition, several studies have shown that customers are not always explicit in their intentions or choices. The adoption of conventional marketing methods such as questionnaires and interviews to evaluate customer choices, purchase intentions and needs, may lead to incorrect or biased conclusions. Likewise, a verbal statement of choice can give rise to unconscious or conscious bias. It is difficult to find the customer's choice directly on the basis of preferences, due to ethical issues, time availability and high cost of the product. These factors represent a contradiction in the view of the customer during the evaluation of the usefulness of the product and their real views, senses, and feelings about the product. Advanced neuroimaging techniques-based marketing studies provide excellent customer behaviour assessment. These techniques eventually help sellers to evaluate the customer's neuronal activities and to understand the subconscious process that provides failed or successful marketing information. This information is obtained by overcoming the basic challenge in conventional marketing research.

BCIs is widely used neuro-imaging technique in the consumer neuroscience field. BCIs techniques permit the researcher to connect the brain with the computer efficiently. Generally, the BCI system does not require muscle interference or external devices to generate commands. In addition, the BCI uses freely produced brain signals to control devices capable of interacting or communicating with the environment. EEG signals analysis is the fundamental technique used to test neuronal processes. The EEG-based method is portable, affordable, versatile, and practical.

Table 20.1 Widely used locations of sensors in consumer neuroscience field for choice detection

Authors	Electrode channel	Waves	Device system
Agarwal et al. [3]	C4, Cz, Fz, Pz, P3, C3, P3 & F3	N200	Muse InteraXon
Touchette et al. [1]	F4 & Fp3	α	NeXus-10 biofeedback system Mind Media Inc
Vecchiato et al. [5]	F1/F2, F7/F8, AF7/AF8 & Fp1/Fp2	α	BrainAmp, Brainproducts GmbH
Vecchiato et al. [5]	F1/F2, F7/F8, AF7/AF8 & Fp1/Fp2	θ	BrainAmp, Brainproducts GmbH
Boksem et al. [4]	CP5, C4, C1, C3, C5, FC4, FC2, FCz, FC1, FC3, F4, F2 & F1	θ	64 active Ag–AgCl electrodes (Biosemi ActiveTwo
Boksem et al. [4]	FCz, FC1, F2 & AFz,	β	64 active Ag–AgCl electrodes (Biosemi ActiveTwo

Noninvasive EEG signals can provide continuous real-time neuronal activities with excellent temporal resolution. Thus, in this chapter of the book, the authors focused on EEG signal based BCI system studies for neuromarketing applications.

By employing some classifiers, BCI can provide excellent analysis to extract choice patterns from neuronal activity, and to convert acquired patterns into a brand-selling response. In addition, the relevant literature on DNN-based choice classification has been extensively explored and assessed. The key objective of this work was to carry out a comprehensive survey of EEG-based Choice Classification studies for the consumer neuroscience application in order to increase the understanding of classifier forecasting performance by comparing the results of DL and other traditional classification techniques, including k-nearest neighbours (KNN), random forest (RF) and vector support machines (SVM).

The typical electrode positions used in the choice identification studies are summarized in Table 20.1. Benjamin et al. [1] has shown that the α wave from positions F4 and F3 relates to the involuntary response of the customer to the items attractions. Vecchiato et al. [2] has shown that asymmetric increments of the α and θ waves are associated with pleasant advertising in the left brain region at positions F1, F7, AF7, FP1. Likewise same author [2] also showed that asymmetric increments of the α and θ waves are linked to the viewing of displeasing advertising observed in the right brain region at positions F2, F8, AF8 and FP2. The spectral power of the α waves increases significantly for pleasant advertising at position F1 and for the unpleasant advertising at positions AF4 and AF8. Whereas for θ waves, enhanced neuronal activity develops at positions F3, AF8, and F2, for unpleasant advertising, and at electrode Fp1 for pleasant advertising. Mohit et al. [3] related choice characteristics such as liking, emotions, attention, to position C4, C3, Cz, Fz, Pz, P3, F3. M.A.S Boskem et al. [4] have shown that the single choice is related to the activity of β in the (15–18) Hz span on the middle-frontal region on position FCz, FC1, F2, and AFz. Besides, they demonstrated that the population choice is related to the (60–105) Hz θ span on the front-central region of electrodes CP5, C5, C4, C3, C1, FC4, FC2, FCz, FC4, FC1, F4, F2, and F1.

20.2 EEG Signal Indices

Based on our review of previously reported studies, we have identified the following EEG features that have been reported for the examination of individual responses to marketing stimulus: (1) valence. (2) choice index, (3) effort index; and (4) the approach–withdrawal (AW) drive index and (5) Task Engagement Index. These indices help advertisers to understand customer reactions to products.

20.2.1 Valence

Frontal electrode asymmetry is related to choice. Right and left frontal stimulation are linked to negative and positive valence, respectively. Several reports have backed the hypothesis that frontal asymmetry in EEG signals is a valence indication.

20.2.2 Choice Index

It is defined with respect to asymmetries of frontal γ and β waves often linked to the actual stage of choice selection. It is defined in relation to the asymmetries of the frontal γ and β waves which are often linked to the choice. In addition, it is the most closely linked component of the willingness-to-pay action to determine the customer's choice, particularly using γ waves. Smaller magnitudes of γ and β waves are correlated with comparatively stronger right brain activation, and larger magnitudes show stronger left prefrontal activation. Michael et al. [6] individually measured the choice index for each wave (β and γ) based on the position of AF4 and AF3, as per Eq. (20.1):

$$Choice\ Index = \frac{\ln(AF3) - \ln(AF4)}{ln(AF3) + \ln(AF4)} \tag{20.1}$$

20.2.3 Effort Index

The effort index is described as the activities of the frontal θ in the prefrontal region. In the this region, increased θ power is linked to a higher degree of work complexity and difficulty. This index functions as an indicator of mental processing due to cognitive fatigue and has been extensively investigated in neuromarketing studies. This element demonstrates the significance of both negative and positive cognitive processing for the development of strong memory tracks during marketing.

20.2.4 AW Index

The approach–withdrawal drive indices, specifies drive, motivation, and approach-avoidance. The frontal asymmetry hypothesis describes that the right hemisphere of the brain is responsible for the withdrawal (negative) drive and the left hemisphere is responsible for (positive emotions. Its negative magnitude results in the prevention behaviour (negative drive) expressed by increased stimulation of the left frontal brain region. Positive AW magnitude, on the other hand, leads to approach behaviour (positive motivation) as expressed by the upper-right frontal cortex stimulation. Several scientists have shown the reliability and dependability of the frontal α asymmetry in neuromarketing studies.

$$AW = \frac{\alpha(\mathrm{F4}) - \alpha(F3)}{a(F4) + a(F3)} \tag{20.2}$$

20.2.5 Task Engagement Index

The Task Engagement Index tests how a person is cognitively engaged in a task. The task Engagement index could be used as a feature for the cognitive task classification. Variations in the demand for tasks can lead to varying degrees of mental Engagement that can be calculated using the task Engagement Index.

$$TEI = \frac{\beta \text{ power}}{a\ power + \theta \text{ power}}$$

20.3 Neuromarketing

Conventional marketing techniques include focus groups, questionnaires, interviews and surveys, in which customers consciously and freely document their views and experiences. However such conventional techniques cannot determine the unconscious part of the behaviour of the customer. EEG signal-based systems have the potential to detect unconscious drives that affect the act of building choices. It has been well documented that almost 91% of the information is being treated subconsciously in the brain. Studies involving consumer neuroscience and neuromarketing, the examination of subconscious actions reveals the real choices of customers relatively more correctly than conventional marketing techniques do. Mainly because subconscious views play a leading role in making customer choices. On the other hand, conventional market research techniques have been found to be lacking in the examination of subconscious actions in the customer's mind, which make the difference between the results of conventional marketing studies and the customer's behaviour at baying point. The word 'neuromarketing' is made up of two words, 'neuro' and 'marketing.' Combination of two broad fields of research: marketing and neuroscience. Neuroscience is a study of the mind from a psychological and biological point of view. In addition, neuroscience research has exceptionally strengthened marketing research, and integration between these fields helps to understand customer behaviour. The word neuromarketing has begun to appear around 2003. Initially, a few companies, such as SalesBrain and Brighthouse, began providing research findings and consultations on neuromarketing, promoting and encouraging neuroscience in the field of marketing. Over the last few years, a number of large corporations have started using the neuromarketing method to examine promotional materials before delivering products to customers. Such a method has been consistently supported by product managers in large companies, including Combell's and Coca-Cola.

Neuromarketing scientists strive to employ a neuroscientific process to understand customer behaviour (preferences, needs, requests) when customers purchase products. This element shows the basic interest of the scientist in assessing the mental and sensorimotor response of the customer to advertisements and products using a number of techniques. Numerous scientific experimental marketing techniques are available other than BCI, such as facial electromyography, facial coding, skin conductivity, galvanic skin output, eye-tracking and EEG. Over the last few years, EEG-based analysis signals have often been used in combination with brands and promotional stimulation. The neuroscience method for assessing customer behaviour has been progressively used by marketing professionals, in particular consumer brand managers, to estimate the memorization of a TV advertisement, thus offering a different advertising assessment tool. See Table 20.2 about a summary of advertising-or buyer research based on EEG signals. Vecchiato et al. [7] clarified the advantages of using the EEG assessment for commercial and political television advertising. The EEG signal analysis evaluates the interest generated by the advertising activity [8]. Vecchiato [9] on the other hand, has identified the influence of another cognitive mechanism in the neuromarketing area.

20.3.1 Ethical Considerations in Neuromarketing

The aim of neuromarketing is to provide evidence of how the brains of customers function in the presence of various market stimuli. The advantage of this measuring system is that the evidence obtained is free from customers' perceptions and their inability to disclose the facts. There is clearly no question of the use of these methods in consumer evaluation, but the great challenge is that researchers can observe beyond the ethical limits. That is why the use of neuromarketing causes ethical apprehensions that could be divided into two categories: the protection of the rights of customers and the protection of other parties that could be influenced or exploited by these experiments [30].

Table 20.2 List of advertising articles that used EEG signals

Author	Marketing substance and stimuli	Techniques	Details
Young [10]	TV advertisement	Waves	Engagement Index = β/(α þ θ)
Venkatraman, et al. [11]	TV ad	Waves	occipital α
Vecchiato et al. [2]	TV advertisement	Waves	θ and α (power spectral density)
Vecchiato et al. [12]	TV advertisement	Waves	θ and gamma
Ohme et al. [13]	TV advertisement	Waves	α (emotional response)
Ohme et al. [14]	TV advertisement	Waves	α band frontal asymmetry
Khushaba, Rami. et al. [14]		Waves	Product α, β, θ, gamma, delta
Daugherty,, et al. [15]	TV advertisement is dense-	Waves	3 epochs: 200–350, 350–500, 500–800
Cohen, Jill, et al. [16]	Print advertisement	Waves	delta and β
Boksem, Maarten et al. [4]	Movie trailers	Waves	β and gamma oscillations
Barnett et al. [17]	Movie trailer	Waves	α oscillation (cross-brain correlations)
Telpaz, Webb and [18]	Product	Waves/field potentials	θ/N200, FRN
Schaefer et al. [19]	Price	Field potentials	FRN (feedback related negativity); P300
Xiaoyi, Wang et al. [20]	Names of brands	Field potentials	N400
Jing and Han [11]	Products	Field potentials	P300
Pohzaharliev et al. [21]	Brands	Field potentials	P2, P3, LPP
Ma et al. [22]	Brands and images	Field potentials	N200, P300
Ma et al. [23]	Brands and product	Field potentials	P300
Ma et al. [24]	Brands	Field potentials	N270
Lin et al. [25]	Scent	Field potentials	LPP
Jones et al. [26]	Price	Field potentials	FN400, LPC, P200
Jin et al. [27]	Brands	Field potentials	N400
Guo et al. [28]	Online recommendations	Field potentials	N200, P300
Chen et al. [29]	Online book reviews	Field potentials	LPP (prolonged P300

Four key areas of neuromarketing ethics can be described in this field: the safety of test subjects, dominant niche audience exploitation, responsible company advertising and public representation [30]. The identity of persons responding to neuroscience trials held in medical institutions is generally protected by regulations. However these rules lose control as neuroscience is used outside medical research facilities, which basically means that the privacy of research objects is based on the researcher's ethical principles.

Misuse of niche populations is the second ethical challenge of neuromarketing. In this situation, children, older people, people with psychiatric disabilities, etc. are more sensitive audiences. Neuroscience research would be able to supply advertisers with vital knowledge of vulnerable customers, and would be able to manipulate these niche markets even more effectively than others. For example, in order to produce new games, video game companies are using children for such testing [31].

Responsible company advertising suggests that neuromarketing research firms must be particularly careful about advertising their services. The fact that human brains are the world's most complex biological organs, with 10 billion cells attached to each other, is scarcely remembered by representatives of major organisations. The method of understanding how brain functions is extremely complex and there is still a chance of making mistakes. However, color-coded brain photographs, complemented by neuroscientific explanations, often appear to be sufficiently conclusive to generate organizational neurotype. This adds to situations under which advertisers are generally keen to support studies into neuroscience without taking into account the relevance of their findings. And if neuroscience-based advertising does not meet requirements, businesses will incur damages.

Another legal issue of neuromarketing is civic representation. The lack of research on neuromarketing data induces distrust and gossip in society. Conversely, new, anonymous tools, along with fear and uncertainty, create more harmful results than constructive ones. And those stories are not repudiated and do not vanish with the unavailability of knowledge. A huge amount of criticism is received for the use of neuroscience in consumer studies. However, as this approach is comparatively new in the whole world of marketing science, very little has been done to study the topic or to establish a better understanding of it. It is also a part of marketing research and of the general marketing area, despite being an evolving field. That is why it is possible to extend prevailing ideas of marketing ethics to bring neuromarketing to the more ethical limelight.

Several theories have so far addressed ethical dilemmas from multiple perspectives and proposed alternatives to a wide range of ethical issues, helping advertisers to further understand ethical challenges and their conceivable remedies. It also helps corporate executives really understand the legal dilemmas and potential solutions. In addition, company choices involve multiple players in the real corporate world with different opinions on what is legal and what is not, through the prism of numerous ethical philosophies, the discussion of challenges makes it possible to consider multiple points of view and find a path to the right solution in individual situations [32].

20.4 EEG Based Choice Detection System

This section discusses the EEG-based recognition of choices, in particular cognitive correlations of choice, diagnostic characteristics of choice, and choice classification techniques. Choice can be described as an attitude towards a range of aspects that can be freely reflected in the judgement. This feature may also be an assessment of the manner in which the product is dis-liked or liked. The potential of computing unconscious and conscious neural activity in advertising, understanding the customer's processing of advertising materials, emotional state, and mental workload cannot be overlooked. The concept of a 'buy button' in the head may be overstated, but scientific attempts to use neural methods to track customer cognitive mechanisms are not trivial. Learning the neural mechanism behind decision-making, emotion and choice could increase the predictability of customer purchase choices and tests. Moreover, neuromarketing has the exact objective of assessing the implicit choices of customers.

Several reports have been published which show that EEG signals can be used to evaluate the customer's choice. In order to use EEG signals effectively in neuroscience studies, the mental processes that cause the choice of action should also b.e identified. This section discusses the neuromarketing experiment method for predicting customer choices and preferences. First we need to put the BCI headset on the head of the customer. Next, neuromarketing scientists are asking the customer to see the product. EEG signals are collected simultaneously during the recording process while customers are viewing the brand. Later seeing each brand, the customer is instructed to choose a specific grade scale of a certain number against the product. Specific grades must be manually labelled as like or dis-like groups when viewing all products. Next, pre-processing and feature extraction operations are performed on the EEG signal.

The classification system is formulated on the basis of the gold standard executed by the choice of customer grades. Figure 20.1 describes a generic BCI choice identification system consisting of three main steps: (1) pre-processing (2) selection/extraction of features and (3) classification process.

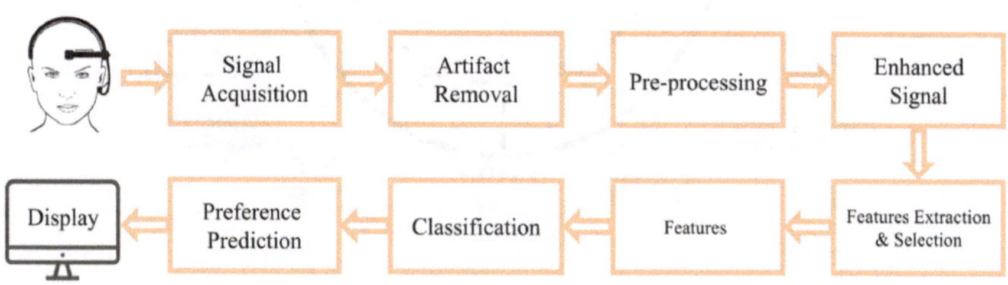

Fig. 20.1 EEG-based choice detection system

20.4.1 The Cognitive Choice Correlations

This section discusses cognitive factors related to choice. Different regions of the cortex are responsible for different mental and cognitive roles. In order to correctly specify the location of the EEG electrodes, it is essential to locate the underlying brain locations responsible for the treatment of choice. Research findings have shown that the choice is related to the front portion of the brain, typically the mid-front orbital region, and the pre-frontal cerebral cortex. Knutson et al. [3, 33] have linked the nucleus accumbens to the choice forecast. Whenever the customer sees the product, a greater stimulation of this area specifies the increased likelihood of the purchase of the item by the customer. In addition, Kirk et al. [34] have demonstrated a link between the medial orbitofrontal cortex and the contextual choice; greater stimulation in this location is associated with a higher degree of choice. EEG signals correlated with specific requirement may be obtained via associated brain region. Key electrode locations and related neural activities based on 10–20 standards are shown in Fig. 20.2. The choice feature is responsible for the medial-frontal cortex.

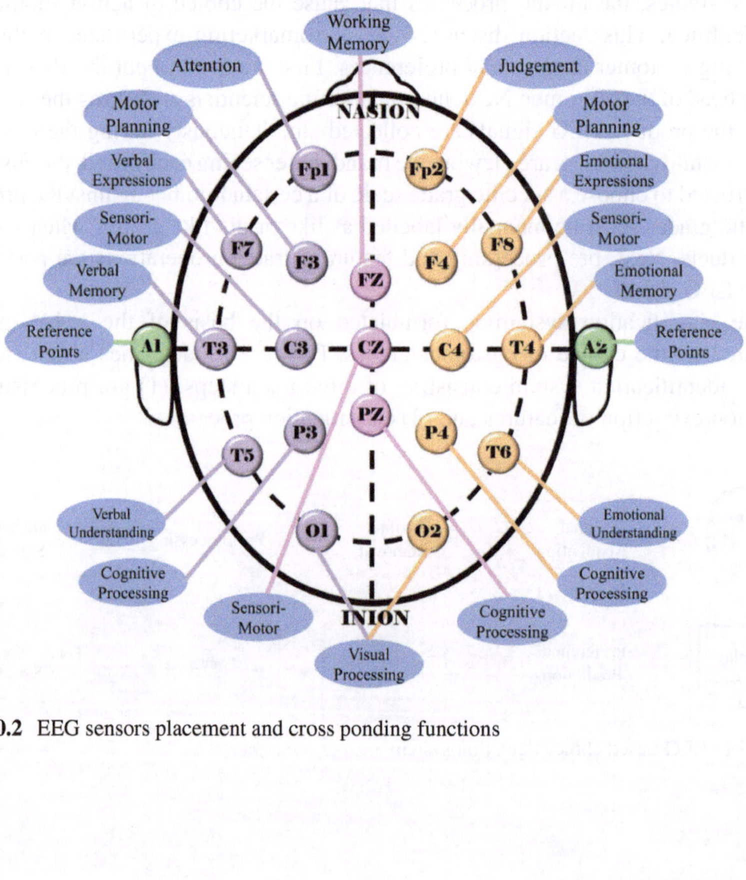

Fig. 20.2 EEG sensors placement and cross ponding functions

20.5 Choice Classification Techniques

While significant developments have been made in linking neural activities to the customer's choice, there is still little evidence that neural evaluation can be truly useful in predicting marketing success. Neural examinations will greatly improve predictive capability above and beyond conventional examinations. Mainly because neural examinations are superior predictors compared to self-reported assessments, the ability of neuromarketing techniques to predict choice in real scenarios has promising results for advertisers. The first study report was published in 2007, and it was found that pre-choice related stimulation could be used in the specific brain region to predict the resulting choices [35]. Since then, similar findings have been presented in other neuromarketing studies. Recently, the use of multivariate techniques, such as pattern classification for predicting preferences, has become a common standard procedure. For example, classification techniques may be used to predict preferences for various products. The resulting models, which are based on neuromarketing techniques, are more efficient in predicting new settings and states than conventional marketing techniques, such as questionnaire forms and focus groups. In addition, these techniques are more likely to be scalable, giving advertisers greater insight into customers and significant financial outcomes.

The system of detection choices based on statistical techniques could be divided into three typical types: time, frequency and their blend. Detection systems based on the time signal uses the recording of the ERPs; on the other hand, the frequency-based modelling is achieved by the recognition of the typical characteristics of the γ, β, α, θ, and δ waves, PSD and other methods of analysis. In addition various frequency-based attribute selection techniques could also be used; that is, spectral filters and conventional spatial patterns have been used in the song sound choice classification based on SVM techniques, with an accuracy of 68% and 75%, respectively [36]. In another study, Fast Fourier Transform (FFT) was used as a feature selection technique based on SVM, which recorded 82% accuracy. Similarly, in another report, the scientist employed the FFT based on Radial SVM for the choice classification and recorded 75% accuracy [37]. The final choice model combines time and frequency to analyse brain signals by measuring time intervals and power spectrum for post-stimulus duration. Numerous conventional data-mining techniques were used to identify choices, and the use of a specific time frequency method has been proposed to track customer music choices. KNN-based techniques have resulted in 83 and 86% accuracy with specific TF techniques, namely spectrogram and Hilbert–Huang Spectrum (HHS) respectively [38]. In their more extended work, Hadjidimitriou et al. [39] achieved a significantly higher accuracy of 91% on the basis of a similar experimental design. A similar study for the classification of musical choice based on the TF technique, namely DFT-KNN, obtained a 98% accurate value. Similarly, a similar accuracy of 97% was obtained on the basis of a quadratic discriminant analysis (QDA) for music choice classification.

The majority of scientists used variations in their prediction models from traditional regression techniques. However, several methods and techniques have been proposed for the processing of EEG based on classification techniques for the identification of customer choice. Table 20.3 provides a comprehensive review of

Table 20.3 Various techniques for the evolution of choice

Authors	Technique	Device/system	
Agarwal et al. [3]	Linear regression (Lasso, Ridge)	Muse InteraXon	Classification/regression
Yadava [44]	Hidden Markov model	14 Channel Emotiv EPOC Inc. Headset	
Teo et al. [41]	Random forest	9 electrode advanced brain monitoring B-Alert X10	
Chew et al. [40]	Deep neural network	Advanced brain monitoring (ABM), the B-alert X10 headset	
Hadjidimitriou et al. [38]	k-nearest neighbor	14 Channel Emotiv EPOC Inc. Headset	
Djamal et al. [45]	Support vector machine	Not mentioned	
Telpaz et al. [18]	Univariate linear regression analysis	ActiveTwo Biosemi system	
Jones et al. [26]	Correlation analysis	Neuroscan, Inc.	Feature selection extraction
Ohme et al. [14]	Frequency domain	16-channel contact precision instruments/neurobehavioral systems	
Khushaba et al. [46]	Wavelet domain	14 Channel Emotiv EPOC Inc. Headset	
Hadjidimitriou et al. [38]	Time–frequency analysis	14 Channel Emotiv EPOC Inc. Headset	
Boksem et al. [4]	Statistical features	64 active Ag–AgCl electrodes Biosemi ActiveTwo	
Telpaz et al. [18]	Average	ActiveTwo Biosemi system	Preprocessing
Hakim et al. [47]	Downsampling	8-electrode EEG system Neuroelectrics	
Hakim et al. [48]	Band-pass filters	8-electrode EEG system Neuroelectrics	
Yadava et al. [49]	Low-pass filter	14 Channel Emotiv EPOC Inc. Headset	

some neuromarketing studies and a comparison of computational techniques. A few choices reported studies are based on two classification techniques to find efficient classifiers for a specific set of features. Chew et al. [40] used EEG signals to determine the customer's choices by using virtual three-dimensional structures. The neuromarketing scientist used EEG wave characteristics to classify the EEG signal into two types that were liked and dis-liked product on the basis of SVM and KNN and obtained 75% and 80% classification accuracy, respectively. However, these findings cannot be considered to be credible, since the authors used a relatively small dataset of five participants. However the authors increased the number of participants to 16 in their extended study [41, 42], but improved accuracy was not achieved. By integrating the EEG signal analysis with the questionnaire tests, another research report [43] obtained 68% value of accuracy based on SVM to determine the most and least preferred items.

Combinations of classifiers, such as stacking, voting or boosting, may be used to form different types of classifiers, which involve training to complement each other and joining their outputs to improve their accuracy. The selection of classification techniques in a neuromarketing study is often dependent on both the neural activity measured and the application where it is intended to be used. However, SVM and LDA are the most commonly used classification techniques, and approximately more than half of the EEG-based neuromarketing studies are based on these techniques. The study of classification techniques was divided into four types: transfer learning, matrix and tensor classifiers, adaptive classifiers, and deep learning. The transfer learning technique seeks to enhance the performance of a learning algorithm by using the knowledge received when learning from another dataset or task. Matrix and tensor classifiers are also known as multi-way arrays, map data with appropriate steps directly to a specific space, and do not involve feature selections and filters. In adaptive classifiers, their factors, including weights of features, are constantly reviewed and updated on the basis of recent neuronal activities.

Loosely inspired by human nervous systems, DNNs are powerful algorithms. Only basic calculations can be made by nodes in a common neural-network. However, DNNs work by linking the node to create multiple layers, where almost every calculation is applied by evaluating the performance of the previous layer. In this way, the DNN could even achieve results that are far more complicated than what a single node could do. The 'deep' in the DNNs is a reference to the fact that the number of layers can be very large. MLP is the most common form of DNN used for BCI applications. Other forms of DNN have been less commonly studied, such as neural networks of learning vector quantification or neural networks of Gaussian Classifier [42]. Teo et al. [42, 43] suggested deep learning techniques to identify the choice through the use of rotating objects in 3D. The findings suggest that the use of deep networks can be more effective than other ML techniques, including KNN, SVM and RF. Teo et al. [50] enhanced the accuracy of the results based on deep network configuration plus dropout and 80% stimulation tanh.

20.6 Forecasting Choice Features

This section of the book chapter discusses research findings involving features of prediction techniques that could help advertisers predict customer choices, as shown in Table 20.4. Several studies have been conducted that have used traditional regression differentiation analysis in their forecasting models. The authors categorized these predictive features according to the type of EEG signals: (1) transient behaviour and (2) Brainwaves as features.

20.6.1 Transient Behaviors as Features

Numerous research reports have investigated the ERP elements P300, N200, and N400 in the choice-related neurological functions, each of which can be explained as follows.

20.6.2 P300

Recently, customers' choices for expanded brand labels have been identified using the higher amplitude value of P300. For specific product features, P300 was used by Wang and Han [11] as a criterion for the choice of the customer.

Other scientists [11, 26] considered factors that had an impact on purchasing decisions. The authors [26, 61] used the active BCI to evaluate the roles of discount promotions, pricing, gender and mathematical ability in customer purchase decisions and strongly linked them to ERP components such as P300 and P200. In addition, in order to recognise the choice, the significance of the different products was also assessed on the basis of θ, β, and α waves of EEG signals. A combination of EEG signals and eye-tracking was also used by the neuromarketing scientist to select the chosen product. Michael et al. [6] looked at the emotional reactions of tourism choices based on video, photos, and words as stimulus. The authors found that the photos showed higher affective reactions than those of words formed by an unconscious bias in travel decision-making. The authors [44] built a prediction model based on the EEG signal dataset for customer product preferences, and further analysed the roles of age and gender in the customer choice process with respect to like and dislike. One more experimental report included an inductive analytical technique to examine three preferred and three different advertisements based on a large EEG dataset. The findings indicate that there were statistically significant differences in ERP between the advertisement that was liked and the advertisement that was disliked [15].

Table 20.4 EEG-based neuromarketing studies

Authors	Method of recording	Type of signal	EEG/ERP component	Marketing element and stimuli	Device/system
Michael et al. [51]	BCI & Galvanic Skin Response	EEG	γ, β, α, θ, & δ	Observing tourism pictures, videos & words	ABM X-10 EEG brain scanner, /iMotions v5.1
Agarwal et al. [52]	BCI	ERP, ERSP	θ, β & N200	Observing product pictures	Muse InteraXon
Modica et al. [53]	BCI	EEG	α & θ	Observing & touching products	21-channel portable BEmicro, EBneuro
Caratù, Myriam et al. [54]	BCI	EEG	α	Tasting drinks	BrainVision LiveAmp Brain Products GmbH
Ramsøy et al. [55]	BCI	EEG	θ	Observing products & prices	14 Channel Emotiv EPOC Inc Headset
Hakim et al. [48]	BCI	EEG	δ, θ, α, β, & γ	Observing advertisement of food products	8-electrode EEG system Neuroelectrics
Touchette et al. [1]	BCI, Galvanic Skin Response Heart Rate	EEG	α	Observing product pictures	NeXus-10 biofeedback system manufactured by Mind Media Inc
Yadava et al. [49]	BCI	EEG	δ, θ, α, β, & γ	Observing product pictures	14 Channel Emotiv EPOC Inc Headset
Cartocci, Giulia et al. [56]	BCI	EEG	θ & α	Viewing advertisement videos	BrainVision Engineering, Neuromarketing Headset
Goto, Nobuhiko et al. [57]	BCI	ERP	N200, LPP, & PSW	Observing product pictures	ASALAB amplifier ANT Neuro
Bosshard et al. [58]	BCI	ERP	LPP	Observing brand names	BioSemi ActiveTwo EEG system
Terry Daugherty et al. [15]	BCI	EEG	Three epochs: 200–350, 350–500, & 500–800	Viewing television advertisement	256-channel Hydrocel Geodesic Sensor Net Electrical Geodesics Inc

(continued)

Table 20.4 (continued)

Authors	Method of recording	Type of signal	EEG/ERP component	Marketing element and stimuli	Device/system
Boksem et al. [4]	BCI, GSR Heart Rate	EEG	β & γ oscillations	Viewing advertisement (movie trailers)	64 active Ag–AgCl electrodes Biosemi ActiveTwo
Telpaz et al. [18]	BCI	ERSP, ERP	θ, N200, & FRN	Observing products	ActiveTwo Biosemi system
Giovanni Vecchiato et al. [59]	BCI	EEG	θ & α	Viewing television advertisement	KT88-2400 device (Contec Medical Systems Co
Wang et al. [11]	BCI	ERP	P300	Observing products	64-channel HydroCel Geodesic Sensor Net
Khushaba et al. [46]	BCI, Eye Tracking	EEG	α, β, θ, γ, & δ	Observing products	14 Channel Emotiv EPOC Inc Headset
Jones et al. [26]	BCI	ERP	FN400, LPC, & P200	Observing products & prices	Neuroscan, Inc
Wang et al. [20]	BCI, EOG	ERP	N400	Observing brand names	Neuroscan synamp Neurosoft Labs, Inc
Vecchiato et al. [5]	BCI	EEG	Asymmetrical increase in θ & α in PSD	Viewing television advertisement	BrainAmp, Brainproducts GmbH
Vecchiato et al. [60]	BCI, Galvanic Skin Response, Heart Rate	EEG	θ & γ	Viewing television advertisement	BE and Galileo EBneuro
Ohme et al. [14]	BCI	EEG	A band frontal asymmetry	Viewing television advertisement	16-channel Contact Precision Instruments/ neurobehavioral systems
Ma, Qingguo et al. [22]	BCI	ERP	N200 & P300	Observing brands & pictures	Neuroscan Synamp2 Neurosoft Labs, Inc
Ohme et al. [13]	BCI, Galvanic Skin Response, EMG	EEG	α	Viewing television advertisement	16-channel Contact Precision Instruments/ neurobehavioral systems

(continued)

Table 20.4 (continued)

Authors	Method of recording	Type of signal	EEG/ERP component	Marketing element and stimuli	Device/system
Ma, Qingguo et al. [23]	BCI	ERP	P300	Observing brands & products	Neuroscan Synamp2 Neurosoft Labs, Inc

20.6.3 N200

Ma et al. [22] found that the magnitude of the N200 revealed a correlation between the categories of brand extension and emotional state. However this correlation only existed with a slight extension of the brand and with negative emotions. Goto et al. indicated that the N200 value can show choices for the product as calculated by the spontaneous process, while PSW and LPP can show choices for the product as calculated by conscious cognitive processes [57]. The Cerebro integrated system ERSP, N200 minima and N200 mean that products are graded on the basis of consumer choice [3]. In the same way, Telpaz et al. [18] combined ERP and ERSP techniques to predict drug choices by analysing FNR, N200, and delta waves.

20.6.4 N400

Several scientists have shown that the N400 value can predict the extension of the brand [62]. Strong correlations with brand titles and amplified N400 values have been observed, predicting increased customer preference. Similarly, another brand extension research work has shown that the value of N400 is correlated with an unconscious response to the band and product classification, but not to conscious evaluations [20].

20.6.5 Brainwaves as Features

The γ and β waves from customers who viewed the film clips were used to estimate the recall and box office sales [4, 63]. These elements were also used as an indicator of the desire to purchase in the determination of customer choice and preference [64] as well as used to calculate cognitive willingness and predict ticket purchases [4].

In addition, frontal asymmetry has been associated with customers' unconscious reactions to the product's desirability. Likewise, Enrica M et al. [65] associated higher-frequency bands with desirable home meals as well as food from foreign nations. In addition, initiatives that have won awards in the public sector anti-smoking message have been linked to higher α frequencies. Relatively low θ frequency bands have been associated with poor product selection outcomes. In addition, these frequencies were associated with effective anti-smoking social care advertisements and imported brands compared to local brands. Table 20.5 shows some neuromarketing findings using various classification techniques to achieve the most reliable performance in predicting customer choice. In order to improve the accuracy of the forecast results, the study emphasises the need to add more functionality to the classification process.

20.7 Conclusions and Future Recommendations

The last decade has provided a solid foundation for the emergence of neuromarketing and consumer neuroscience as a scientific discipline for research and study. There are several interdisciplinary groups across the globe that pursue this scientific field and are constantly working on more challenging marketing issues. In this book chapter previous reports on the use of EEG signals based choice classification have been reviewed to analyse customer issues in advertising and collaborative areas and show their benefits over conventional approaches in the study of consumer-relevant behaviours. This work is one of the first to study previously published EEG literature and to provide a methodological context for marketing researchers. In order to understand the EEG-based choice classification, we started with an outline of the widely studied emotional and cognitive mechanisms in neuromarketing, with related ERP and EEG wave aspects. Considerations, technical information and ethical considerations in neuromarketing were also discussed. The key objective of this chapter of the book is to close void between conventional market analysis focused on explicit customer response and a neuromarketing analysis focused on implicit consumer response. The EEG based detection of choice in marketing research has been thoroughly investigated. One More void in marketing studies is the absence of comprehensive data-mining techniques for estimating and classifying customer choice. In order to find customer choice based on EEG data using power spectral density and valence characteristics, etc., deep-learning approaches need to be implemented for effective marketing. We concluded that the KNN + Fourier transform algorithm demonstrated better performance, recall and accuracy, particularly in comparison to the KNN, SVM and RF classifiers. Suggestions for future research could well involve the assessment of DNNs across the transfer learning framework for choice prediction.

Table 20.5 Different classifier used for the detection of choice

Authors	Class	Waves	Location	Classification algorithm	Accuracy (%)	
Telpaz et al. [18]	Liked/disliked	δ, θ, α, β & γ	AF3, F7, F3, FC5, T7, P7, O1, O2, P8, T8, FC6, F4, F8, & AF4	Hidden Markov Model	70.33	ActiveTwo Biosemi system
				Random Forest	68.41	
				Support Vector Machine	62.85	
Hakim et al. [48]	Most favored/least favored	δ, θ, α, β & γ	F7, Fp1, Fpz, Fp2, F8, Fz, Cz, & Pz	Decision trees with Adaboost M1	63.34	8-electrode EEG system Neuroelectrics
				k-Nearest Neighbor	59.98	
				Logistic Regression	67.32	
Abdul Kader et al. [66]	Liked/disliked	δ, θ, α, β & γ	POz, Fz, Cz, C3, C4, F3, F4, P3, & P4	DNN with dropout & tanh	74.38	9 Electrode ABM B-Alert X10 Headset
				Support Vector Machine	60.19	
				Deep Neural Network	63.99	
Moon et al. [39]	Most preferred Preferred Less preferred Least preferred	θ, α, β, γ	AF3, AF4, F3, F4, F8, F7, T8, T7, P7, P8, O1, & O2	k-Nearest Neighbor + Fourier transform	97.99	14 Channel Emotiv EPOC Inc. Headset
		θ, α, β & γ		quadratic discriminant analysis + Fourier transform	97.39	
Yaozhang et al. [36]	Liked/disliked	δ, θ, α, β & γ	FP1 & FP2	Support Vector Machine + CSP	74.77	2 EEG electrodes NeuroSky-MindBand

(continued)

Table 20.5 (continued)

Authors	Class	Waves	Location	Classification algorithm	Accuracy (%)	
				Support Vector Machine + Spectral filter	68.22	
Hadjidimitriou et al. [38, 67]	Liked/Disliked	β & γ	AF4, AF3, F8, F7, F4, F3, FC5, T8,T7, P8, P7, O2, O1, & FC6	k-Nearest Neighbor + time–frequency analysis	91.02	14 Channel Emotiv EPOC Inc. Headset
		δ, θ, α, β & γ			86.5	
Ma, Qingguo et al. [22]	Preferred image/unnoticed image	α	F8 & T8	Support vector machine	83.64	Neuroscan Synamp2 Neurosoft Labs, Inc.
		α, β, θ, & γ	AF4, AF3, F7, F3, FC5, T8, T7, P8, P7, O2, O1, FC6, F8, & F4,		75.44	
Chew et al. [40]	Liked/disliked	α, θ & δ	Fz, F3, F4	KNN	80	advanced brain monitoring (ABM), the B-alert X10 headset
		α θ	**F3 & F4**	**Support vector machine**	**75**	

References

1. Touchette, B., & Lee, S.-E. (2017). Measuring neural responses to apparel product attractiveness: An application of frontal asymmetry theory. *Clothing and Textiles Research Journal, 35*(1), 3–15.
2. Vecchiato, G., Toppi, J., Astolfi, L., Fallani, F. D. V., Cincotti, F., Mattia, D., Bez, F., & Babiloni, F. (2011). Spectral EEG frontal asymmetries correlate with the experienced pleasantness of TV commercial advertisements. *Medical & Biological Engineering & Computing, 49*(5), 579–583.
3. Agarwal, M., & Sivakumar, R. (2019). Cerebro: A wearable solution to detect and track user preferences using brainwaves. In *The 5th ACM Workshop on Wearable Systems and Applications* (pp. 47–52).
4. Boksem, M. A., & Smidts, A. (2015). Brain responses to movie trailers predict individual preferences for movies and their population-wide commercial success. *Journal of Marketing Research, 52*(4), 482–492.
5. Vecchiato, G., Toppi, J., Astolfi, L., De Vico Fallani, F., Cincotti, F., Mattia, D., Bez, F., & Babiloni, F. (2011). Spectral EEG frontal asymmetries correlate with the experienced pleasantness of TV commercial advertisements. *Medical & Biological Engineering & Computing, 49*(5), 579–583.
6. Michael, I., Ramsoy, T., Stephens, M., & Kotsi, F. (2019). A study of unconscious emotional and cognitive responses to tourism images using a neuroscience method. *Journal of Islamic Marketing.*
7. Vecchiato, G., Astolfi, L., Fallani, F. D. V., Salinari, S., Cincotti, F., Aloise, F., Mattia, D., Marciani, M. G., Bianchi, L., & Soranzo, R. (2009). The study of brain activity during the observation of commercial advertising by using high resolution EEG techniques. In *2009 Annual International Conference of the IEEE Engineering in Medicine and Biology Society* (pp. 57–60). IEEE.
8. Yoshioka, M., Inoue, T., & Ozawa, J. (2012). Brain signal pattern of engrossed subjects using near infrared spectroscopy (NIRS) and its application to TV commercial evaluation In *The 2012 International Joint Conference on Neural Networks (IJCNN)* (pp. 1–6), IEEE.
9. Vecchiato, G., Babiloni, F., Astolfi, L., Toppi, J., Cherubino, P., Dai, J., Kong, W., Wei, D. (2011). Enhance of theta EEG spectral activity related to the memorization of commercial advertisings in Chinese and Italian subjects. In *2011 4th International Conference on Biomedical Engineering and Informatics (BMEI)* (pp. 1491–1494). IEEE.
10. Young, C. (2002). Brain waves, picture sorts®, and branding moments. *Journal of Advertising Research, 42*(4), 42–53.
11. Wang, J., & Han, W. (2014). The impact of perceived quality on online buying decisions: An event-related potentials perspective. *NeuroReport, 25*(14), 1091–1098.
12. Vecchiato, G., Astolfi, L., Fallani, F. D. V., Cincotti, F., Mattia, D., Salinari, S., Soranzo, R., & Babiloni, F. (2010). Changes in brain activity during the observation of TV commercials by using EEG, GSR and HR measurements. *Brain Topography, 23*(2), 165–179.
13. Ohme, R., Reykowska, D., Wiener, D., & Choromanska, A. (2009). Analysis of neurophysiological reactions to advertising stimuli by means of EEG and galvanic skin response measures. *Journal of Neuroscience, Psychology, and Economics, 2*(1), 21.
14. Ohme, R., Reykowska, D., Wiener, D., & Choromanska, A. (2010). Application of frontal EEG asymmetry to advertising research. *Journal of Economic Psychology, 31*(5), 785–793.
15. Daugherty, T., Hoffman, E., & Kennedy, K. (2016). Research in reverse: Ad testing using an inductive consumer neuroscience approach. *Journal of Business Research, 69*(8), 3168–3176.
16. Cohen, J., & Polich, J. (1997). On the number of trials needed for P300. *International Journal of Psychophysiology, 25*(3), 249–255.
17. Barnett, S. B., & Cerf, M. (2017). A ticket for your thoughts: Method for predicting content recall and sales using neural similarity of moviegoers. *Journal of Consumer Research, 44*(1), 160–181.
18. Telpaz, A., Webb, R., & Levy, D. J. (2015). Using EEG to predict consumers' future choices. *Journal of Marketing Research, 52*(4), 511–529.

19. Schaefer, A., Buratto, L. G., Goto, N., & Brotherhood, E. V. (2016). The feedback-related negativity and the P300 brain potential are sensitive to price expectation violations in a virtual shopping task. *PloS One, 11*(9), e0163150.
20. Wang, X., Ma, Q., & Wang, C. (2012). N400 as an index of uncontrolled categorization processing in brand extension. *Neuroscience Letters, 525*(1), 76–81.
21. Pozharliev, R., Verbeke, W. J., Van Strien, J. W., & Bagozzi, R. P. (2015). Merely being with you increases my attention to luxury products: Using EEG to understand consumers' emotional experience with luxury branded products. *Journal of Marketing Research, 52*(4), 546–558.
22. Ma, Q., Wang, K., Wang, X., Wang, C., & Wang, L. (2010). The influence of negative emotion on brand extension as reflected by the change of N2: A preliminary study. *Neuroscience Letters, 485*(3), 237–240.
23. Ma, Q., Wang, X., Shu, L., & Dai, S. (2008). P300 and categorization in brand extension. *Neuroscience Letters, 431*(1), 57–61.
24. Ma, Q., Wang, X., Dai, S., & Shu, L. (2007). Event-related potential N270 correlates of brand extension. *NeuroReport, 18*(10), 1031–1034.
25. Lin, M.-H. J., Cross, S. N., Childers, T. L. (2018). Understanding olfaction and emotions and the moderating role of individual differences. *European Journal of Marketing*.
26. Jones, W. J., Childers, T. L., & Jiang, Y. (2012). The shopping brain: Math anxiety modulates brain responses to buying decisions. *Biological Psychology, 89*(1), 201–213.
27. Jin, J., Wang, C., Yu, L., & Ma, Q. (2015). Extending or creating a new brand: Evidence from a study on event-related potentials. *NeuroReport, 26*(10), 572–577.
28. Guo, F., Zhang, X., Ding, Y., & Wang, X. (2016). Recommendation influence: Differential neural responses of consumers during shopping online. *Journal of Neuroscience, Psychology, and Economics, 9*(1), 29.
29. Chen, M., Ma, Q., Li, M., Dai, S., Wang, X., & Shu, L. (2010). The neural and psychological basis of herding in purchasing books online: An event-related potential study. *Cyberpsychology, Behavior, and Social Networking, 13*(3), 321–328.
30. Murphy, E. R., Illes, J., & Reiner, P. B. (2008). Neuroethics of neuromarketing. *Journal of Consumer Behaviour: An International Research Review, 7*(4–5), 293–302.
31. Acuff, D. (2005). Taking the guesswork out of responsible marketing. *Young Consumers: Insight and Ideas for Responsible Marketers, 6*(4), 68–71.
32. Hahn, R. (2009). The ethical rational of business for the poor–integrating the concepts bottom of the pyramid, sustainable development, and corporate citizenship. *Journal of Business Ethics, 84*(3), 313–324.
33. Knutson, B., Rick, S., Wimmer, G. E., Prelec, D., & Loewenstein, G. (2007). Neural predictors of purchases. *Neuron, 53*(1), 147–156.
34. Kirk, U., Skov, M., Hulme, O., Christensen, M. S., & Zeki, S. (2009). Modulation of aesthetic value by semantic context: An fMRI study. *NeuroImage, 44*(3), 1125–1132.
35. Plassmann, H., Venkatraman, V., Huettel, S., & Yoon, C. (2015). Consumer neuroscience: Applications, challenges, and possible solutions. *Journal of Marketing Research, 52*(4), 427–435.
36. Pan, Y., Guan, C., Yu, J., Ang, K. K., Chan, T. E. (2013). Common frequency pattern for music preference identification using frontal EEG. In *2013 6th International IEEE/Embs Conference on Neural Engineering (Ner)* (pp 505–508). IEEE.
37. Kim, Y., Kang, K., Lee, H. & Bae, C. (2015) Preference measurement using user response electroencephalogram. In *Computer Science and Its Applications* (pp. 1315–1324). Springer.
38. Hadjidimitriou, S. K., & Hadjileontiadis, L. J. (2012). Toward an EEG-based recognition of music liking using time-frequency analysis. *IEEE Transactions on Biomedical Engineering, 59*(12), 3498–3510.
39. Moon, J., Kim, Y., Lee, H., Bae, C., & Yoon, W. C. (2013). Extraction of user preference for video stimuli using EEG-based user responses. *ETRI Journal, 35*(6), 1105–1114.
40. Chew, L. H., Teo, J., & Mountstephens, J. (2016). Aesthetic preference recognition of 3D shapes using EEG. *Cognitive Neurodynamics, 10*(2), 165–173.

41. Teo, J., Hou, C. L., & Mountstephens, J. (2017). Deep learning for EEG-based preference classification. In *AIP Conference Proceedings* (p. 020141). AIP Publishing LLC.
42. Teo, J., Hou, C. L., & Mountstephens, J. (2018). Preference classification using electroencephalography (EEG) and deep learning. *Journal of Telecommunication, Electronic and Computer Engineering (JTEC), 10*(1–11), 87–91.
43. Hakim, A., Klorfeld, S., Sela, T., Friedman, D., Shabat-Simon, M., & Levy, D. J. (2018). Pathways to consumers minds: Using machine learning and multiple EEG metrics to increase preference prediction above and beyond traditional measurements. *bioRxiv, 2018*, 317073.
44. Yadava, M., Kumar, P., Saini, R., Roy, P. P., & Dogra, D. P. (2017). Analysis of EEG signals and its application to neuromarketing. *Multimedia Tools and Applications, 76*(18), 19087–19111.
45. Djamal, E. C., Indrawan, R., Pratama, J., Renaldi, F. (2017). EEG based neuropsychology of advertising video using fast fourier transform and support vector machine. *Journal of Telecommunication, Electronic and Computer Engineering (JTEC), 9*(3–7), 105–109.
46. Khushaba, R. N., Wise, C., Kodagoda, S., Louviere, J., Kahn, B. E., & Townsend, C. (2013). Consumer neuroscience: Assessing the brain response to marketing stimuli using electroencephalogram (EEG) and eye tracking. *Expert Systems with Applications, 40*(9), 3803–3812.
47. Hakim, A., Klorfeld, S., Sela, T., Friedman, D., Shabat-Simon, M., & Levy, D. J. (2018). Pathways to consumers' minds: Using machine learning and multiple EEG metrics to increase preference prediction above and beyond traditional measurements. *bioRxiv, 2018*, 317073.
48. Pathways to consumers' minds: Using machine learning and multiple EEG metrics to increase preference prediction above and beyond traditional measurements.
49. Yadava, M., Kumar, P., Saini, R., Roy, P. P., & Prosad Dogra, D. (2017). Analysis of EEG signals and its application to neuromarketing. *Multimedia Tools and Applications, 76*(18), 19087–19111.
50. Teo, J., Chew, L. H., Chia, J. T., & Mountstephens, J. (2018). Classification of affective states via EEG and deep learning. *International Journal of Advanced Computer Science and Applications, 9*(5), 132–142.
51. Michael, I., Ramsoy, T., Stephens, M., & Kotsi, F. (2019). A study of unconscious emotional and cognitive responses to tourism images using a neuroscience method. *Journal of Islamic Marketing, 10*(2), 543–564.
52. Agarwal, M., & Sivakumar, R. (2019). Cerebro: A wearable solution to detect and track user preferences using brainwaves. In *The 5th ACM Workshop on Wearable Systems and Applications* (pp. 47–52). Seoul, Republic of Korea: Association for Computing Machinery.
53. Modica, E., Cartocci, G., Rossi, D., Martinez Levy, A. C., Cherubino, P., Maglione, A. G., Di Flumeri, G., Mancini, M., Montanari, M., Perrotta, D., Di Feo, P., Vozzi, A., Ronca, V., Aricò, P., & Babiloni, F. (2018). Neurophysiological responses to different product experiences. *Computational Intelligence and Neuroscience, 2018*, 9616301.
54. Caratù, M., Cherubino, P., & Mattiacci, A. (2018) Application of neuro-marketing techniques to the wine tasting experience. In: Proceedings of the 11th annual conference of the euromed academy of business 2018 Sep.
55. Ramsøy, T. Z., Skov, M., Christensen, M. K., & Stahlhut, C. (2018). Frontal brain asymmetry and willingness to pay. *Frontiers in Neuroscience, 12*(138).
56. Au-Cartocci, G., Au-Caratù, M., Au-Modica, E., Au-Maglione, A. G., Au-Rossi, D., Au-Cherubino, P., Au-Babiloni, F. (2017). Electroencephalographic, heart rate, and galvanic skin response assessment for an advertising perception study: Application to antismoking public service announcements. *JoVE*, 2017(126), e55872.
57. Goto, N., Mushtaq, F., Shee, D., Lim, X. L., Mortazavi, M., Watabe, M., & Schaefer, A. (2017). Neural signals of selective attention are modulated by subjective preferences and buying decisions in a virtual shopping task. *Biological Psychology, 128*, 11–20.
58. Bosshard, S. S., Bourke, J. D., Kunaharan, S., Koller, M., & Walla, P. (2016). Established liked versus disliked brands: Brain activity, implicit associations and explicit responses. *Cogent Psychology, 3*(1), 1176691.

59. Neurophysiological tools to investigate consumer's gender differences during the observation of TV commercials.
60. Changes in brain activity during the observation of TV commercials by using EEG, GSR and HR measurements.
61. Ramsøy, T. Z., Friis-Olivarius, M., Jacobsen, C., Jensen, S. B., & Skov, M. (2012). Effects of perceptual uncertainty on arousal and preference across different visual domains. *Journal of Neuroscience, Psychology, and Economics, 5*(4), 212.
62. Lin, M.-H. J., Cross, S. N., Jones, W. J., Childers, T. L. (2018). Applying EEG in consumer neuroscience. *European Journal of Marketing, 2018.*
63. Boksem, M., Chan, H.-Y., Schoots, V., Sanfey, A., & Smidts, A. (2016). Emotional responses to movie-trailers predict individual preferences for movies and their population-wide commercial success. *ACR North American Advances, 2016.*
64. Ramsøy, T. Z., Skov, M., Christensen, M. K., & Stahlhut, C. (2018). Frontal brain asymmetry and willingness to pay. *Frontiers in Neuroscience, 12*, 138.
65. Modica, E., Cartocci, G., Rossi, D., Martinez Levy, A. C., Cherubino, P., Maglione, A. G., Di Flumeri, G., Mancini, M., Montanari, M., Perrotta, D. (2018). Neurophysiological responses to different product experiences. *Computational Intelligence and Neuroscience, 2018.*
66. Classification of affective states via EEG and deep learning.
67. Hadjidimitriou, S. K., & Hadjileontiadis, L. J. (2013). EEG-based classification of music appraisal responses using time-frequency analysis and familiarity ratings. *IEEE Transactions on Affective Computing, 4*(2), 161–172.

Chapter 21
A Study on Artificial Intelligence for Economic Renaissance in India

Meda Srinivasa Rao, Venkateswararao Podile, Durgaprasad Navvula, and Bharath Samishetti

Abstract India emerged as an IT/ITEs hub in the world. With the same spirit, we would like to put all our efforts into becoming artificial Intelligence technologies destination to the world. It requires provided proper initiatives to all stakeholders. Artificial intelligence is buzzword worldwide; it replaces the human element with machines and performs similar to human beings. As stated by former President A. P. J. Abdul Kalam, India has been turned into five areas: cultivation and food manufacturing; education and health care; information and communication technology; infrastructure growth and self-reliance on vital innovations. Technology is the critical driver of progress in any country; recently, Artificial Intelligence has been the buzzword worldwide and spurred heated debate globally. AI Products are slowly penetrating in every corner of human life with the advent of the Internet of Things, Cyber Security, Blockchain and others. In the above environment, the present conceptual paper highlighted the issues like AI in the world at a glance, AI potentiality to Progress India, AI Adoption in India, AI and its impact on select industries/domains in India, Barriers to Adopt AI technologies in India, Challenges for Adoption of AI technologies in India, Way out to overcome Challenges and finally Conclusion.

Keywords Artificial Intelligence · Machine learning Language · Adoption · NITI Aayog · Cyber security

M. S. Rao · V. Podile · D. Navvula
KL Business School, Koneru Lakshmaiah Education, Foundation, Vaddeswaram, Andhra Pradesh 522502, India
e-mail: msrinivasarao@kluniversity.in

V. Podile
e-mail: vraopodile@kluniversity.in

D. Navvula
e-mail: ndurgaprasad@kluniversity.in

B. Samishetti (✉)
Koneru Lakshmaiah Education Foundation, Vaddeswaram, Vijayawada, India

Sree Chaitanya College of Engineering, Karimnagar, Telangana, India

P. Kumar et al. (eds.), *A Fusion of Artificial Intelligence and Internet of Things for Emerging Cyber Systems*, Intelligent Systems Reference Library 210,
https://doi.org/10.1007/978-3-030-76653-5_21

21.1 Introduction

India's Vision 2020 document was prepared by the then Prime Minister P.V. Narasimha Rao. It was granted his replacement, late Atal Bihari Vajpayee, who declared that India would become a financially developed country by 2020 in the legislature and one of their Independent Day messages. To realize our great visionaries' dreams, we have to transform India in five areas where India has core competencies that include Agriculture and food handling; schooling and health care; ICT; production of facilities, and self-reliance on vital innovations. Technology is a crucial driver for any country's progress; recently, Artificial Intelligence is a buzzword throughout the world and spurred heated debate globally. AI Products are slowly infiltrating home and workplaces. The focus is not only a specific area or industry spread across functionalities but also to any particular sector and country spread across the world. Artificial intelligence is the machines applied designed and programmed in such a manner that they think and act like a human. Till now, the devices have been doing what human beings were told to do, but with Artificial Intelligence, they will be able to think and behave like human beings [1]. AI is an 'Umbrella' A definition that consists of various subfields. One such sub-field is machine learning, which focuses on creating programmers that will train itself to read, comprehend, interpret, prepare, and act when introduced to new data in the correct amounts. India's digital transformation journey started very long back but moving a slow pace compared to other nations. India's competitive advantage over other emerging markets for capturing significant market share in IT/ITES, similarly India has every possibility to take advantage of AI technology provided multi-stakeholders with a proper strategy towards their destination relating to AI. In terms of flagship initiatives such as Modern India and Ability India, the government has also introduced a new technology surge. Every stakeholder community should be proud of the efforts and accomplishments in developing the world's largest rural broadband network, target direct value transmission system, the most extensive digital identity system and my government-citizen interaction portal. Numerous government programmer provides the framework for injecting the next generation of artificial intelligence technology solutions [2]. India's potential to emerge as a pioneer in the production of goods and services in this field is evident. Recognizing the tremendous potential that AI has for India, it is necessary to highlight the need to establish an efficient programme and governance mechanism to handle smart systems-facilitated development. Skills, reskilling and capacity development, job organization and overhaul, standardization and interoperability, regulatory system work, data protection and public technology would continue to be the subject of Artificial Intelligence projects. A proactive attitude can help initiate work in this direction and step on. Artificial Intelligence (AI) relates to human intelligence emulation of computers designed to feel like humans and imitate their behaviour. The word can also refer to any computer that displays human mind-related characteristics such as learning and problem-solving. Artificial intelligence's perfect attribute is the capacity to rationalize and take action with the most significant probability of reaching a particular target.

21.1.1 State of Artificial Intelligence in the World

It is no wonder that artificial intelligence has evolved more than ever since 2010; it has grown at an annual average growth pace of almost 60% (Burno Jacobson). But we have to think about what AI will generate this, could imply for our future. What were the nations that are driving the AI uprising [3]? The USA ranks as the top country with AI applications, companies like IBM, Microsoft, Google, Face book and Amazon invest heavily in this area. USA's pool of scientific knowledge combined with its market power will allow it to stay on top in Fig. 21.1.

Next to the USA, the Chinese government was very ambitious to concentrate on AI technologies. The government announced its intention to become "a principal world centre of Artificial Intelligence Innovation" by 2030. There are companies like Tencent, Alibaba and Baidu avail AI in their success. Japanese Prime minister Abe has urged the businesses to invest more into AI technologies to overcome an ageing population's problems and decrease workforce. With plenty of research into AI, a decreasing force and a high automation potential, Japan is likely to continue right at the top [4]. Companies like Soft Bank and Toyota started their effort to tap AI technologies. The UK also put its effort to become global in AI technologies; companies like Deep Mind technologies are the worldwide AI programming leader.

Fig. 21.1 AI-based economic analysis

Germany also plans to become a leading hub for Artificial Intelligence. Companies like Porsche, Daimler and Bosch are critical players for further promoting AI technologies in their respective domains.

21.2 Literature Survey

21.2.1 AI Potentiality to Progress India

Accenture said that AI could raise India's annual Gross Value-Added growth rate by 1.3% points, increasing its GDP by 15% in 2035. A study entitled 'Rewire for Prosperity' predicted that by improving the essence of work to produce positive outcomes for companies and society, AI could add $957 billion to the Indian economy. The study was performed by comparing two situations to estimate the economic efficiency of AI. Second, under existing projections about the future, the projected average economic growth rate was the second one, the AI example, which indicates anticipated economic growth after the influence of AI has been incorporated into the economy. Compared to other G20 economies, the study also examines the country's AI innovation ecosystem's power through five main pillars: colleges, startups, big companies, decision-makers, and collaborations with multiple stakeholders [5].

21.2.2 Universities

As of 2016, India had produced 26 million graduates involved in STEM (Science, Technology, Engineering and Mathematics) subjects. The problems of these graduates have a lie in their practical proficiency levels and therefore their employability. This is where universities would have to set up and match international standards by investing in research and providing quality support to graduates trying to produce AI technologies.

21.2.3 Startups

According to a Zinnov Consulting study, India is the fourth largest cluster after the US, China and the UK with around 250 AI Startups. The problems here are that the funding Indian startups receive substantially low than the other three key players in the global market. As per the 2017 Inc42 Data Labs report, the industry has picked up 100% sales, with Bangalore becoming India's most significant breeding ground for AI startups. Via 58 deals in the AI room, around $87.85 million was generated.

Big Data Firm Qubole raised $25 million and bagged the largest amount of financing raised by an Indian AI business in 2017.

21.2.4 Large Companies

The leading Indian Banks and Tata Motors are beginning to experiment with AI. Accenture Study Report reveals that 80% of Indian companies believed that AI transforms their business wholly or significantly in the next three years, with an average growth of 68%. If this is to be the case, and if large companies in India decided to invest in AI seriously, we could very well have an Indian AI revolution on our hands [6].

21.2.5 Policymakers

So far, The Ministry of Electronics and Information Technology has created a task force group to work out a road map for new technologies such as AI and Blockchain. The Ministry of Commerce and Industry has initiated a task force for AI, to work out how it can develop manufacturing and services. This is just a very start; there is much more to do in terms of AI policy to take off. The country needs a comprehensive, well thought out and well-funded set of guidelines for an AI revolution.

21.2.6 Multi-Stakeholders Partnerships

Europe is probably the model to follow. The PPP, where the European Commission is funding up to 700 million Euros, and the Private Investments is over 2 billion Euros, is the most serious effort in Research into AI. The situation in India so far a partnership between Government bodies and private institutions are very infantry stage. Need for multi stakeholder's interaction and association to move further to grab the potentiality [7].

21.2.7 AI Adoption Policies in India

India is already on the road of a digital transformation, and the next move is to make wise decisions using the big data produced. The ASSOCHAM-PWC joint research on 'Artificial Technology and Robotics' 2017 report illustrated the opportunity for businesses to set up AI-focused Innovation Centers in India with initiatives such as Digital India and Make in India to build advantageous initiatives such as Digital

India and Make in India, thereby needing strong cooperation between academia, the private domain and the public sector to understand and resolve problems holistically. More than 36% of major financial companies have invested in these innovations currently. The study also stated that AI, Artificial Language, and Automation have a wide variety of financial information security, but their ability in India has not been wholly realized [8]. It also noted that AI could leverage to defend the infrastructure sector, such as airlines and power plants susceptible to threats, sensor devices' usage, and the identification of trends. AI security framework uses sensors to conduct dangerous tasks for humans-recovery of weapons, identifying combat activities on borders in neighbouring countries. Another field in which computer defence programs depend on historical data from cyber-attacks and artificial intelligence is predictive intelligence to defend and identify potential risks expected to exist in the future. Accessibility for people with various conditions is an untapped opportunity in India [9].

In conjunction with other designs such as 3D printing and IoT, AI has great potential to drive widespread usability, affordability, and design viability in India. AI environmental studies technologies have still not acquired substantially in India [10]. The opportunity here involves AI integrated smart energy grids for power generation, precision development for reduced waste in landfills in Fig. 21.2.

Fig. 21.2 AI-based innovations

21.3 Methodology

21.3.1 Impact of Artificial Intelligence on the Indian Economy in Select Industries

1. **Agriculture**: India named the agricultural economy and its contribution to GDP is around 16% and provides food security to 130 crore people. Nearly 50% of the workforce depend on this sector. The government prioritized doubling framers income by 2022 As a strategy for public relations. One side of spectacular progress and the other side of government focus, depending on volatile factors, has a supply chain and efficiency week. Globally, modern and AI innovations are helping to address pressing challenges around the value chain of agriculture. India has 30 million mobile growers, projected to rise threefold by 2020, and 315 million rural Indians are projected to use the internet by 2020. Accenture research highlighted that 70 million Indian farmers would be impacted by digital farming and linked farm services by 2020, contributing USD 9 billion to farmers' profits. Startups such as Intelllo Labs, Aibono, Trithi Robotics, Saturn are at the forefront of tech-based agro startups earned USD 313 million in 2016, and startups are extensively active this field in the immediate future. Any areas worth noting are soil quality monitoring and regeneration, crop monitoring systems and offering real-time guidance to growers, the productivity of field mechanization, and AI innovations can have a significant effect on various stages of the value chain. AI solutions can still solve poor market management, shortage of guaranteed irrigation, and fertilizer and pesticide overuse/abuse [11].
2. **Manufacturing**: It is anticipated that the manufacturing sector will be one of the main recipients of AI-based solutions, allowing 'Factory of the Future' to automate processes and machinery through versatile and adaptable technological systems to react to new or unpredictable circumstances through making smart decisions. The 2018 Automotive Report showed that 92% of senior manufacturing executives agree that emerging technology, like AI, would enable them to improve their efficiency levels and encourage workers to work smarter. A new Boston Advisory Group study showed that a large difference exists between the 'ambition and implementation' of an enterprise, with only one in five organizations integrating AI into one or more of their processes. Similarly, Forrester, a multinational consulting company, claims that 58% of industry and technology experts are studying AI solutions, but only 12% are currently implementing them. We can see sensors detecting flaws on the production line in the case of manufacturing. The data is fed to the cloud to ensure that the faulty component is automatically extracted from the bar, and a new al is requested when measuring just-in-time timetables in real-time. Applying AI technology and instruments can impact numerous manufacturing fields, including engineering (AI for R&D efforts), supply chain management (demand forecasting), manufacturing (AI

will decrease costs and improve efficiency), monitoring (predictive maintenance and improved usage of assets), quality control, in-plant manufacturing and material handling, and more [12].

3. **Health Care**: From the current USD 100 billion, this sector will expand to USD 280 billion by 2020. However, it presents enormous efficiency, usability and affordability problems for broad segments of individuals. Applying AI to health care will help resolve challenges such as obstacles to utilization to health care services. Computer vision and micro-bank imaging initiatives can aim to reliably and precisely diagnose the disease at a very early level. An imitative by the private sector contributes to creating a handheld computer called '3Nethra' between Microsoft and Indian startup Forums Health, which can track primary eye disorders and complicated conditions such as diabetic retinopathy. There is a range of large-scale measures by the government to resolve health care issues. Ayushman Bharat's mission is to encourage e-Health, where AI can play a key role in its development. Health care options focused on AI will help move from 'sick medicine' to 'real medical care' to focus on preventive techniques. Healthcare implementation of AI will help tackle high barriers to healthcare services, especially in rural areas with low connectivity and restricted availability of healthcare practitioners. This can be done by utilizing cases such as AI-driven diagnostics, customized counselling, early detection of possible pandemics, and diagnostic devices, amongst other [13].
4. **Education**: A thriving education sector can change the county through human capital growth and efficiency enhancement. Estimates suggest that over half of the country's population is actually under the age of 25. To provide quality schooling and instruction, the use of automated means of data collection is successfully leveraged. With many AI resources, problems including low graduation rates in rural schools and weak learning performance can effectively be solved. Adaptive coaching platforms for personalized learning are the key goal, and insightful and collaborative tutoring solutions are some of the resources in this area to solve very long-term problems. Organizational India began its efforts by developing personalized learning videos for companies such as Content Technologies Inc (CTI) and AI research & design Company. In Andhra Pradesh, Micro-Soft helps to forecast dropouts. Pearson's writing utilizes NLP-based technologies to provide students with direct reviews to develop their writing skills for learning apps. To rationalize teacher resources and establish tailored formal training to recognize and fulfil expertise and ability differences, AI technology and instruments would be beneficial [14].
5. **Smart cities and infrastructure**: As the metropolitan population proportion is rising exponentially and by 2050, it is projected to hit about 60%. Unplanned urbanization would generate many problems, such as congestion, noise, high crime rates, low housing conditions, which will theoretically place massive pressure on current Indian cities' infrastructure, which operational needs. By fostering AI-powered service delivery based on citizen data, rationalizing admin workers based on predicted service demand and transition pattern research, and AI grievance and resolution by chatbots, AI technology would address the

problem. Crowd control, smart defence networks, cyber threat protection are only a couple more fields that discuss AI technology and resources [15].

6. **Smart mobility and Transportation**: The NITI Aayog study stresses the focus on AI introduction to resolve connectivity and connectivity concerns. The usage of AI technology such as traffic congestion and road crashes, high rates of traffic collisions, lack of mass transit facilities and others can quickly solve some of the problems in this field. Many of these problematic areas can be solved through AI Supported smart technology such as Supported Car, Green Field Infrastructure, Automated Trucking, Intelligent Transportation Networks, Ride Path & Flow optimization and community-based parking. In this area, future use cases involve autonomous ride-sharing vehicles, semiautonomous applications such as driver assist, and tracking and management of predictive engines. Autonomous trucking and shipping and better traffic control are other fields that AI will influence [16].
7. **Energy**: AI is expected to make lasting changes to industries globally, and the power sector is no exception. In terms of the implementation, it is still early days. However, there are various use cases for AI in the energy market, including modelling and forecasting energy system structures to minimize unpredictability and maximize power balance and usage performance. In renewable energy resources, AI will allow power generation by smart meter-enabled intelligent grids and enhance photovoltaic energy's efficiency and affordability. AI can even be implemented for the predictive management of continue, analogous to the industrial industry [17].
8. **Retail**: The retail sector in India seems to be on the right path as huge investments and technological advancements lead it. These funds are now being used to create solutions with new emerging technologies like blockchain, Chatbots, AI and Machine Learning. With implementations such as optimizing user interface by offering customized tips, preference-based searching and image-based product quest, it has become one of the early investors of AI solutions. Other usage cases involve an expectation of consumer interest, increased inventory control and proactive fulfilment control..AI Magnify is an Artificial Intelligence and Computer Vision-based company based in New Delhi started operations in 2017 by Neeraj Kumar and Devarshi Mishra with an essence of AI, computer vision, Facial recognition, Machine Learning, Deep Learning and Data Analytics. Its aims to target the new combination of the retail sector with the power of AI [18, 19].
9. **Banking and Financial Services (BFSI)**: Artificial Intelligence to play a significant role in the Banking segment in the area of customer support, efficiency and security and compliance, according to a report. The report by CyberMedia Research and NASSCOM on AI for Banking and Financial Services (BFSI) sector based on a survey of IT decision-makers states that many see a strong need for AI in four major business areas. First one is Proactive and personal customer care. The second one is the Automation of back-end business processes to reduce human errors and improve turnaround time. The third one uses technology to track consumer behaviour for offering customized products. The last

one is to monitor processes and data for regulatory compliance, anti-money laundering, and risk management [20, 21].

21.3.2 Barriers to Adopt AI Technologies in India

The adoption of AI by different industries was affected, among other variables, by technological and regulatory difficulties, but the most significant determinant was the commercial consequences [22]. Although technical viability, organized data availability, legislative hurdles, privacy concerns, ethical problems, human interaction choice have all played their part in deciding a sector's readiness for large-scale AI adoption; compelling business use cases (e.g. improved performance, precision, distance [23–25], forecasting and efficient decision-making) that have a direct influence on rev As seen in the AI adoption and usage survey of the McKinsey Global Institute, sectors leading the adoption of AI today also plan to increase their expenditure in AI as often as possible, thus further enhancing the varying degrees of AI adoption across industries [26, 27].

A task force appointed by Ministry of Commerce and Industry to address three crucial policy questions one is what are the areas the government should play a role the second one is how AI can improve the quality of life and solve the problems of Indian citizens and third, what are the sectors that can generate employment and growth by the use of AI technology. The committee also highlighted what the policies needed to turn India into an AI research powerhouse and not only mere a user of AI solutions [28, 29].

21.3.3 Challenges for Adoption of AI Technologies in India

- The country produced second-largest Science, Technology, Engineering and Management graduate some 27 million annually next to china in size. Still, in terms of the research contribution, it was a distant fifth behind China, The US, Japan, the UK, and just ahead of Germany and France.
- Private Sector tech companies have not developed AI technologies more competitively like other major players in the world. The private sector's contribution in the US and China is very aggressive in AI-related areas like design, develop and commercialize the AI tools and technologies, that was missing here.
- Shortage of expertise is another challenge, for instance, demand for AI and Machine Learning specialists will likely rise by 60% just this year, but the supply is not matching the request. The country could face a demand–supply gap of 2 lakh data Analytics professionals by 2020.
- Absence of enabling data ecosystems and access to intelligent data.
- Lack of suitable approaches to protect startups and small ones from big players in this field. The natural competitive advantage that larger companies like Google

and Face book have over smaller companies is based on the amount of accessible data.

- With the diversity and amount of data outlets, the insecurity of data protection grows. The more data gathered from websites, the harder it is to determine whether the data source is harmful. With the adoption of cloud-based systems, the risks of breaching data access and privacy have multi-folded.
- The funds required to set up and implement AI are very high; thus, not every business owner or organization can invest.
- Lack of collaborative efforts between significant stakeholders, mainly Corporate India and the government, is an essential deterrent in promoting India's AI technologies.
- Lack of awareness about how AI is supported to address many critical problems still unsolved.
- Lack of supporting efforts from academia and institutions to promote AI-centric programs and training.
- Potential but Untapped and non explored SMEs segment to implement AI technologies and tools is absolutely at a nascent stage.
- We are integrating efforts between corporate India and global Research agencies and forerunners of MNCs.
- Limited efforts by state governments, they are behind in line with central government towards AI strategy and implementation.

21.3.4 Way Out to Overcome Challenges

- NITI Aayog study report proposes two tired framework for promoting AI research involving the creation of Centers of Research Excellence in AI and International Centers for Transformational Artificial Intelligence will provide the ecosystem for application-based technology development and deployment.
- To resolve data ecosystem related problems encourage decentralized data market place based on blockchain technology, which aims to enable data providers to share data they would usually hesitate to, given trust and control issues. A data annotation done via crowd-sourced, untrained, or non-expert anonymous annotators could very viable for long term
- The data security and privacy concerns brought by latest technologies like IoT, Big data and cloud can be addressed by Advanced algorithms and data models to identify potential data threats and eliminate them before confidential data is compromised.
- Public–Private Partnership based. Research activities will help move further and make India the best source of AI technologies globally.
- Multifold strategy for uplifting awareness among stakeholders about AI technologies and its benefits by using Mass Media and social networks.

- Need for a particular emphasis on facilitating the creation of large foundational annotated data sets that will enable and accelerate AI solutions by startups and other AI research organizations.
- For promoting and developing AI tools and the adoption of AI in different fields, partnership and collaborative approach involving various stakeholders and the government certainly help us.
- Supporting AI startups financially as well as other means like establishing National Artificial Intelligence Market place (NAIM) and Data Market place.

21.4 Conclusion

India needs to focus on establishing cooperative partnerships with some of the front runners in AI to become a global influence participant, such as Japan, the United Kingdom, Germany and Singapore, Israel, and China, to build responses to tactical social and economic problems that can support and promote strategy formulation. The immediate need to create excellence centres to facilitate interdisciplinary study in law, health, engineering, management, and social sciences. To build flexible innovations that can be quickly replicated in most developed and emerging countries, AI can change the overall quality of life and create an ideal sandbox for companies and organizations globally. The paradigm for Artificial Intelligence as a Service (AIaaS) may be 'Solved in India', and it is a fresh idea to move forward. Moreover, national initiatives need to be centred on creating a long-term strategy and action plan that unites the environment and tackles strategic and ethical problems when they emerge, in cooperation with policymakers, civil society and the private sector.

References

1. McCarthy, J. http://Jmc.stanford.edu/ArtificialIntelligence/What-is-AI/index.html
2. TCS AI Global trend study Report. http://sites.tcs.com/AI/keyfindings
3. *Ai based detection of diabetic retinopathy*. http://blog.google/topics/machine-learning/detcti 9ng-diabetic-eye-disease-machine-learning/
4. *Artificial intelligence effect: Five years from now*. http://www.businesstoday.in/current/eco nomy-politics/AI-future-jobs-ficci-job-study-nasscom/story/265976.html
5. *Artificial intelligence imperils India Inc jobs*. http://www.thehindu.com/business/industry/art ificialintelligence-imperils-India-inc-jobs/article/19529813.ece
6. *Study-state of AI in India 2017*. https://analyticsindiamag.com/study-state-of-artificia-Intellige nce-in-india-2017
7. Raju, M., & Venkateswara Rao, P. (2020). Financial analysis of selected IT companies in India. *Test Engineering & Management, 83*, 13356–13364.
8. *Future of jobs in India-A 2022 perspectives*. http://www.ey.com/publicaion/vwLUAssets/ey-future-of-jobsin-india/$file/ey-future-ofjobs-in-India.Pdf
9. *AI, Automation will cost 7 lakh jobs by 2022*. http://www.businesstoday.in/sectors/it/ai-and-aut omation-to-cost-7lakh-it-jobs-by-2022-Savs-report/story/259880.html
10. Sailaja, V. N., & Madhavi, N. B. (2016). Relationship between retirement wealth and householders' 'personal financial and investment behavior.' *Man in India, 96*(5), 1521–1529.

11. NASSCOM Knowledge Center. *Ai, beyond the myth & the hype*. Retrieved from http://www.nasscom.in/knowledgecenter/publications/ai-beyond-myth-hype
12. *Indian industries that will be impacted by AI: The dawn of an era of tectonic change VishalKrishna*. https://yourstory.com/2017/10/7-indian-industries-that-will-be-impacted
13. Artificial Intelligence has the potential to add 1 trillion to India's economy in 2035
14. By: The Financial Express. Published 9 June 2018
15. Jacobsen, B. (2018) *Countries leading the Way in AI*; 8 Jan 2018. https://www.futuresplatform.com/.../5-countries-leading-way-ai-artificial-intelligence
16. Datta, D., Mishra, S., & Rajest, S. S. (2020). Quantification of tolerance limits of engineering system using uncertainty modeling for sustainable energy. *International Journal of Intelligent Networks, 1*, 1–8. https://doi.org/10.1016/j.ijin.2020.05.006
17. Faggella, D. (2018). *Machine learning in finance—Present and future applications*. http://www.techemergence.com/machine-learning-in-finance/. 1 June 2018
18. Ravi, S., & West, D. M.: Artificial intelligence and data analytics in India. https://www.brookings.edu/.../2018/.../17/artificial-intelligence-and-data-analytics-in-i
19. Sekhara Rao, K. S., Madhav, V. V., & Udaya Shankar, A. (2019). Financing methods and risks—Impact of risk on the selection of financing methods in power projects: A study of indian power sector. *International Journal of Innovative Technology and Exploring Engineering, 8*(9), 242–245
20. Balaji, C., Rao, P. V. D., Rekha, Y. C., & Basheerunisa, S. (2019). Technological relationship between banknifty and other selected sectorial indices of NSE. *International Journal of Recent Technology and Engineering, 8*(3), 2314–2318.
21. Balaji, C., Venkateswara Kumar, K. S., Lalitha Devi, G., Pravallika, S., & Harika, K. (2019). Impact of acquisitions on financial performance of selected softwarecompanies in India. *International Journal of Innovative Technology and Exploring Engineering*, *8*(9), 2415–2419.
22. Ames, J. G., & Tabuada, P. (2014). Control barrier function based quadratic programs with application to adaptive cruise control. In *Proceedings of IEEE Conference on Decision and Control (CDC)* (pp. 6271–6278).
23. Sharma, S., & Obaid, A. J. (2020). Mathematical modelling, analysis and design of fuzzy logic controller for the control of ventilation systems using MATLAB fuzzy logic toolbox. *Journal of Interdisciplinary Mathematics, 23*(4), 843–849.
24. Sharma, S., & Obaid, A. J. (2020). Contact-mechanics and dynamics analysis of three-different ellipsoidal raceway geometries for deep Groove ball bearing using Abaqus 6.13 version FEA simulation for high load-bearing as well as speed-rotating applications. *International Research Journal of Multidisciplinary Science and Technology, 3*(5), 36–43.
25. Obaid, A. J., & Sharma, S. (2020). Recent trends and development of heuristic artificial intelligence approach in mechanical system and engineering product design. *Saudi Journal of Engineering and Technology, 5*(2), 86–93.
26. Sekhara Rao, K. S., & Sahyaja, C. (2018). Financial performance—A study on Indian Banks. *International Journal of Engineering & Technology, 7*(1), 177–181.
27. Alkaam, N.O., Mohammed, D. M. Q., & Obaid,A. J. (2018). A hybrid technique for object detection and recognition using local features algorithms. *Journal of Advance Research in Dynamical and Control Systems, 10*(2), 2330–2343.
28. Pradeep Kumar Patnaik, R., Balaji, C., & Narayana, M. S. (2019). Technological relationship between capital structure and financial efficiency in ACC Ltd. *International Journal of Innovative Technology and Exploring Engineering, 8*(11), 2009–2013.
29. Al-Yasiri, D. S. Q., & Obaid, A. J. (2018). A new approach for object detection, recognition and retrieving in painting images. *Journal of Advance Research in Dynamic and Control System, 10*(2), 2345–2359.

Chapter 22
A Review of Resource Allocation and Management Methods in IoT

Aykut Karakaya and Sedat Akleylek

Abstract IoT systems are one of the most important areas of developing technology. IoT application solutions are becoming widespread and their usage areas are expanding. Therefore, studies to develop IoT technologies are also increasing. Although the benefits of developing technology are enormous, it includes some difficulties. One of the most important challenges in IoT systems is resource allocation and management. Cloud, fog, or edge computing methods are used for storage and computing processes in IoT applications. Data perceived from resource-constrained devices reach these computing nodes. Resource allocation and management must be made in the cloud, fog, or edge nodes for computing and storage. The correct and complete resource allocation and management are very important for the performance of the system. Numerous methods are proposed for this. Artificial intelligence-based methods are one of them. This study examines IoT resource allocation and management.

22.1 Introduction

Internet of Things (IoT) is a network structure that connects detection devices such as sensors and end devices such as actuators to the Internet and communicates these devices according to the specified protocols to perform tasks such as monitoring, recognition, and management [1]. By using IoT systems, effective applications can be developed in many areas such as smart cities, health, industry, agriculture, military. IoT-based approaches can be examined on a wide scale from small wearable devices to large industrial systems. In IoT systems, personal area networks, wireless sensor networks, wireless body area networks are generally created by using wearable sensor devices and specialized sensors on various points such as people, animals, machines,

A. Karakaya (✉)
Department of Computer Technologies, Bulent Ecevit University, Zonguldak, Turkey
e-mail: aykut.karakaya@bil.omu.edu.tr

S. Akleylek
Department of Computer Engineering, Ondokuz Mayis University, Samsun, Turkey
e-mail: sedat.akleylek@bil.omu.edu.tr

P. Kumar et al. (eds.), *A Fusion of Artificial Intelligence and Internet of Things for Emerging Cyber Systems*, Intelligent Systems Reference Library 210,
https://doi.org/10.1007/978-3-030-76653-5_22

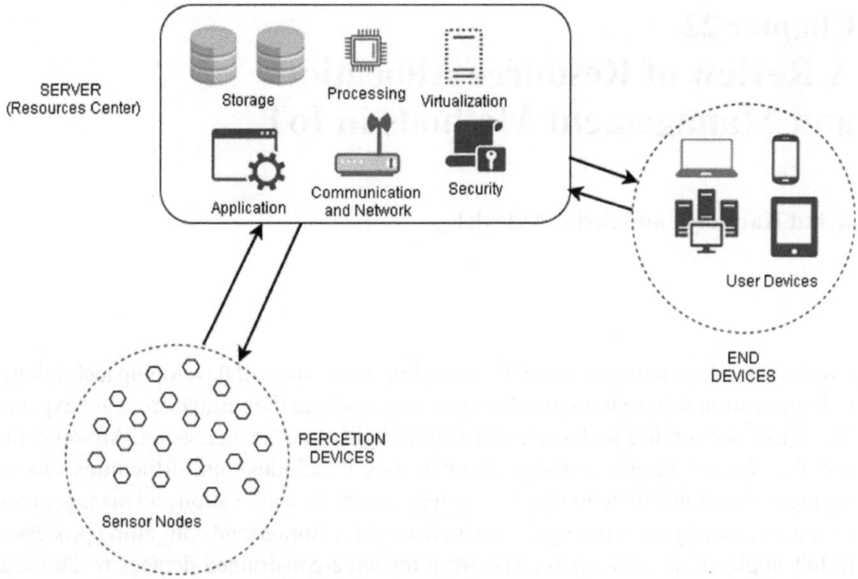

Fig. 22.1 Shared resources and processes in IoT systems

underground, underwater, forest areas. Some data is detected from the sensor devices and sent to a data processing and storage device at the upper layer. The results obtained in this layer are sent to the actuator devices in the monitoring position.

High-quality, fast, reliable, smart, and modern applications can be developed with IoT systems. Therefore, it is preferred in various fields. However, IoT systems bring many challenges as well as these advantages. The need for security, non-existent software, sensors affected by external factors (such as wearable devices), the need for power and energy saving, efficient storage, and data communication are some of these. In this study, solution approaches for the efficient use of different resources such as storage, processor, network structure. in IoT systems and allocation of these resources to tasks are examined.

The resources and processes shared by the tasks are shown on the general structure of IoT systems in Fig. 22.1. These are generally storage, processing, virtualization, application, network management, security, distributed structure.

22.1.1 Motivation

Generally, cloud, fog, and edge systems are used alone or together for data storage and processing in IoT. The resources in these systems have to be shared by the different tasks. Processes such as sharing resources, distributed data processing, and time scheduling are often referred to as resource allocation, resource management, load balancing. Also, energy saving, long battery life, low computation load processes are often referred to as offloading. With offloading approaches, system performance can be increased by providing efficient resource usage, high process capacity and low response time [2].

In this study, limited and shared resources in IoT systems are examined and solution approaches are evaluated for efficient sharing of resources. The study aims to determine the trend in recent years in both system infrastructure and artificial intelligence and optimization methods in this field and to direct and contribute to future studies.

This study is the extended version of [43] in which resource management and security in fog computing are explained.

22.1.2 *Organization*

In the following sections of the paper, fog, cloud, and edge computing systems that provide resources for IoT applications are examined in Sect. 22.2. Resource management processes are classified and resource allocation approaches are examined and compared in Sect. 22.3. Open problems for resource allocation and management in IoT systems are given in Sect. 22.4. The conclusion in the last section.

22.2 Cloud, Fog and Edge Computing Systems

In this section, fog, cloud, edge computing systems that provide data storage and processing resources to IoT systems are examined. In the upper layer of the IoT architecture, these systems process data, store it, and transmit the results to the actuator devices. Fog computing and edge computing provide a local solution to data processing and storage. Fog computing extends the storage, networking, and computing possibilities offered by cloud computing towards the edge of the IoT network. Therefore, the fog computing method is a factor that increases the performance and efficiency of the system [3]. Edge computing and fog computing are both localized. Unlike fog computing in edge computing, the computing load is on the end devices such as sensors, and mobile devices [4]. In fog computing, the computing load is on the fog nodes in the fog layer. In most models using one of these two methods, cloud systems (fog-cloud or edge-cloud) are used for permanent storage.

Three-layer architecture is commonly used in fog computing. The first layer consists of end devices such as sensor nodes, smart devices, IoT-enabled devices. The second layer consists of fog nodes such as routers, gateways, switches, and access points. The fog nodes carry out storage and computing activities. The third layer consists of remote cloud servers. Also, this layer provides high storage and computing service [5]. Data obtained from resource-constrained end devices are processed by the fog nodes and results are transmitted to the cloud for permanent storage or

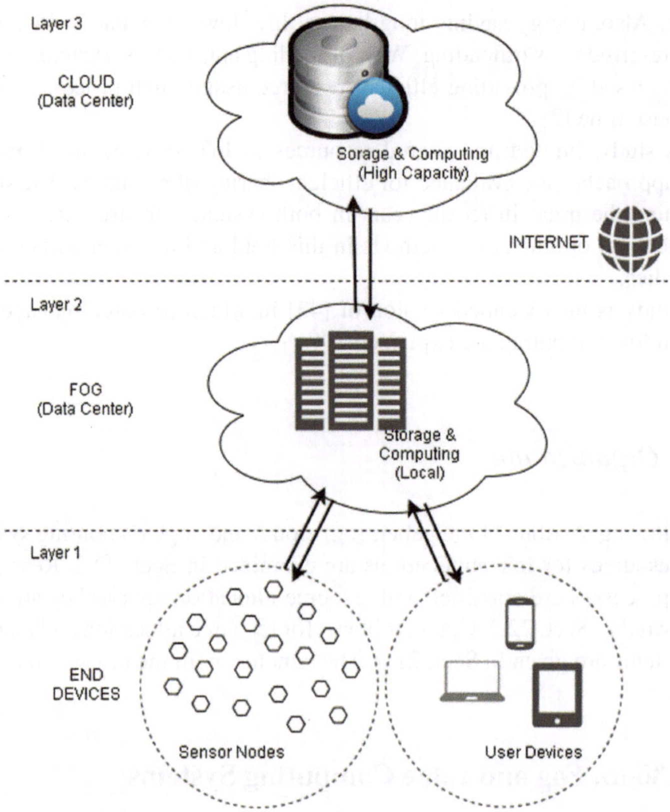

Fig. 22.2 Fog computing

to mobile end devices as a response. Thus, latency is reduced and data transmitted to the cloud is protected. The processing, storage, and transmission of data in fog computing are shown in Fig. 22.2.

While cloud computing optimizes and manages data through the remote server, fog computing optimizes and manages the local environment. The fog systems have features such as low latency, creating security steps, location awareness, localization. The cloud systems, on the other hand, have features such as high latency, security steps not defined, globalization.

Edge and fog computing both provide local solutions to data management. The difference the edge computing is that the computing load is on the end devices, sensors, and mobile devices [4]. The edge is fully localized. The fog extends system resources towards the edges of the network [5]. In other words, the fog is a layer between end devices and the cloud. In edge computing systems, time-sensitive data is processed and stored on smart devices, other data is sent to the cloud. They allow the cloud to spread over a wide area. They have many challenges because mobile end devices are resource-constrained. However, their advantage is that they can be

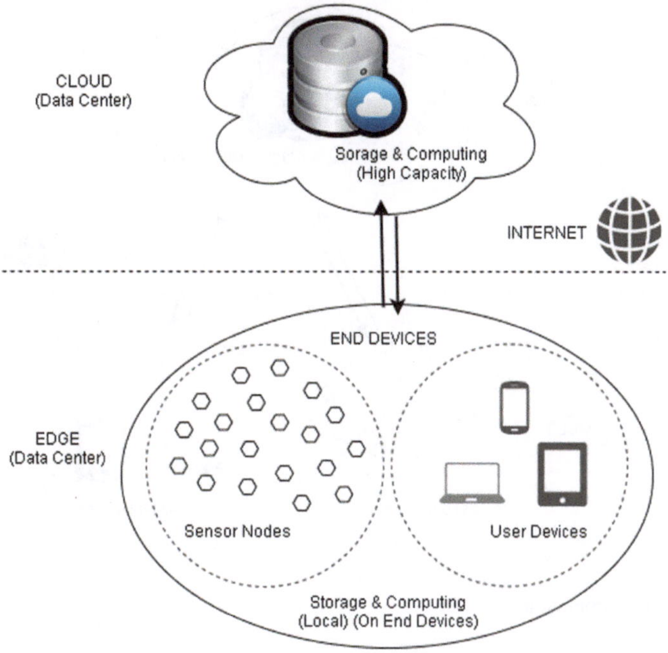

Fig. 22.3 Edge computing

spread over large geographical areas than the fog systems. Fog computing can be used in environments where end nodes are close to the data center (such as hospital, sports field, industrial companies). Edge computing can be used in remote and wide geographic environments (such as automobile systems, smart cities). Both of these computing structures extend the cloud to end devices and support the optimization of the cloud. The processing, storage, and transmission of data in edge computing are shown in Fig. 22.3.

In cloud computing, data is sent directly to the cloud server for storage and processing. Data security risks may occur as data is sent to the cloud without being encrypted via The Internet. Data security processes can be performed on end devices. However, the cost is high because end devices are resource-limited. The cloud computing systems alone are sufficient in IoT applications where the maximum waiting tolerance of tasks is higher than a certain threshold value. However, in IoT applications where the waiting tolerance must be lower than a certain threshold, the use of cloud computing systems together with fog computing or edge computing increases the performance of the applications. Data processing, storage, and transmission in cloud computing are shown in Fig. 22.4.

Differences in the fog, edge, and cloud computing used for processing and storing data obtained from sensors are shown in Table 22.1 [6].

In IoT-based applications, fog or edge computing methods can be used with cloud systems. Thus, they support increasing the performance of these applica-

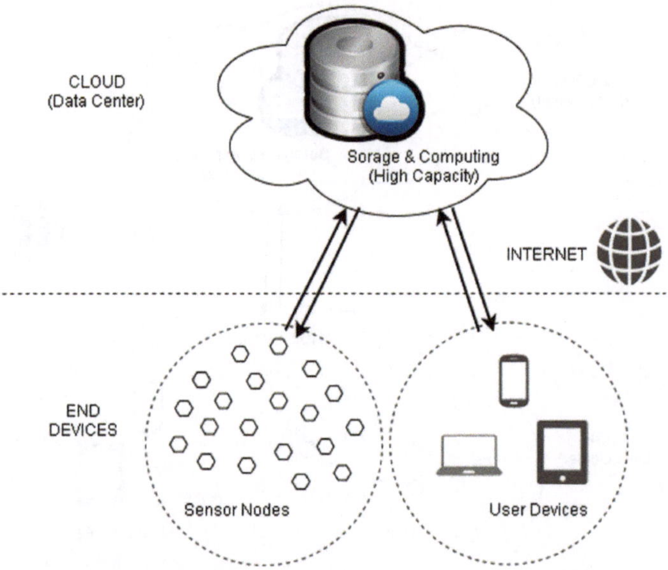

Fig. 22.4 Cloud computing

Table 22.1 The cloud, fog and edge computing comparison

Features	Cloud computing	Fog computing	Edge computing
Latency	High	Low	Low
Response time	High	Low	High
Data processing and storage process	On the device accessed via the Internet	On devices in the fog layer	On edge devices
Position	Global	Local	Local
Mode of operation	Internet connected	At the edge of the network	At the edge of the network
Storage and data processing capacity	High	High	Low
Mobility	Constricted	Supported	Supported

tions and ensuring secure data communication. With the development of hardware, data processing and storage, task management, resource allocation, and security in IoT systems, low-cost, fast, and reliable results are obtained. Therefore, IoT-based approaches are preferred and developed in many areas. These three computing structures have some challenges to develop IoT-based applications. Resource allocation is the most important of these.

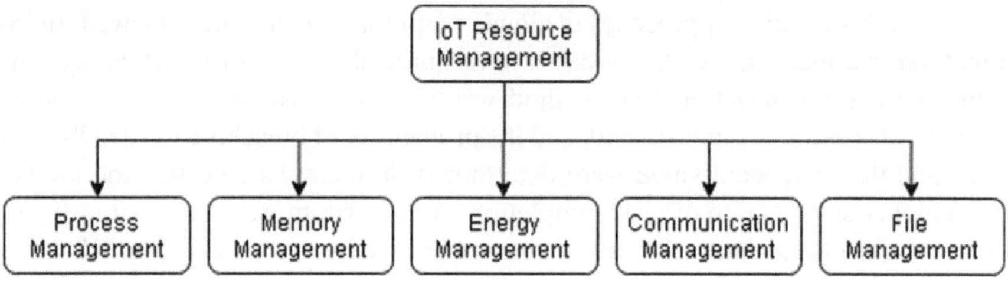

Fig. 22.5 Classification of the resource management process [7]

22.3 Resource Management in IoT Systems

In this section, resource management and allocation problems in IoT systems and the recent studies on the solution of these problems are examined. In IoT systems, almost every resource used by many tasks and requests such as server resources, transmission lines, wireless resources must be shared. It is very important to design this sharing process efficiently and effectively. The most obvious problems and challenges on resource management are high computational complexity, high transmission latency, narrowing of scope, allocation of limited resources, low energy efficiency, Quality of Service (QoS) requirements, security issues that can be caused by inefficient resource allocation.

Resource management classification is shown in Fig. 22.5 [7]. In recent studies, some issues are very popular. These issues: "task scheduling" and "task offloading" for the process management, "resource allocation" for the memory management, "energy efficiency" for the energy management, "load balancing" for the communication management, and "virtualization" and "storage" for the file management.

Resource management and allocation issues are important for resource-constrained systems such as standard IoT, the narrowband IoT (NB-IoT), the cognitive IoT, machine to machine (M2M). In recent studies for resource allocation in these systems, researchers tend to use artificial intelligence (AI), machine learning (ML), and optimization approaches for low-latency and high-performance frameworks that serve a large number of users, enable full use of resources. The challenges in IoT systems are handled in different studies and effective methods are developed to overcome these challenges.

22.3.1 Resource Allocation Approaches in IoT Systems

In IoT applications, efficient resource allocation is performed on server resources such as the cloud, fog, and edge computing, as well as frequency channels, spectrum band, and transmission bandwidth. The methods developed to solve the resource allocation problem are examined and analyzed according to many parameters.

Resource allocation approaches in cloud computing systems are reviewed. In [8], a method that meets the QoS standards in resource allocation of the cloud systems using the combinatorial auction method was proposed. The urgency for efficient resource allocation was considered, and the probability of breaching the deadline for tasks and their expected value were determined. It is emphasized that the method has a higher success rate of task completion. An algorithm was proposed in [9] for different objects to share joint resources in the cloud systems. Using a ML-based algorithm, an artificial intelligence controller provided centralized management with the information collected from the network. Also, algorithms were designed for the message queuing telemetry transport (MQTT) protocol to consume a lower bandwidth and to establish inter-machine relationships. It has been stated that the method has higher performance and usability than traditional IoT communication systems.

In [10], a method for estimating tasks and allocating available resources in the WebIDE cloud system with task pre-planning was proposed. The tasks were grouped into three classes according to requirements. Then, transition probability between tasks was calculated with the Markov state matrix and a prediction model was designed. Finally, resource allocation in the cloud was achieved by pre-planning with the ant colony algorithm. Thus, it was emphasized that the method reduces task response time and provides efficient resource use. In [11], a whale optimization algorithm-based model was proposed to reduce the cost and processing time of tasks and to allocate the most appropriate resources for the tasks, in cloud systems. In the model, some algorithms were redesigned. These are the distance algorithm between two whales, the spiral algorithm for a whale to surround its prey according to the generated value, the shrinking algorithm to move a whale towards its prey, and the search prey algorithm to move a whale to another randomly selected whale. Thus, it was reported that resource allocation problems were detected and the most appropriate optimization process was carried out.

There are many resource allocation studies specific to fog computing systems in the literature. The trend in this area is to develop frameworks that allocate resources intelligently and efficiently, make maximum use of resources in the fog layer, and are supported by algorithms and mathematical models. In [12], a method of user association and resource allocation was proposed, considering the QoS requirements for reliable and low-latency communication and mobile broadband. With the analytical hierarchy process (AHP), the priorities of the QoS requirement of IoT applications were determined. Then, a bilateral comparison matrix was created in the fog networks coordinator (FNC) to establish relationships in IoT devices and bilateral matching was performed. In this matching, the most appropriate resource allocation approach was implemented for externalities caused by the delay. It has been stated that this method provides useful and efficient resource allocation and user relationship stability. In [13], radio or computational resource allocation problems were examined to increase system optimization and user satisfaction. For this reason, the student-project allocation (SPA) game algorithm for the most appropriate management of resources and in addition to this, the user-oriented collaboration (UoC) method was proposed to eliminate the conflict caused by external influences. Thus, it is emphasized that the externalities are eliminated with the most appropriate resource allocation policy and the method is more efficient than similar studies.

Markov decision process (MDP) was used for the fog radio access network (Fog-RAN) resource allocation, in [14]. The method of discovering the most optimum policy was proposed by using reinforcement methods such as state - action - reward - state - action (SARSA), expected SARSA, Q learning, and Monte Carlo to solve the problems. It has been reported that with this method, tasks get the maximum benefit from resources in fog nodes and high performance for low latency. In [15], a multi-criteria decision making (MCDM) based load balancing method was proposed to prevent congestion in gateways. These congestions can lead to security problems. Load balancing operation is solved with the AHP approach, which works best in Mixed Integer Linear Programming as a multi-objective optimization problem. It is stated that reliable and fast results are achieved for users by preventing bottlenecks and providing load balancing. In [16], a resource allocation scheme was proposed that minimizes the fog computing overhead and provides QoS requirements. This scheme covered three methods. The AHP-based method was developed to determine QoS parameters and task priorities. A resource block allocation method was designed to determine the allocated resources according to parameters such as priority, satisfaction, and trust. For the relationship optimization, a bilateral matching game method was established. It was emphasized that with this resource allocation scheme, efficient load balancing can be provided, QoS can be provided to users and additional loads can be reduced.

Edge computing is one of the computing approaches that require efficient resource allocation due to its resource-constrained nature. End systems can perform higher data processing by being supported by end servers. Similar to methods in the fog and cloud, edge studies also are often in the form of efficient resource management, load balancing, priority tasks. In [17], a resource management scheme was proposed between edge servers (ES) and edge gateways (EG) based on the Lagrangian and the Karush-Kuhn-Tucker (KKT) structure. This scheme includes that an algorithm determines the most efficient resource allocation for edge servers and gateways and a heuristic algorithm that considers the workload of gateways and plans among them. It is stated that the scheme provides workload justice and efficient resource allocation. In [18], a resource allocation algorithm that allows finding the best policy to comply with the service-level agreement (SLA) in terms of reliability and an MDP-based model that allows the adaptation of resources to services was proposed. Reinforcement learning (RL) was created with these methods and a high-level resource allocation was provided for the reliability. It is stated in the study that the method is more efficient than similar approaches. In [19], a deep reinforcement learning (DRL) based resource scheduling scheme was proposed for transmission power, resource allocation, load balancing, and low latency. In the scheme, the related and regularized stacked autoencoder (2r-SAE) algorithm was used to compress high-dimensional data. Later, adaptive simulated annealing (ASA) customized structure was used and DRL was enabled to search efficiently. As a result, the DRL structure was provided to train the network and find the optimal load balance. In the study, it is emphasized that this scheme reaches optimum performance and reduces the processing time.

In [20], RL-based joint communicational and computational resource allocation algorithm (RJCC) was proposed to optimize energy and resource-constrained processing latency in the mobile edge computing (MEC) systems in smart city applications. To optimize the calculation cost, a Q learning-based offloading algorithm was developed in multiple segments and a Lagrange-based migration algorithm in end platforms. In this way, it has been stated that low latency is maintained, better performance is achieved in energy consumption and offloading problems, and QoS is met. In [21], algorithms for dynamic resource allocation in edge systems using ML approaches were proposed. The users were grouped according to the k-means algorithm, considering the priority parameters. The highest priority group performed the computing processes on the edge server, and the lowest priority group performed locally. Others were processed according to the MDP model. Deep Q network (DQN) method was used to control high-dimensional data and find the most appropriate policy. It has been explained that these algorithms show the same performance with a lower cost than similar approaches. In [22], a scheme was proposed for efficient energy consumption, optimum resource allocation, and load calculation management in heterogeneous networks in IoT systems using MEC. This scheme was divided into sub-problems. These are the optimum CPU frequency solution, successive convex approximation (SCA) based power management algorithm, and a distributed load calculation and resource allocation. Algorithms were developed to eliminate these problems. With these algorithms, it has been reported that processes are performed with low latency and efficient computational complexity.

In [23], a cooperative edge computing-based task allocation scheme was proposed that reduces task completion time. In this scheme, the two-edge-node cooperative task allocation based on improved particle swarm optimization (TCA-IPSO) algorithm was designed to improve the optimization process with the crossover and mutation in the genetic algorithm. In this way, it has been stated that it reduces the average task completion time according to a similar algorithm and can be applied to large IoT systems. In [24], an efficient resource allocation policy aimed at minimizing the average deadline and average resource requirement was proposed. The problem was formulated with the MDP and a DQN-based algorithm was designed to solve the problem. It has been declared that this approach provides more efficient resource allocation with less resource usage and less task completion time compared to the standard DQN algorithm and similar studies.

In [25], task offloading and resource allocation approach was proposed to optimize network energy efficiency in the cloud radio access network (C-RAN) structures integrated with the MEC. Four sub-problems were created for the problem examined with mixed-integer nonlinear programming. These are Lyapunov optimization theory, convex decomposition methods for task offloading and resource scheduling, original problem, and matching game for radio resource allocation. In the study, it is emphasized that the methods give great results in terms of energy efficiency and latency.

The use of machine learning approaches is restricted due to data privacy problems in IoT systems used in industries. Therefore, a distributed federated learning (DFL) structure was proposed in [26], which provides resource optimization and protects confidential information in industries. Convex optimization structure was used for relationship and resource allocation, and the rounding technique was used to obtain relationship and resource allocation variables. A lightweight authentication approach was used between small and large base stations with all edge computing. In the study, it has been declared that the DFL-based approach is more efficient and protects privacy compared to random resource allocation and association schemes for smart industries. In [27], a Stackelberg game model was proposed based on the results of a federated learning model transmitted to a coordinator in the cloud and running on the MEC systems. The data detected from IoT sensors were trained locally with the MEC system and the results were transmitted to the cloud coordinator. The cloud system created a global market behavior model with these results. In addition, the Stackelberg equilibrium was created between satisfaction and cost. It has been emphasized that this approach maximizes the benefit for all market participants.

In [28], optimization models that provide optimum deployment, optimum resource allocation, and optimum routing of gateways in a network function virtualization (NFV) supported IoT system was proposed. These are gateway placement and multihop routing algorithm (GMA) for gateway deployment and multihop routing, service placement algorithm 1 and 2 (SP1A and SP2A) for service placement. These algorithms were developed based on the optimization models of similar operations (GMO, SP1O, SP2O (O means optimization)). In the study, it is emphasized that there are differences between SP1A and SP2A in terms of calculation and energy costs, and by using these models, results close to optimum are obtained with less calculation time.

Resource allocation studies that can be used in all computing systems such as fog, cloud, and edge are also performed. In [29], resource allocation and transmission speed improvement models were proposed to provide quality of experience (QoE) in green IoT. In this context, shortest path tree (SPT), SPT with centrality (SPTC), and DQN algorithms were used for resource allocation. In content-centric IoT systems, three algorithms were used together for heterogeneous resources instead of a single algorithm. Thus, in the study, it has been reported that high QoE performance is achieved with efficient resource allocation and high transmission speed in these IoT systems. In [30], a Lyapunov-based scheme was proposed by developing two time-scale resource allocation algorithms to optimize energy costs in green IoT systems. This scheme was created in three parts: energy management, data collection control, and Cross Entropy-based channel planning. It was aimed to provide energy and spectrum sources over time. It is stated in the study that the efficiency of the analysis and algorithm is verified.

In [31], non-orthogonal multiple access (NOMA) based and energy-efficient resource allocation was proposed for heterogeneous IoT systems. A gradual resource allocation was planned for mobile users and IoT users in cognitive radio networks, and this plan was resolved with a deep recurrent neural network (RNN) based algorithm. It has been stated that this algorithm produces the optimum solution with low computational complexity and is more efficient in terms of power and energy than similar approaches. Time offset and repetition operations are used in NB-IoT systems to reduce computational complexity and to achieve a wide scope. In [32], a resource allocation scheme that is sensitive to interference and maximizes speed was proposed for NB-IoT systems. Besides, a cooperative and iterative algorithm was developed to find the optimum solution for resource allocation. It has been stated that the data rate is 89.2 Kbps for downlink and 92 Kbps for uplink, and there is also an improvement and energy saving in the system. In [33], a Stackelberg game theory-based resource allocation scheme was proposed in an IoT system consisting of a hybrit access point (HAP) and a large number of sensors. This scheme was designed to minimize transmission costs and energy transfer costs. The Steckelberg game model was used for the analysis of the problem and the mean-field game approach was used to create the goal function. It is emphasized that the method provides optimum power control for HAP and sensors and reduces costs.

In [34], an automatic resource allocation model for IoT networks was proposed to avoid interferences. Spreading-time technology and orthogonal frequency division multiplexing (OFDM) method were used in the model to prevent low power, bandwidth limits, and interference. Additionally, a joint recursive algorithm was developed to allocate resources without jamming in the model. In the study, it is stated that the model is effective and can be applied to resource-limited IoT systems. In [35], a distributed and Q learning-based algorithm for slot planning in clustered IoT networks was proposed. With this algorithm, it was aimed to prevent interference and improve the signal noise ratio (SIR) in multi-channel systems. In the study, it is reported that channel allocation was made at an appropriate SIR level for each communication device and the IoT network was approximated to a transmission without collision. In [36], a low-complexity framework for efficient allocation of limited resources such as spectrum and energy in wireless IoT networks was proposed. In the framework, users were grouped for channels and designed as a multiple-to-one matchmaking game. In this context, an MDP-based power allocation algorithm and a channel allocation algorithm were developed. It is stated that the algorithms are efficient, at a higher level than similar approaches, and have low complexity.

In [37], an alternating direction method of multipliers (ADMM) based resource allocation approach was proposed for large-scale resource allocation problems. Dinkelbach algorithm was used for optimum energy efficiency and Jacobian and ADMM parallel algorithm was used to optimize controller network latency. Besides, a two-layers iterative resource allocation algorithm based on the Steckelberg game approach was used to resolve the target conflict between the controller and the regional IoT servers. In the study, it is declared that these algorithms are efficient in terms of scalability and approach to the optimum. In [38], a multi-agent DRL-based resource allocation scheme was proposed to optimize shared channel allocation and transmission power. Prioritized experience replay (PER) was used to increase network performance and transmission efficiency and to ensure collaboration in a distributed environment. It has been emphasized that this scheme improves resource management performance by providing energy savings and QoS requirements.

In [39], a graphical-based algorithm and methodology were proposed for allocating mobile resources in shared communication over a difficult scenario with high attempts in the spectrum band. This scenario was designed to solve problems in ultra-reliable low-latency communication for the IoT systems in a smart factory environment. As a result, it is reported that the spectral efficiency of the method is much higher than similar approaches and provides fast resource allocation in dense networks. A resource management approach for the satellite-terrestrial network (STN) was proposed in [40]. The matching design considers the resource cube (MCRC) algorithm, which uses a matching considered preferences (MCPR) algorithm to maximize resource utilization, and Markov approximation analysis to minimize total system latency was developed. Layered architecture and the queuing method were used to manage resources. It has been stated that these algorithms slightly increase resource utilization and greatly reduce the total latency. In [44], a priority queue-based task management algorithm was developed for fog nodes. High performance and low latency were analyzed with simulations. Besides, security policies were ensured with blockchain and lightweight authentication approaches.

22.3.2 Comparison of Resource Allocation Approaches in IoT

The application areas and the data storing and processing structures of recent studies for resource allocation in the IoT systems are shown in Table 22.2. The term "general" refers to all resource supplier types in Table 22.2A. Besides, a more detailed version of the reviewed studies is shown in Table 22.3.

Table 22.2 Solution approaches for resource allocation problem in recent years (Data storage and processing structure (A) and usage area (B))

(A)

Data storage and processing structure	Papers
Cloud	[8–11]
Fog	[12, 15, 16, 44]
Edge	[17–19, 21–24, 26]
Cloud-Fog	[13, 14]
Cloud-edge	[20, 25, 27, 28]
General	[29–40]

(B)

Usage area	Papers
Smart industry	[26]
Market analysis	[27]
Smart city	[20]
Green IoT system	[29–31]
NB-IoT system	[28, 32, 34, 36]
Cognitive IoT system	[38, 40]
Many IoT systems	[8–19, 21–25, 33, 35, 37, 39, 44]

Table 22.3 Solution approaches for resource allocation problem in recent years (Extended details)

Paper	Method description	Advantage	Developed and used algorithm(s)
[8]	Task management approach using combinatorial auction	High task completion and QoS support	No algorithm
[9]	Resource sharing and self-management algorithm of IoT devices	Higher performance and communication capacity	Control into IoT gateway, M2M relations establishment
[10]	Resource allocation approach for WebIDE cloud server, which makes task pre-planning with ant colony optimization	Less task response time and efficient resource use	No algorithm
[11]	IoT resource allocation and optimization with whale optimization based heuristic approach	Problem discovery and optimum solution in resource allocation	Whales distance alg., Spiral alg., Shrinking alg. Searchprey alg.
[12]	User association and resource allocation scheme for IoT applications with high QoS needs	User relationship stability and efficient resource allocation	Pair-wise comparison matrix creation at FNC, Global weight vector at each fog devices, Matching algorithm with externalitier, Best fit resource allocation
[13]	Resource allocation scheme with SPA and UOC approaches	High performance	SPA modelling alg., UOC strategy
[14]	Solution approach for Fog-RAN resource allocation with reinforcement learning methods	Most efficient use for limited resources	Learning optimum policy using Monte Carlo, Learning optimum policy using Q-learning, E-SARSA and SARSA
[15]	Load balancing approach for multiple gateways with MCDM method using AHP	Congestion prevention, reliable and fast result	AHP based load association algorithm
[16]	Common task offloading and QoS-supported resource management in IoT networks	QoS support, low additional load, load balancing	AHP based QoS evaluation alg., Research block allocation alg., Bilateral matching game alg.
[17]	Resource allocation and interference management in the MEC	Performance: 42% total efficiency, 59% latency, 37% data retrieval rate, 40% load balance	Optimized research allocation scheme for ES and EG, Scheduling algorithm between EGs
[18]	RL-based dynamic resource allocation that ensures SLA in terms of reliability in the edge systems	21.72% higher performance	DeraDE (dynamic service resource allocation for distributed edges) algorithm

(continued)

Table 22.3 (continued)

Paper	Method description	Advantage	Developed and used algorithm(s)
[19]	DRL based resource scheduling scheme	Near-optimum performance, low latency, real time	2r-SAE alg., ASA alg., DRL with ASA and 2r-SAE
[21]	Resource allocation in the edge systems with the ML approaches	Lower cost, same performance with the other approaches	Priority driven clustering based on k-means, DQN training for computation task offloading, DQN based computation task offloading scheme
[22]	Efficient resource allocation, energy consumption and load-balancing scheme in heterogeneous structures in the MEC-based systems	Low latency, efficient computational complexity	Iterative joint subchannel and power allocation alg., Successive convex approximation based power allocation alg. based on logarithmic approximation, Joint computation offloading and resource allocation alg.
[23]	Optimum task allocation approach with TCA-IPSO algorithm developed based on genetic algorithm	53.8% and 36% reduction in average task completion time from two similar algorithms	TCA-IPSO algorithm
[24]	DRL-based efficient resource allocation framework	Low task completion time, less resource usage	DQN with multiple replay memories
[25]	Optimum task offloading and resource allocation approach in C-RAN and MEC integrated systems	59% reduction in energy consumption, 57% reduction in average task delay	Matching game based subchannel allocation alg., Dynamic task offloading and resource allocation alg.
[26]	DFL-based resource optimization approach in smart industries	High efficiency, data privacy protection	Proposed decomposition relaxation based algorithm
[27]	Stackelberg game model running on the MEC systems for market analysis and using DFL results in the cloud	Maximizing benefit for all market participants	No algorithm
[28]	Optimization model for gateway placement, resource allocation and routing in NFV-enabled IoT systems	Near-optimum results with low calculation time	Neighborhood generation of gateway placement and multihop routing, Neighborhood generation of service placement

(continued)

Table 22.3 (continued)

Paper	Method description	Advantage	Developed and used algorithm(s)
[20]	RL based joint communucation and computational resource allocation mechanism	Reduce in energy consumption and processing time, QoS support	RJCC, Q-Offloading, L-Migration
[29]	Transmission speed and resource allocation model in Green IoT	High QoE performance	Resource allocation based on SPT, Resource allocation based on SPTC, Resource allocation based on DQN
[31]	RNN-based resource allocation model using NOMA technology for heterogeneous IoT	Low computational complexity and energy, high power	Stepwise peeling alg. for the subchannel assignment of the primary users, Reweighted message passing alg. for the subchannel assignment of the secondary, The learning alg. for training tied parameter
[32]	Maximum speed resource allocation scheme for NB-IoT systems	High speed, energy saving, and overall improvement	Iterative efficient and reliable radio resource management procedure
[33]	Steckelberg-based resource allocation scheme in IoT systems with many sensors and HAP	High power, less cost for HAP and sensors	Mean field control algorithm for the HAP and sensor nodes
[30]	Development of online and two time-scale resource allocation algorithm and Lyapunov based scheme	Energy and spectrum optimization, efficient resource allocation	Channel allocation alg., Two timescale resource allocation alg.
[34]	OFDM-based resource allocation model that avoids interference and jams	Appropriate and effective for resource constrained IoT	Subcarriers assignment of multi-user OFDM system in IoT network, The joint code length and power allocation
[35]	Q learning based algorithm for anti-interference and proper channel allocation in multi-channel systems	Channel allocation at the proper SIR level, non-collision transmission	SIR based slot allocation
[36]	MDP-based resource allocation approach for channel and power allocation	Better performance, low computing and long network life	Efficient channel allocation alg., Matching initialization alg., MDP based power allocation alg.

(continued)

Table 22.3 (continued)

Paper	Method description	Advantage	Developed and used algorithm(s)
[37]	Large-scale resource allocation approach with Dinkelbach, Jacobian-ADMM algorithms	Scalable, optimum energy consumption and latency	ADMM and Dinkelbach based resource allocation alg., Jacobian-ADMM based resource allocation parallel alg., Game and Jacobian-ADMM based two layer iterative resource allocation alg.
[38]	DRL-based efficient resource allocation scheme	Energy saving, QoS support, efficient resource allocation	Resource management with coordinated multi-agent DRL-PER
[40]	Markov approximation based resource management scheme in satellite-terrestrial networks	Increase resource usage, decrease total latency	MCPR alg., MCRC alg.
[39]	Graphically based resource allocation approach for ultra reliable low latency communication	Balanced traffic, spectrum efficiency, fast resource allocation	No algorithm
[44]	Priority queue based task management algorithm and lightweight IoT application scheme	High performance, low latency, secure communication	Priority assignment algorithm, management of fog nodes

22.4 Open Problems

In this section, open problems and future directions for resource management and allocation in the IoT systems are offered. Many studies are carried out in this area. However, ML approaches are just beginning to be integrated. Therefore, open problems for the future works on the issue of resource management and allocation of the resource-constrained systems can be listed as follows:

- IoT systems have become more effective with ML approaches. For this reason, resource allocation methods that are supported by IoT systems and machine learning cooperation and are more effective than traditional methods can be developed.
- Many different devices can work together in IoT applications. Therefore, interoperability can be increased by developing an effective resource allocation method in the IoT systems that share heterogeneous resources to heterogeneous devices.
- In IoT applications with continuous data flow from end devices, some tasks are likely to wait for very long periods. To solve this problem, ML and optimization methods can be combined to provide a balanced and fair resource sharing between devices.

- In IoT systems where spectrum and transmission channels are shared to end devices, there may be an overload on a certain channel. To overcome this challenge, by providing fair channel allocation, overflows that may occur in channels and insecure transmissions can be prevented.
- Computational complexity can be reached to optimum levels when designing ML and optimization-based approaches for more efficient resource allocation in resource-constrained systems. Thus, by integrating these methods developed with existing resource-constrained systems, performance can be increased with lower computational costs.
- Priority-based resource allocation and task management approaches can be developed for tasks transmitted from IoT edge devices to the fog, cloud, or edge computing systems. Thus, urgent processes can be prioritized.
- In IoT systems, resource allocation algorithms can be developed by considering network congestion because data privacy problems may occur in cases of congestion and overflow.
- ML-based resource allocation approaches can be developed by extracting the features for field-specific processes in IoT systems developed specifically for health, industry, smart city, smart home systems, sports, military, agriculture.
- DRL-based approaches can be improved for resource allocation and management in IoT systems, which have been widely used in recent years.
- Similar to [44], high-performance resource management algorithms can be developed in IoT systems.
- Multivariate quadratic and cubic polynomial based identification approaches [41] and key exchange approaches [42] are proposed for quantum resistive IoT applications. Similarly, methods based on post-quantum cryptography can be developed for IoT systems.

22.5 Conclusion

Today, IoT technologies continue to be developed and adopted rapidly. IoT infrastructures provide effective results in applications such as remote control of systems, monitoring an object, and reporting status. In IoT networks that have many heterogeneous resources and serve heterogeneous devices, the sharing, management, and allocation of resources are extremely important for the overall performance of the system. In IoT applications, there are different shared resources such as channels, spectrum, transmission media as well as server resources such as the cloud, fog, edge computing.

In this study, the cloud, fog, and edge computing approaches that provide server resources to IoT systems and have a very important place in these systems are examined. Resource management mechanisms in the IoT systems are classified and resource allocation studies are reviewed. Also, these studies are analyzed in detail and future work suggestions for open problems are offered.

Considering the studies in recent years, the results of the researchers' studies on resource management and allocation in IoT systems are as follows:

- It is observed that in IoT systems, they tend to focus more on MEC structures as a server resource center and try to find solutions to open problems in this area.
- It can be said that they focus on systems with low energy, low latency, and low computational complexity, high efficiency, and supplying QoS requirements in the allocation of transmission environments and server resources.
- It is concluded that they tend to machine learning and artificial intelligence methods such as DQN, DRL, and optimization methods such as game-based approaches, MDP, genetic algorithm, and Lyapunov.

References

1. Patel, K. K., & Patel, S. M. (2016). Internet of things-IOT: Definition, characteristics, architecture, enabling technologies, application & future challenges. *International Journal of Engineering Science and Computing*, *6*(5).
2. Min, M., Xiao, L., Chen, Y., Cheng, P., Wu, D., & Zhuang, W. (2019). Learning-based computation offloading for IoT devices with energy harvesting. IEEE Transactions on Vehicular Technology, 68(2), 1930–1941.
3. Bader, A., Ghazzai, H., Kadri, A., & Alouini, M. S. (2016). Front-end intelligence for large-scale application-oriented internet-of-things. IEEE Access, 4, 3257–3272.
4. Varghese, B., Wang, N., Barbhuiya, S., Kilpatrick, P., & Nikolopoulos, D. S. (2016, November). Challenges and opportunities in edge computing. In *2016 IEEE International Conference on Smart Cloud (SmartCloud)* (pp. 20–26). IEEE.
5. Mukherjee, M., Matam, R., Shu, L., Maglaras, L., Ferrag, M. A., Choudhury, N., & Kumar, V. (2017). Security and privacy in fog computing: Challenges. IEEE Access, 5, 19293–19304.
6. Roman, R., Lopez, J., & Mambo, M. (2018). Mobile edge computing, fog et al.: A survey and analysis of security threats and challenges. *Future Generation Computer Systems*, *78*, 680–698.
7. Musaddiq, A., Zikria, Y. B., Hahm, O., Yu, H., Bashir, A. K., & Kim, S. W. (2018). A survey on resource management in IoT operating systems. IEEE Access, 6, 8459–8482.
8. Choi, Y., & Lim, Y. (2016). Optimization approach for resource allocation on cloud computing for iot. International Journal of Distributed Sensor Networks, 12(3), 3479247.
9. RamíÂrez, P. L. G., Taha, M., Lloret, J., & Tomás, J. (2019). An intelligent algorithm for resource sharing and self-management of wireless-IoT-gateway. *IEEE Access*, *8*, 3159–3170.
10. Wang, H., Li, J., Tian, J., & Wang, K. (2019, November). WebIDE cloud server resource allocation with task pre-scheduling in IoT application development. In *2019 IEEE International Conference on Industrial Internet (ICII)*
11. Sangaiah, A. K., Hosseinabadi, A. A. R., Shareh, M. B., Bozorgi Rad, S. Y., Zolfagharian, A., & Chilamkurti, N. (2020). IoT resource allocation and optimization based on heuristic algorithm. Sensors, 20(2), 539.
12. Abedin, S. F., Alam, M. G. R., Kazmi, S. A., Tran, N. H., Niyato, D., & Hong, C. S. (2018). Resource allocation for ultra-reliable and enhanced mobile broadband IoT applications in fog network. *IEEE Transactions on Communications*, *67*(1), 489–502.
13. Gu, Y., Chang, Z., Pan, M., Song, L., & Han, Z. (2018). Joint radio and computational resource allocation in IoT fog computing. *IEEE Transactions on Vehicular Technology*, *67*(8), 7475–7484.

14. Nassar, A., & Yilmaz, Y. (2019). Reinforcement Learning for Adaptive Resource Allocation in Fog RAN for IoT With Heterogeneous Latency Requirements. *IEEE Access*, *7*, 128014–128025.
15. Banaie, F., Yaghmaee, M. H., Hosseini, A., & Tashtarian, F. (2020). Load-balancing algorithm for multiple gateways in Fog-based Internet of Things. IEEE Internet of Things Journal.
16. Huang, X., Cui, Y., Chen, Q., & Zhang, J. (2020). Joint task offloading and QoS-aware resource allocation in Fog-enabled Internet of Things networks. IEEE Internet of Things Journal.
17. Na, W., Jang, S., Lee, Y., Park, L., Dao, N. N., & Cho, S. (2018). Frequency resource allocation and interference management in mobile edge computing for an Internet of Things system. IEEE Internet of Things Journal, 6(3), 4910–4920.
18. Deng, S., Xiang, Z., Zhao, P., Taheri, J., Gao, H., Yin, J., et al. (2020). Dynamical Resource Allocation in Edge for Trustable Internet-of-Things Systems: A Reinforcement Learning Method. *IEEE Transactions on Industrial Informatics*, *16*(9), 6103–6113.
19. Jiang, F., Wang, K., Dong, L., Pan, C., & Yang, K. (2020). Stacked auto encoder based deep reinforcement learning for online resource scheduling in large-scale MEC networks. IEEE Internet of Things Journal.
20. Xu, S., Liu, Q., Gong, B., Qi, F., Guo, S., Qiu, X., et al. (2020). RJCC: Reinforcement-Learning-Based Joint Communicational-and-Computational Resource Allocation Mechanism for Smart City IoT. *IEEE Internet of Things Journal*, *7*(9), 8059–8076.
21. Liu, X., Yu, J., Wang, J., & Gao, Y. (2020). Resource Allocation With Edge Computing in IoT Networks via Machine Learning. *IEEE Internet of Things Journal*, *7*(4), 3415–3426.
22. Tang, L., & Hu, H. (2020). Computation Offloading and Resource Allocation for the Internet of Things in Energy-Constrained MEC-Enabled HetNets. *IEEE Access*, *8*, 47509–47521.
23. Wang, Q., Shao, S., Guo, S., Qiu, X., & Wang, Z. (2020). Task Allocation Mechanism of Power Internet of Things Based on Cooperative Edge Computing. *IEEE Access*, *8*, 158488–158501.
24. Xiong, X., Zheng, K., Lei, L., & Hou, L. (2020). Resource Allocation Based on Deep Reinforcement Learning in IoT Edge Computing. *IEEE Journal on Selected Areas in Communications*, *38*(6), 1133–1146.
25. Zhang, Q., Gui, L., Hou, F., Chen, J., Zhu, S., & Tian, F. (2020). Dynamic Task Offloading and Resource Allocation for Mobile-Edge Computing in Dense Cloud RAN. *IEEE Internet of Things Journal*, *7*(4), 3282–3299.
26. Khan, L. U., Alsenwi, M., Yaqoob, I., Imran, M., Han, Z., & Hong, C. S. (2020). Resource Optimized Federated Learning-Enabled Cognitive Internet of Things for Smart Industries. *IEEE Access*, *8*, 168854–168864.
27. Lee, J., Kim, D. J., & Niyato, D. (2020). Market analysis of distributed learning resource management for Internet of Things: A game theoretic approach. IEEE Internet of Things Journal.
28. Pham, T. M. (2020). Optimization of Resource Management for NFV-Enabled IoT Systems in Edge Cloud Computing. *IEEE Access*, *8*, 178217–178229.
29. He, X., Wang, K., Huang, H., Miyazaki, T., Wang, Y., & Guo, S. (2018). Green resource allocation based on deep reinforcement learning in content-centric IoT. IEEE Transactions on Emerging Topics in Computing.
30. Zhang, D., Qiao, Y., She, L., Shen, R., Ren, J., & Zhang, Y. (2018). Two time-scale resource management for green Internet of Things networks. IEEE Internet of Things Journal, 6(1), 545–556.
31. Liu, M., Song, T., & Gui, G. (2018). Deep cognitive perspective: Resource allocation for NOMA-based heterogeneous IoT with imperfect SIC. IEEE Internet of Things Journal, 6(2), 2885–2894.
32. Malik, H., Pervaiz, H., Alam, M. M., Le Moullec, Y., Kuusik, A., & Imran, M. A. (2018). Radio resource management scheme in NB-IoT systems. IEEE Access, 6, 15051–15064.
33. Su, J., Xu, H., Xin, N., Cao, G., & Zhou, X. (2018). Resource allocation in wireless powered iot system: a mean field stackelberg game-based approach. Sensors, 18(10), 3173.
34. Dou, Z., Si, G., Lin, Y., & Wang, M. (2019). An adaptive resource allocation model with anti-jamming in IoT network. IEEE Access, 7, 93250–93258.

35. Hussain, F., Hussain, R., Anpalagan, A., & Benslimane, A. (2020). A New Block-Based Reinforcement Learning Approach for Distributed Resource Allocation in Clustered IoT Networks. *IEEE Transactions on Vehicular Technology*, *69*(3), 2891–2904.
36. Liu, X., Qin, Z., Gao, Y., & McCann, J. A. (2019). Resource allocation in wireless powered IoT networks. *IEEE Internet of Things Journal*, *6*(3), 4935–4945.
37. He, Y., Zhang, S., Tang, L., & Ren, Y. (2020). Large Scale Resource Allocation for the Internet of Things Network Based on ADMM. *IEEE Access*, *8*, 57192–57203.
38. Yang, H., Zhong, W. D., Chen, C., Alphones, A., & Xie, X. (2020). Deep reinforcement learning based energy-efficient resource management for social and cognitive Internet of Things. IEEE Internet of Things Journal.
39. Librino, F., & Santi, P. (2020). Resource Allocation and Sharing in URLLC for IoT Applications Using Shareability Graphs. *IEEE Internet of Things Journal*, *7*(10), 10511–10526.
40. Chen, D., Yang, C., Gong, P., Chang, L., Shao, J., Ni, Q., et al. (2020). Resource Cube: Multi-Virtual Resource Management for Integrated Satellite-Terrestrial Industrial IoT Networks. *IEEE Transactions on Vehicular Technology*, *69*(10), 11963–11974.
41. Akleylek, S., Soysaldi, M., Boubiche, D. E., & Toral-Cruz, H. (2019). A novel method for polar form of any degree of multivariate polynomials with applications in IoT. *Sensors*, *19*(4), 903.
42. Akleylek, S., & Seyhan, K. (2020). A probably secure bi-GISIS based modified AKE scheme with reusable keys. IEEE Access, 8, 26210–26222.
43. Akleylek, S. & Karakaya, A. (2020). Data security in fog computing and applications. In S. Sagiroglu, & S. Akleylek (Eds.), *Cyber security and defense: Biometric and cryptographic applications*. Turkey: Nobel Academic Publishing. ISBN:978-625-439-024-1.
44. Karakaya, A., & Akleylek, S. (2021). A novel IoT-based health and tactical analysis model with fog computing. *PeerJ Computer Science*, *7*. https://doi.org/10.7717/peerj-cs.342

Chapter 23
An Empirical Evaluation of Artificial Intelligence Algorithm for Hand Posture Classification

Asad Hussain, Syed Sajjad Hussain, Muhammad Moin Uddin, Muhammad Zubair, Pardeep Kumar, and Muhammad Umair

Abstract During the past decade, an intensive growth of Human–Computer Interaction (HCI) has been evolved. It includes, but is not limited to, Virtual Reality, Augmented Reality, Voice Control Systems, EEG-based systems, etc. Primarily, HCI is the hybridization of Information & Communication Technology and Biometric. Besides, the biometric inputs like Human Face, Voice, Physical actions, EEG signals, etc. are cascaded with the computing module for robust and intelligent decision making. The hand postures are one of the most common biometric for system automation & feedback. In this study, exhaustive empirical research of the machine learning algorithm for hand posture classification has been established. In this connection, a recent dataset, 'Mocap Hand Postures Data Set,' has been opted to employ the different variants of the machine learning algorithm. To the best of the knowledge, the exhaustive comparative study on the said dataset is found to be deficient in the literature. Primarily the principal focus is to identify the best candidate for real-time hand posture classification. Moreover, the performance of each

A. Hussain
Institute of Business Management, Karachi, Pakistan

S. S. Hussain (✉)
Department of Computer Science, SZABIST, Karachi, Pakistan
e-mail: dr.sajjad@szabist.edu.pk

M. M. Uddin · M. Zubair
IQRA University, Karachi, Pakistan

P. Kumar
Department of Software Engineering, Quaid-e-Awam University of Engineering, Science and Technology (QUEST), Nawabshah, Pakistan
e-mail: pardeep.kumar@quest.edu.pk

M. Umair
Department of Computer and Information Sciences, High-Performance Cloud Computing Center (HPC3), Universiti Teknologi PETRONAS, Seri Iskandar, Malaysia

P. Kumar et al. (eds.), *A Fusion of Artificial Intelligence and Internet of Things for Emerging Cyber Systems*, Intelligent Systems Reference Library 210,
https://doi.org/10.1007/978-3-030-76653-5_23

method has been rigorously measuring as a function of training accuracy, testing accuracy, prediction speed, and training time. Besides, the percentage recognition of each corresponding class is illustrated using the respective confusion matrix.

Keywords Artificial intelligence · Machine learning · Classification · Prediction · Virtual reality · Human–computer interaction · Augmented reality

23.1 Introduction

The recent growth of information and communication technology (ICT) has played a catalytic role in a variety of applications like education, business, government, everyday life activities [1]. This also has up- lifted the scope of HCI in the realities of application areas such as security, medical, entertainment, education, etc. [2]. The nascent trend of HCI in the modern-day world has motivated researchers to devise robust dataset and classification models for controlling and automating real-time applications [3]. In this connection, the different dataset has been designed for the verities of forms [4]. HCI with hand gestures plays a vital role as human sometimes relies on their hands to communicate and interact with their environment. So, hand-gesture-based models are prominent from other approaches by providing an easy and natural way of communication and interaction. Many studies have been conducted to evaluate hand- gesture-based techniques, constraints, and how their effectiveness can be increased [5].

MoCap Hand Postures Data Set is one of the recent datasets for hand posture classification. In the literature, very few works have been reported on this dataset to identify the best classification algorithm. This has created the pressing need to devise a comprehensive study to advocate the best candidate classifier for this dataset. The old-style two-dimensional Human-Computer Interaction (HCI), which is keyboard and mouse, cannot handle complications of the latest Virtual Environment (VE) applications. Instead, those applications require the utilization of technologies and different modalities and their integration to create an immersive user experience [6]. There are many difficulties and constraints in the recognition of hand gestures accurately from images as the human hand structure is very complex, and they have versatile shapes [7]. Once the hand is located, hand gesture recognition (HGR), focuses on inferring predetermined gestures to independent signs which are mostly static and can be classified by machine learning pattern classifiers algorithms i-e K-Nearest Neighbors (KNN), Support Vector Machine, Hidden Markov Models, Principal Component Analysis (PCA) [8, 9].

23.2 Literature Review

Hand gesture recognition is accomplished in two ways, glove based electromechanical approach (used in this study) and the other one is a vision-based approach [9, 10]. Motion capture (MoCap) generally refers to techniques and ways for modeling the movement of a body during an activity. This normally involves the coordinated action of the subject's face, hands, and other body parts [11].

Humans use gestures to interact with their surroundings in the early stages of development. The human also uses gestures as facial expressions, body movement, pointing finger, etc. gesture patterns that are used to control applications should be easy to perform and memorize [12]. A large number of innovations are done through human gestures and human gesture recognition machines. Body language is an essential way of communicating between humans, especially in cases of dumb and deaf peoples. In only India, almost 63 million peoples are suffering from hearing loss [13].

Motion capture or Mo-cap is a technique to capture and analyze human expressions. Mo-cap has been used widely to animate graphics figures in motion pictures. Most motion capture camera systems are expensive, cumbersome, time-consuming, and intrusive. These drawbacks not only avoid Mocap data from easy to use, but it might make it impractical for other applications [14]. Human-Computer Interface (HCI) can be helpful for the entertainment industry as well where the control of personalized computer graphic model is based on appearances, shapes, and motions which are captured. It is developing the end products more believable [15]. Control Applications where the transduced motion or pose is used to govern something. Games interfaces are the best examples of this. HCI can also be used entertainment industry where the controller of personalized computer graphics models is based on appearances, motion, and shapes, and motion, which is making the products more believable [15].

A static hand posture is the one in which the hand remains still in a certain position such as a closed fist, holding, etc. whereas a dynamic gesture involves motion of the hand, fingers, and arms, e.g., waving, shaking, etc. [16]. There are three non-vision-based procedures for monitoring orientation and hand position, which are inertial tracking, magnetic and acoustic. A transmitter device releases an alow- frequency magnetic field. From that magnetic field, the receiver, which is a small sensor device determines its orientation and position, which is relative to a magnetic source [17]. In recent times, the interaction of humans with computing devices has become so advanced that it has now become a necessity, and we cannot even live without it. We have embedded technology in our daily lives that we are now using for our daily routine tasks like marking attendance, communication, and even for the sake of entertainment, etc [18].

Primarily there are three main methods in analyzing hand gestures vision-based, glove-based, and analysis by drawing gestures. The glove-based approach works by sensors that can be optical and can also be mechanical which are attached to the glove. The transducer converts finger flexions into electrical signals, which are used to determine hand posture [19]. Human Motion Capture (MoCap) is often used to conduct a study on musculoskeletal biomechanics and also clinical problems. It also provides animations to the entertainment industry. For capturing human motions, the most popular technique is using markers that are placed on the skin, even though it has some serious drawbacks [20].

23.3 Classifiers

23.3.1 Decision Tree

A decision tree (DT) is a supervised learning technique. A Decision Tree (DT) is the representation of various decisions and their possible significances graphically, which include but are not limited to chance like events, utility, and resource cost [21]. It is a machine learning algorithm that is a decision support tool that uses a graph that resembles a tree structure or model of decisions. There are mainly three parts of a decision tree internal node, Branch, and Leaf Node [22]. This algorithm can be used in both classification and regression problems. The variants of a decision tree used in this study are Fine tree, Medium tree, and coarse tree.

23.3.2 Discriminant Analysis

It is an algorithm used in machine learning to find a combination of features that characterize two or more classes of events. The result combination can be used as a linear classifier. It can also be used in regression [23]. The variants of discriminant analysis used in this study are linear discriminant and Quadratic Discriminant.

23.3.3 Support Vector Machine

SVM is a discriminative classifier that separates two linear classes by a hyperplane. In supervised learning, when labeled training data is inputted, SVM outputs an optimal hyperplane, which categorizes new examples [24]. The variants of SVM used in this study are Linear SVM, Quadratic SVM, Cubic SVM, Fine Gaussian SVM, Medium Gaussian SVM, and Coarse Gaussian SVM.

23.3.4 *K-Nearest Neighborhood*

A KNN algorithm is an approach to classify data that estimates that how possible data point is to lie in one group or the other depending on in which group data points close to it are in [25]. The variants of K-NN used in this study are Fine K-NN, Medium K-NN, Coarse K-NN, Cosine K-NN, Cubic K-NN, and Weighted K-NN.

23.3.5 *Ensemble*

Ensemble classifier refers to a group of individual classifiers which are collectively trained in a dataset [26]. The variants of ensemble classifiers used in this study are Boosted trees, Bagged trees, Subspace discriminant, Subspace K-NN, and RUSBoosted trees.

23.4 Dataset

The dataset includes 5 gestures which are static (hand poses) and is captured against ten users by using a Vicon motion capture camera system, and there is a glove on which infrared markers on certain joints are attached. A rigid pattern on gloves' back was used to establish translation and rotation invariant local coordinate system. The gestures which were captured were pointing with one finger, fist, flat hand, pointing with two fingers and fingers curled (grab). Hundreds of instances of each gesture were captured in streams where the user held the gesture for a short time. All the missing values, values in 0, and markers more than 200 mm from the origin and which transformed to local coordinates were all removed [27]. The details of the dataset and its attribute information can be referred from the data source [27].

23.5 Analysis and Results

The main focus and objective of this research are to give a detailed overview of the comparative study of different Artificial intelligence classification algorithms w.r.t to its different parameters to classify Hand posture. For this purpose, we used MATLAB (classification learner App) 2018 as a tool. This app contains various classification algorithms with its different variants which are discussed in this research.

Tested ML algorithms have a set of results based on different classification parameters. The list of tested parameters is "Training Accuracy (%)," "Testing Accuracy (%)," "Prediction Speed (10 K-Obs/s) ", "Training Time (s)." Through these parameters, we find the most competent and capable classification algorithm. In classification

Algorithms Main parameter to check and validate the algorithms is Testing Accuracy (%), as shown in the reading Table. According to the reading table, as seen that Fine Gaussian SVM got the highest testing accuracy, which is then succeeded by Subspace KNN. We select 96% as the minimum threshold so that the total algorithms that lie in this are listed from highest to lowest rank: bagged Trees, Fine KNN, Weighted KNN, Cubic SVM, Medium Gaussian SVM. If we talk about another Measurement parameter second in precedence is Prediction Speed. The top two algorithms w.r.t to Prediction speed are Coarse Tree and Medium Tree. Fine tree secured third place in terms of Prediction speed. The third parameter, which is necessary to check the performance of the Algorithm is Training Speed. According to the reading table, we can state that Cubic SVM and Fine Gaussian SVM are the top two algorithms in terms of Training Speed. Quadratic SVM, Cubic KNN are the further two in rank third and fourth in the list. In the account of Training Accuracy, Fine Gaussian SVM, Fine KNN, Weighted KNN, Bagged Trees, and Subspace KNN all stated have 100% training accuracy. Cubic SVM is in the second rank, with a training accuracy of 98.5%. Concluded from all above that **Fine Gaussian SVM** is the most optimal and efficient Machine Learning algorithm for the classification of hand posturing. Its reading for parameters are, For Training Accuracy 100%, Testing Accuracy 100%, Prediction Speed 0.049, and Training Speed 298.24 all these measurements can be seen visually in the form of a graph.

Furthermore, detailed reading and observation for all 22 tested classification algorithms are shown in Table 23.1. It also illustrates the confusion matrix for each algorithm by which the performance of every candidate can be validated. The table also contains values for True positive and false negative for all algorithms Detail table is stated below

In addition to the empirical evaluation, this study also submits the graphical illustration of the testing result. In the following figure, the testing result of each algorithm is presented with the corresponding label. The testing graph illustrated the relation of True values in blue color on X-Axis (Samples) against the classified value in red color at Y-axis (Output). In these figures, The true value is represented with blue color and the algorithm classified result is shown with red color. The correct recognition is the point where both curves intersect for the corresponding sample. The graphical results also advocate for the Fine Gaussian as the optimal algorithm for this problem domain (Fig. 23.2).

23.6 Conclusion and Future Work

It is evident from the Literature review that detail investigation upon the selected dataset is witnessed to be scarce in the past. As stated above the main focus of this study is to figure out the best algorithm based on several Measurement parameters, testing accuracy (%), Prediction Speed (Obs/sec), Training Time (Obs/Sec), and training accuracy (%) along with its confusion matrix and values of True positive and false negative The same can be visualized in testing graphs and for a detailed

Table 23.1 Comparison results of different parameters concerning training accuracy, testing speed, prediction speed, training time along with confusion matrix, and the values of its true positive and false negative

	Confusion matrix			True class					True positive	False negative	Training accuracy (%)	Testing accuracy (%)	Prediction speed (obs/s)	Training time (s)
				1	2	3	4	5						
Decision tree	FineTree	Predicted class	1	95	1	2	1	2	95	5	76.50%	84.0956	~410,000	15.86
			2	1	76	2	7	14	76	24				
			3	3	2	76	15	4	76	24				
			4	2	11	14	65	8	65	35				
			5	1	15	5	10	69	69	31				
	Medium tree	Predicted class	1	81	4	3	5	8	81	19	62.8	73.2617	~640,000	9.9451
			2	1	65	4	5	26	65	35				
			3	5	6	70	9	10	70	30				
			4	1	20	23	39	16	39	61				
			5	1	20	10	14	56	56	44				
	Coarse Treeh	Predicted class	1	82	6	8	0	4	82	18	46.5	68.3068	~1,200,000	2.3163
			2	2	73	20	0	6	73	27				
			3	24	10	49	0	18	49	51				
			4	7	25	38	0	31	0	100				
			5	4	44	28	0	24	24	76				
Discriminant analysis	Linear	Predicted Class	1	74	0	13	2	11	74	26	59.3	66.2855	~33,0000	3.3856
			2	1	70	9	3	17	70	30				
			3	7	1	71	6	15	71	29				
			4	16	17	18	27	22	27	73				
			5	7	19	8	14	52	52	48				

(continued)

Table 23.1 (continued)

	Confusion matrix			True class 1	2	3	4	5	True positive	False negative	Training accuracy (%)	Testing accuracy (%)	Prediction speed (obs/s)	Training time (s)
	Quadratic	Predicted class	1	97	<1	2	1	1	97	3	83.6	88.681	~280,000	1.575
			2	<1	83	2	8	6	83	17				
			3	1	2	86	6	5	86	14				
			4	<1	5	22	68	5	68	32				
			5	0	11	1	6	81	81	19				
Support vector machine	Linear SVM	Predicted class	1	86	0	6	1	7	86	14	64.2	75.1213	~100,000	205.16
			2	2	68	9	4	17	68	32				
			3	7	1	73	5	15	73	27				
			4	12	12	15	38	23	38	62				
			5	8	18	7	14	52	52	48				
	Quadratic SVM	Predicted class	1	>99		<1	<1	<1	>99	<1	91.3	94.2596	~5100	311.55
			2	<1	88	2	4	6	88	12				
			3	1	<1	95	2	1	95	5				
			4	<1	3	9	86	2	86	14				
			5		9	1	3	87	87	13				
	CubicSVM	Predicted class	1	100	<1	<1	<1	<1	100		98.5	98.1173	~10,000	276.81
			2	<1	98	<1	1	1	98	2				
			3	<1	<1	99	1	<1	99	1				
			4	<1	<1	3	96	<1	96	4				

(continued)

Table 23.1 (continued)

	Confusion matrix			True class 1	2	3	4	5	True positive	False negative	Training accuracy (%)	Testing accuracy (%)	Prediction speed (obs/s)	Training time (s)
			5	<1	1	<1	<1	99	99	1				
	Fine Gaussian	Predicted class	1	100	<1	<1	0	<1	100		100	100	~490	298.24
			2	<1	>99	0	<1	<1	>99	<1				
			3	<1	0	>99	<1	<1	>99	<1				
			4	<1	0	<1	>99	<1	>99	<1				
			5	<1	<1	<1	<1	100	100					
	Medium Gaussian	Predicted class	1	>99	<1	<1	0	<1	>99	<1	95.4	96.0383	~2500	72.419
			2	<1	95	1	2	2	95	5				
			3	1	<1	98	1	<1	98	2				
			4	<1	2	7	90	<1	90	10				
			5	0	4	1	1	94	94	6				
	Coarse Gaussian	Predicted class	1	98	0	2	0	<1	98	2	77	87.3527	~890	156.67
			2	<1	75	6	4	15	75	25				
			3	3	1	82	4	10	82	18				
			4	3	10	14	56	17	56	44				
			5	<1	16	6	6	71	71	29				
KNN	Fine KNN	Predicted class	1	100	<1	<1	<1	<1	100	0	100	98.6602	~1200	79.013
			2	<1	100	<1	<1	<1	100	0				
			3	<1	0	100	0	0	100	0				

(continued)

Table 23.1 (continued)

	Confusion matrix			True class					True positive	False negative	Training accuracy (%)	Testing accuracy (%)	Prediction speed (obs/s)	Training time (s)
				1	2	3	4	5						
			4	<1	0	0	100	0	100	0				
			5	<1	<1	0	0	100	100	0				
	Medium KNN	Predicted class	1	>99	<1	<1	<1	0	>99	<1	91.3	92.6311	~1400	59.345
			2	1	91	2	3	3	91	9				
			3	2	<1	96	2	<1	96	4				
			4	1	3	12	84	<1	84	16				
			5	2	9	2	2	86	86	14				
	Coarse KNN	Predicted class	1	>99	0	<1	<1		>99	<1	80.7	87.5144	~940	82.859
			2	2	80	7	6	6	80	20				
			3	7	<1	87	5	<1	87	13				
			4	3	6	21	70	1	70	30				
			5	9	15	5	5	66	66	34				
	Cosine	Predicted	1	>99	<1	<1	<1	0	>99	<1	91.3	92.7466	~1000	77.136
			2	1	92	2	3	3	92	8				
			3	2	1	86	2	<1	96	4				
			4	1	4	12	82	<1	82	18				
			5	1	9	2	2	86	86	14				
	Cubic KNN	Predicted class	1	>99	<1	<1	<1	0	>99	<1	88.8	90.9217	~41	1968.6
			2	1	89	3	4	3	89	11				

(continued)

Table 23.1 (continued)

	Confusion matrix			True class					True positive	False negative	Training accuracy (%)	Testing accuracy (%)	Prediction speed (obs/s)	Training time (s)
				1	2	3	4	5						
			3	3	<1	95	2	<1	95	5				
			4	1	4	14	80	1	80	20				
			5	3	12	3	2	80	80	20				
	weighted KNN	Predicted class	1	100	<1	0	0	0	100	0	100	98.614	~1100	71.566
			2	<1	100	<1	0	0	100	0				
			3	0	<1	100	<1	0	100	0				
			4	0	0	<1	100	<1	100	0				
			5	0	0	0	<1	100	100	0				
Ensemble classifiers	Boosted trees	Predicted Class	1	85	3	1	2	8	85	15	74.9	77.5121	~50,000	35.059
			2	1	76	3	4	16	76	24				
			3	5	4	77	5	8	77	23				
			4	2	19	18	56	5	56	44				
			5	<1	12	4	6	78	78	22				
	Bagged trees	Predicted class	1	100	0	0	0	0	100	0	100	99.4109	~32,000	19.453
			2	<1	100	0	<1	<1	100	0				
			3	<1	<1	100	<1	<1	100	0				
			4	<1	<1	0	100	0	100	0				
			5	0	0	0	0	100	100	0				
	Subspace	Predicted class	1	80	3	8	1	9	80	20	57.9	68.2144	~16,000	9.8466

(continued)

Table 23.1 (continued)

	Confusion matrix		True class 1	2	3	4	5	True positive	False negative	Training accuracy (%)	Testing accuracy (%)	Prediction speed (obs/s)	Training time (s)
		2	1	71	9	2	17	71	29				
		3	10	5	70	3	13	70	30				
		4	19	18	25	15	23	15	86				
		5	11	24	7	8	50	50	50				
Subspace KNN	Predicted class	1	100	<1	0	0	<1	100	0	100	99.4225	~15,000	11.54
		2	<1	100	<1	0	0	100	0				
		3	0	<1	100	<1	0	100	0				
		4	0	<1	<1	100	<1	100	0				
		5	0	0	0	<1	100	100	0				
RUSboosted trees	Predicted class	1	83	4	2	4	8	83	17	65.2	73.3426	~50,000	32.569
		2	1	78	6	6	10	78	22				
		3	11	6	68	7	7	68	32				
		4	3	23	28	40	6	40	60				
		5	<1	22	9	14	55	55	45				

	Confusion matrix	True class	1	2	3	4	5	True positve	False negative	Training accuracy (%)	Training accuracy (%)	Prediction speed (obs/s)	Training time (s)
Artificial neural	Predicted class	1	80	6	3	4	80	83	17	61.2	42.3426	~23,000	47.569
		2	2	73	8	6	10	73	24				

(continued)

Table 23.1 (continued)

Confusion matrix	True class						True positive	False negative	Training accuracy (%)	Training accuracy (%)	Prediction speed (obs/s)	Training time (s)
		1	2	3	4	5						
	3	19	6	50	7	7	50	42				
	4	6	23	28	36	8	36	60				
	5	<1	22	9	14	55	55	45				

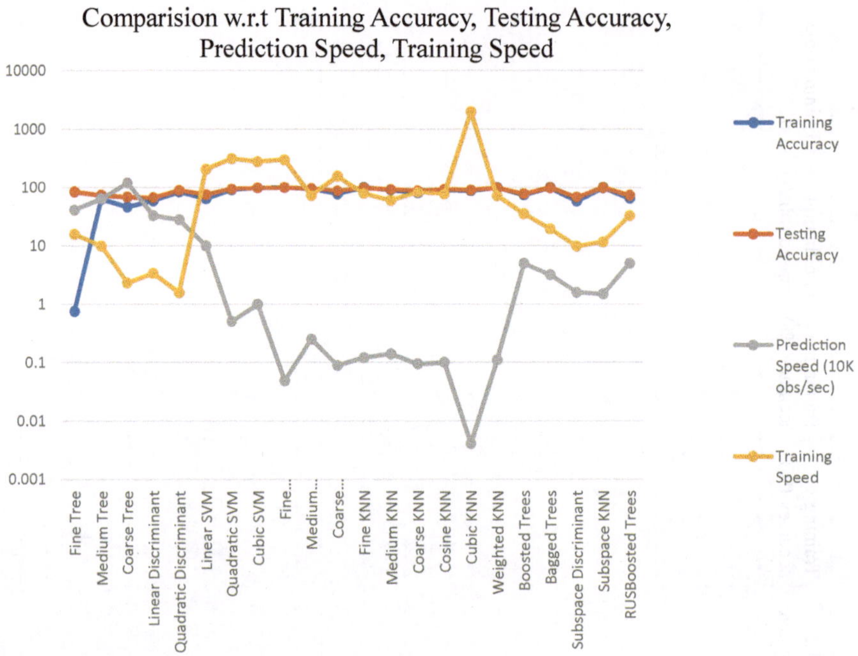

Fig. 23.1 Comparison w.r.t training accuracy, testing accuracy, prediction speed and training speed

view of every calculation is outlined in and Table 23.1. A vital part of verified classification performance of ML algorithms is testing the accuracy of a particular Candidate, as shown in Table 23.1. Fine Gaussian SVM is the most optimal and efficient Machine Learning algorithm for the classification of hand posturing. It's reading for parameters are, For Training Accuracy 100%, Testing Accuracy 100%, Prediction Speed 0.049, and Training Speed 298.24. We can say that Fine Gaussian SVM is the best ML classification Algorithm, and the least feasible algorithm is Linear Discriminant for the said dataset. As a future work new robust variants of Deep learning algorithm are suggest to devise a pre-trained model.

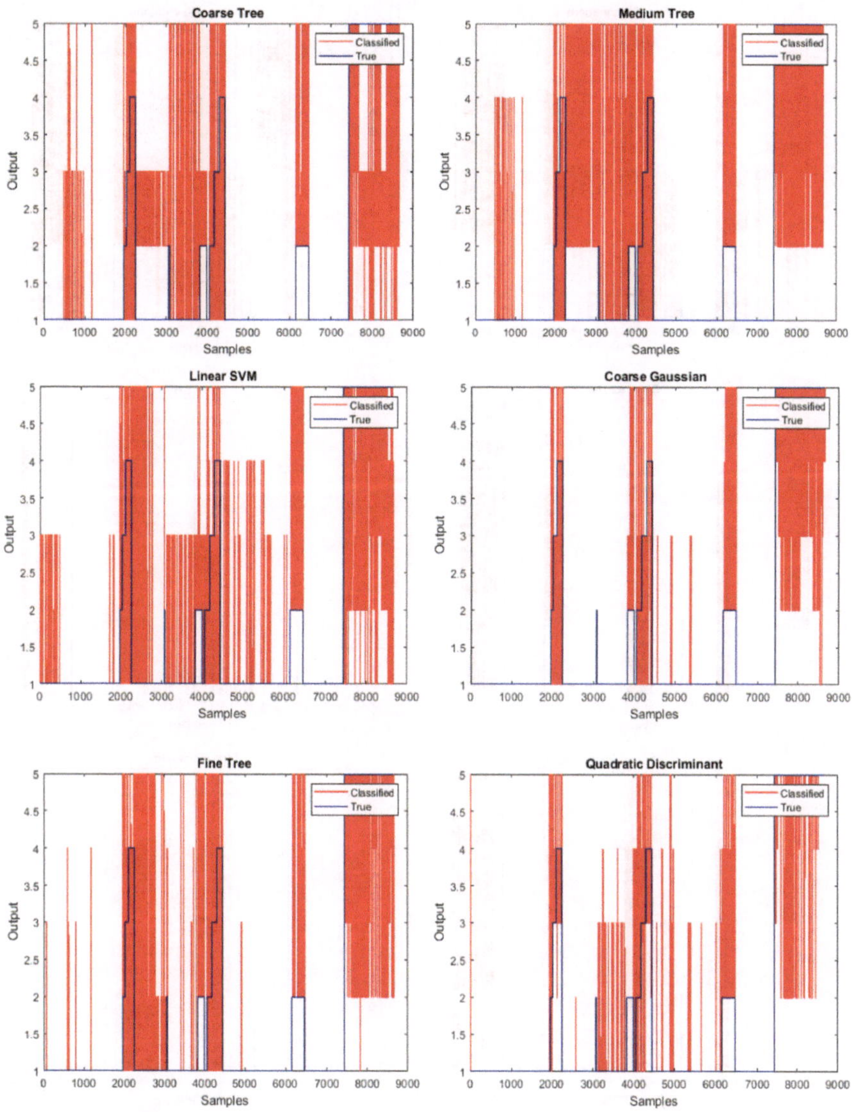

Fig. 23.2 Testing accuracy curve of ML algorithm

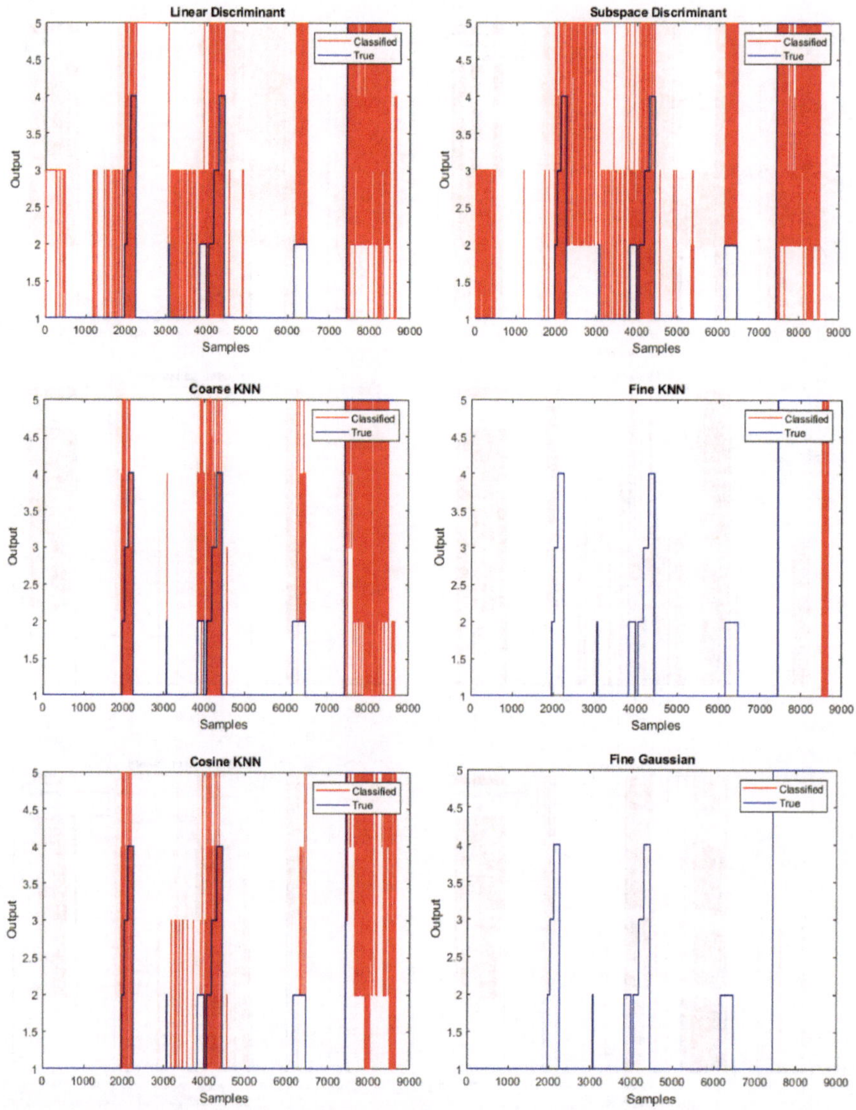

Fig. 23.2 (continued)

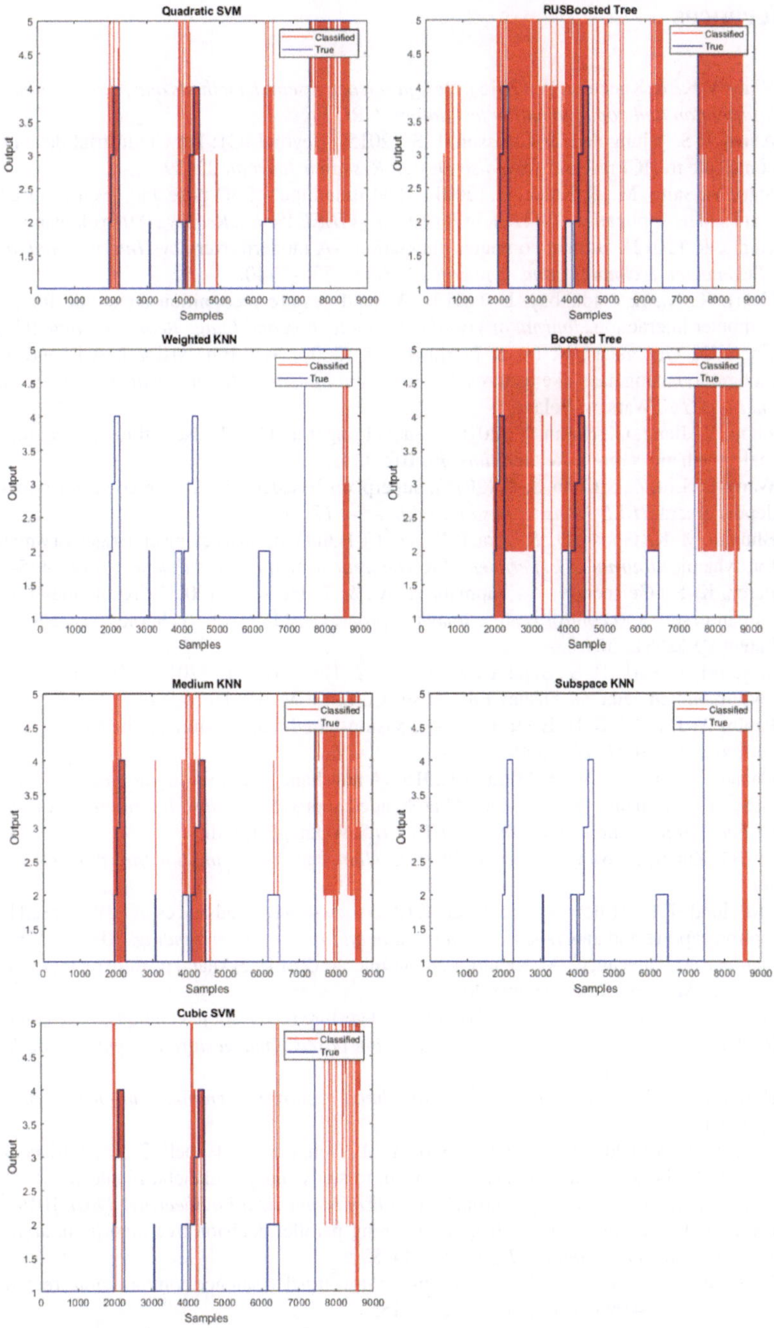

Fig. 23.2 (continued)

References

1. Vijay, V. K., & Kuechler, W. (2015). *Design science research methods and patterns: innovating information and communication technology*. CRC Press.
2. Anna, T. S., Ljungblad, & Claesson, I. S. (2015). Beyond ICT: How industrial design could contribute to HCI research. *Swedish Design Research Journal,* 22–29.
3. Sato, Y., Saito, M., & Koike, H. (2001). Real-time input of 3D pose and gestures of a user's hand and its applications for HCI. In *Proceedings IEEE Virtual Reality 2001*, Yokohama, Japan.
4. Gupta, R. (2012). Human-computer interaction—A modern overview. *International Journal of Computer Technology and Applications, 3*(5), 1736–1740.
5. Çağrı, K. N., & Güdükbay, U. (2015). A hand gesture recognition technique for human-computer interaction. *Journal of Visual Communication and Image Representation,* 97–104.
6. Chen, Q., Georganas, N. D., & Petriu, E. M. (2007) Real-time vision-based hand gesture recognition using haar-like features. In *Instrumentation and Measurement Technology Conference—IMTC*, Warsaw, Poland.
7. Zhou, Y., Jiang, G., & Lin, Y. (2016). A novel finger and hand pose estimation technique for real-time hand. *Pattern Recognition, 49*, 102–114.
8. Wang, C., Liu, Z., & Chan, S.-C. (2015). Superpixel-based hand gesture recognition with kinect depth camera. *IEEE Transactions on Multimedia, 17*(1).
9. Bhuyan, M. K., Ghosh, D., & Bora, P. K. (2006). Hand motion tracking and trajectory matching for dynamic. *Journal of Experimental & Theoretical Artificial Intelligence, 18*(4), 435–447.
10. Perez, K. S., Geisner, K. A., Kipman, A. A., & Tsunoda, K. (2009). Techniques for using human gestures to control gesture unaware programs. Microsoft Technology Licensing LLC Patent US12551370.
11. Shiratori, T., Park, H. S., Sigal, L., Sheikh, Y., & Hodgins, J. K. (2011). Motion capture from body-mounted cameras. Disney Enterprises Carnegie-Mellon University.
12. Pablo, W. J. K. M. S. H. E., & Y. (2011). Vision-based hand-gesture applications. *Communications of the ACM, 54*(2), 60–71.
13. Chanda, K., Ahmed, W., & Mitra, S. (2015). A new hand gesture recognition scheme for similarity measurement in a vision-based barehanded approach. In *Third International Conference on Image Information Processing (ICIIP) 2015*, Waknaghat, India.
14. Liu, G., Zhang, J., Wang, W., & McMillan, L. *Human motion estimation from a reduced marker set*.
15. Moeslund, T. B., Hilton, A., & Kruger, V. (2006). A survey of advances in vision-based human motion capture and analysis. *Computer Vision and Image Understanding, 104*, 90–126.
16. Gardner, A., Duncan, C. A., Kanno, J., & Selmic, R. (2014). 3D hand posture recognition from small. In *ResearchGate*, October 2014.
17. LaViola, J. J., Jr. A survey of hand posture and gesture recognition techniques and technology.
18. Rautaray, S. S., & Agrawal, A. (2012). *Vision-based hand gesture recognition for human.* Springer, Dordrecht.
19. Hasan, H., & Kareem, S. K. (2012). *Static hand gesture recognition using neural networks.* Springer.
20. Corazza, S., Mundermann, L., Chaudhari, A. M., Dematteo, T., Cobell, C., & Andriacchi, T. P. (2006). A Markerless motion capture system to study musculoskeletal biomechanics: Visual hull and simulated annealing approach. *Annals of Biomedical Engineering, 34*(6), 1019–2019.
21. Haim, Y. B., & Tov, E. T. (2010). A streaming parallel decision tree algorithm. *Journal of Machine Learning Research, 11*(2010), 849–872.
22. Gupta, P. (2017). *Towards data science.* https://towardsdatascience.com/decision-trees-in-machine-learning-641b9c4e8052. Accessed 2018.
23. Lachenbruch, P. A., Sneeringer, C., & Revo, L. T. (2007). Robustness of the linear and quadratic discriminant function to certain types of non-normality. *Communications in Statistics,* 39–56.
24. Adankon, M. M., & Cheriet, M. (2015). Support vector machine. S. Z. Li & A. K. Jain (Eds.), *Encyclopedia of biometrics*. Springer.

25. Maillo, J., Ram'ırez, S., Triguero, I., & Herrera, F. (2016). *kNN-IS: An iterative spark-based design of the k- nearest neighbors classifier for big data, knowledge-based systems.*
26. Rokach, L. (2010). Ensemble-based classifiers. *Artificial Intelligence Review, 33*(1–2), 1–39.
27. Gardner, A., Duncan, C. A., Kanno, J., Selmic, R. (2014). 3D hand posture recognition from small. In *IEEE Conference on Computer Vision and Pattern Recognition (CVPR)* (pp. 137–143), June 2014.

Chapter 24
Framework for Mid-Air Traffic Collision Detection Using Data Analytics

Rohail Qamar, Raheela Asif, Pardeep Kumar, and Syed Abbas Ali

Abstract Flight safety is a wide area ranging from human life to aircraft. One of the biggest risks is lack of communication or communication failure between the pilot and air traffic control tower can lead to flight delays and catastrophic accidents not just destroying the aircraft but also losing hundreds of human lives. The purpose of this study is the betterment of aircraft safety. We developed a system using data analytics, which creates an environment in the aircraft's cockpit by providing the pilot with a visual system. It allows identifying the neighboring aircraft that lies in the proximity of one aircraft. Generating the alert and warning that other aircraft comes in that radius. This system tries to ensure the maximum possible safety of the aircraft in a mid-air collision.

Keywords Data analytics · Proximity of aircraft · Exploratory data analysis · Flight safety · Safety toolkit · Aviation accidents

R. Qamar
Hamdard University, Karachi, Pakistan
e-mail: rohail.qamar@hamdard.edu.pk

R. Asif (✉) · S. A. Ali
N.E.D University of Engineering & Technology, Karachi, Pakistan
e-mail: rahmed@neduet.edu.pk

S. A. Ali
e-mail: saaj@neduet.edu.pk

P. Kumar
Department of Software Engineering, Quaid-e-Awam University of Engineering, Science and Technology (QUEST), Nawabshah, Pakistan
e-mail: pardeep.kumar@quest.edu.pk

P. Kumar et al. (eds.), *A Fusion of Artificial Intelligence and Internet of Things for Emerging Cyber Systems*, Intelligent Systems Reference Library 210,
https://doi.org/10.1007/978-3-030-76653-5_24

24.1 Introduction

24.1.1 Research Work Description

The framework is a web-based application that will assist the pilot in making decisions in the emergency in case of communication failure between ATC (Air Traffic Controller) and captain. It takes historical ADS-B (Automatic Dependent Surveillance-Broadcast) raw data and then displays insights, mapping of proximity the surrounding area of the aircraft on the world map, generates alerts in case of other aircraft enters the proximity of the aircraft. Predicts the collision time, so the pilot would easily take the decision to change the position of the aircraft.

Aircraft feeder sends data to ADS-B receiver it will then upload on to ADS-B API (Application Programming Interface) so we can fetch raw data from it. Extraction of meaningful information using data cleaning and preprocessing. Finally, the data are shown in the form of different charts, graphs and map the visual representation on the dashboard.

The goal of the presented framework to develop a dashboard-based web application that provides a second personal eye to the pilot that will allow making decisions on their own in case of communication failure. Determine the threshold values for altitude, latitude, longitude, proximity to avoid a collision.

24.1.2 Research Significance

The framework is intended to develop consists of a visual dashboard that could be integrated into the pilot's cockpit, which will be visualized to the pilot. It will monitor the surroundings such as other aircraft in the proximity by analyzing their coordinates, speed, and direction. The system extracts the required data fields from this API and after data filtering plot the aircraft and depict proximity around the aircraft.

It is a 3-tier proximity level of an aircraft that range from 9 to 28 km (5–15 nm). This proximity level is the area around the aircraft, whether in mid-air or landing, or even taking off. The 3-tier proximity is further discussed in detail in the section methodology. This proximity area around the aircraft will have the following attributes or elements in it.

1. Aircraft in the proximity
2. Speed of the aircraft
3. Position and Direction of the aircrafts.

This mapping is done using the raw data from JSON (JavaScript Object Notation) file. The raw data originate from the ADS-B API. It consists of a count of almost 50+ attributes which were to be brought down to only the attributes that were useful for us in developing the visual system. These attributes that were used to map the aircraft were just around 10 attributes that deemed important. This was done using the cleaning process that will get to in the later sections.

The beneficiaries of the proposed solution are the aviation industry for enhancement and betterment. They will now have the ability to perceive the environment and take decisions accordingly. The web-based application includes the following features.

- Dashboard
- Authentication and authorization
- Admin panel
- Mapping of proximity and aircraft in that vicinity
- Generates alerts and warning.

The goal of this proposed work is to find, identify collision detection and generate the alert so pilots can easily take their own decision if there is a communication failure or gap between ATC [1–3].

24.2 System Description

The proposed work is based on 4 major subsystems that make it up as a whole system. These subsystems are the major building foundation of the whole system. Each of the subsystems has defined roles and jobs that they conduct. These subsystems rely on each other for inputs and outputs to generate the final output of the system. The major subsystems are.

24.2.1 Data Acquisition

Every aircraft in the world either it is in the air or land, feeds out data to the ADS-B system through a sensor placed inside the aircraft [4–7]. This data is recorded every 8 s and is refreshed in an interval of every 8 s. This data contains almost 50+ attributes out of which some of them are missing at times, but the data that is of use to the system consists of so far 20+ attributes that we have used in our system. These attributes are the speed of the aircraft, the latitude, longitude, altitude, aircraft type, ICAO (International Civil Aviation Organization) identifier of the aircraft, engineering companies, and the engine type of the aircraft, etc. The online real-time API has not been available for some time and as of now stale data is accessible, but it is expected to be made available soon.

This data is extracted which then makes use of this data, by extracting the useful features of the data by passing it through cleaning and filtering operations, which yield the useful data, which is then implemented by the application to analyze and generate results on the data. The data comes in the form of the JSON format. Figure 24.1 illustrates the preparation of data in the presented framework.

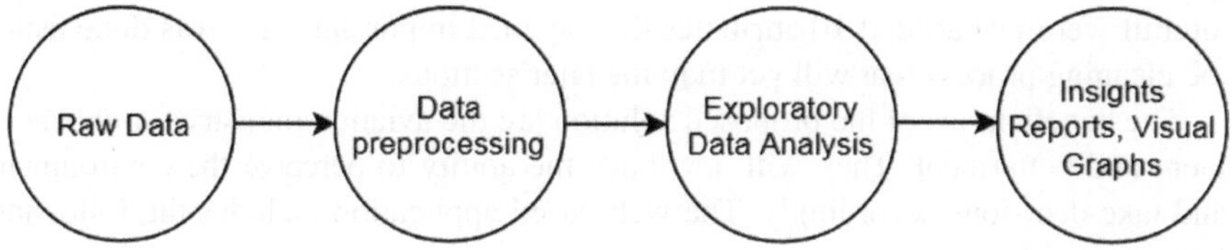

Fig. 24.1 Data preparation

24.2.2 The Safety Toolkit

This subsystem is where all the calculations and the visualization algorithms are deployed in the system. Here, we deal with the proximity visualization algorithm and the alerts calculations, which are all then passed onto the reporting subsystem.

This subsystem first takes input from the Data API as the attributes that are conveyed to the application in from JSON list. The application uses an algorithm to process the attributes to gain insight from them. It performs two major operations as a whole subsystem. These operations are as follows:

- Visualizing the Proximity
- Insights for Alerts.

This visualization is based on a 3-tier proximity level, which will be defined later in the methodology section. This mapping is used to draw the whole map of the world and the aircraft are then placed on the maps with their respective attributes.

This operation concludes by passing on the visualization to the client subsystem and the alert insights to the reporting subsystem which then displays the information to the client. The overall flow is shown in Fig. 24.2.

24.2.3 The Reporting

Figure 24.3 shows the subsystem is responsible for the successful reporting of the alerts and the risks to the client by using certain data transfer techniques. This occurs after the proximity visualization phase by the toolkit subsystem and works on the data provided by the toolkit to analyze the risks for the pilot or the client.

24.2.4 The Client/Pilot

This subsystem occurs in the form of a dashboard. It conveys the necessary information to the pilot or the other concerned authorities such as the ATC, the aircraft

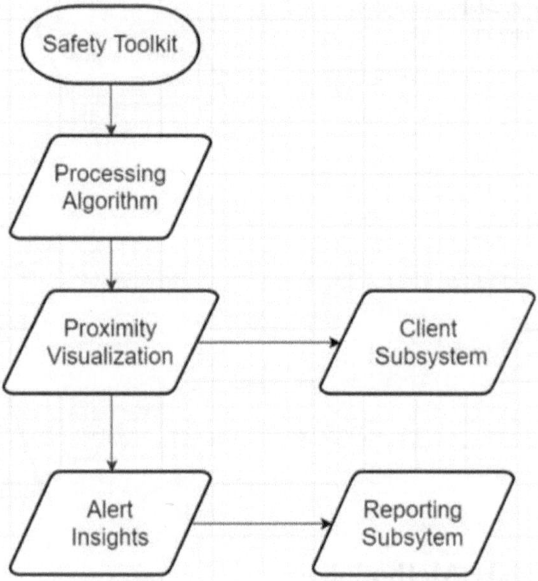

Fig. 24.2 Safety toolkit flow diagram

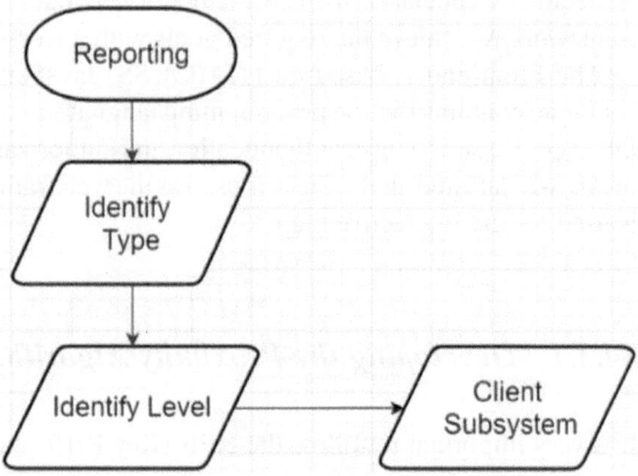

Fig. 24.3 Reporting flow diagram

carriers, or owners. Figure 24.4 shows the client with a visual dashboard that integrates the map and the proximity component from the toolkit subsystem with the reporting insights and alerts from the reporting subsystem.

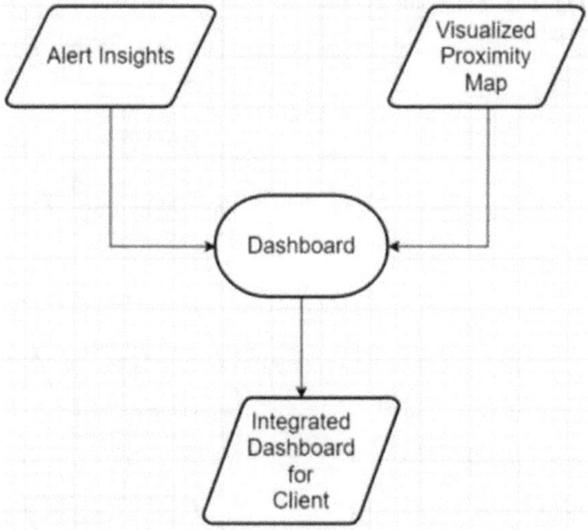

Fig. 24.4 Client flow diagram

24.3 Methodology

A detailed explanation of the methodology that is being used to complete this framework & achieve the required goals with a working prototype.

The Front-end is based on HTML/CSS, JavaScript, and jQuery. Similarly, the back-end contains Python programming language and databases MySQL and Mongo DB. SQL is used for user authentication and authorization and Mongo DB for storage of ADS-B data and alerts. Flask is used as the web framework, while the folium library is used for the interactive map.

24.3.1 Developing the Proximity Algorithm

It is very important to follow the KPIs (Key Performance Indicators) of the Federal Aviation Administration for the aircraft separation. There are two types of separation of aircraft which are to be considered.

1. Vertical separation
2. Horizontal separation.

The consideration in vertical separation is that aircraft flying less than or equal to 29,000 ft (8839 m) must maintain the separation of at least 1000 ft (305 m) vertically unless both aircraft are separated horizontally as illustrated by Fig. 24.5. If aircraft flying above 29,000 ft, then vertical separation will be 2000 ft (610 m).

The horizontal separation standard between aircraft flying at the same altitude is 5 nm (9260 m). This 5 nm is the red zone in the proximity algorithm [8–11]. Three proximity levels are defined.

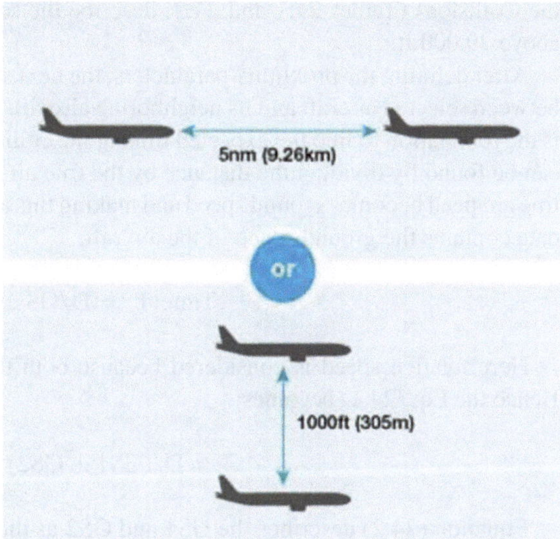

Fig. 24.5 Separation standards

Table 24.1 Separation standards for commercial aircraft at most 29,000 ft

Sr. No.	Latitude and longitude	Altitude	Proximity
1	≤5 nm	<1000 ft	Red zone
2	>5 nm and ≤10 nm	<1000 ft	Warning zone
3	>10 nm	Not concerned	Green zone

Table 24.2 Separation standards for commercial aircraft above 29,000 ft

Sr. No.	Latitude and longitude	Altitude	Proximity
1	<= 5 nm	<2000 ft	Red zone
2	>5 nm and ≤10 nm	<2000 ft	Warning zone
3	>10 nm	Not concerned	Green zone

1. Red region 5 nm
2. Yellow region 10 nm
3. Green region 15 nm.

When the aircraft is in the green region, there will be no alerts as this is a purely safe region. As soon as the airplane reaches the yellow region the application starts a warning alert with a flashing yellow screen sound. Finally, if it hit the worst-case scenario which is the red region the alert becomes more with a red alert flashing screen with the sound. Above all this, it will also give the ETC (Expected Time of the Collision) (Tables 24.1 and 24.2) describe the separation atmost 29,000 ft and above 29,000 ft.

After defining the proximity parameters, the next step is to determine the distance between selected aircraft and its neighboring aircrafts using WGS-84, and Eq. (24.1) is the foundation to find the expected time of the collision between two aircraft. This can be found by dividing the distance by the true airspeed. If there is no wind, then true airspeed becomes ground speed and making this assumption because the ADS-B data contains the ground speed of the aircraft.

$$\mathrm{Time(T)} = \mathrm{D/GS} \tag{24.1}$$

Here relative speed is considered because both the objects are in motion [12]. Hence the Eq. (24.1) becomes;

$$\mathrm{T} = \mathrm{D/(GS1 + GS2)} \tag{24.2}$$

Equation (24.2) describes the GS1 and GS2 as the ground speed of both aircraft that are in contact with each other in the 3-tier proximity level.

24.4 Results and Discussion

After building the proximity of the algorithm, the system is tested on the historical data and also interpolates some dummy aircraft in the proximity to check the working of the algorithm. When an aircraft is in the green zone, then there will be no sound. If the aircraft is in the yellow region (10 nm) it will start giving the flash warning beep sound proximity. If the aircraft comes in the red proximity (5 nm) of the other aircraft, it will start giving the red alert beep. The alert will display the ICAO number of the aircraft its source and destination location as taken from ADS-B API, its coordinates, and the expected time of the collision.

The black color aircraft is the selected one. The algorithm creates the 3-tier proximity and then start analyzing around its surrounding. Figure 24.6 illustrates the three aircraft are in the radius two of them are identified as red because they have horizontal separation less than 5 nm when altitudes are almost identical. The one aircraft is in the green color due to the vertical separation when latitudes and longitudes are the same. The application will start generating the alerts on the web application when the other aircraft reach the yellow or red proximity tier with the screen flashing sound.

Figure 24.7 shows the two aircraft identified as golden in color due to they are in the yellow region (warning). The one aircraft is in the green color due to the vertical separation when latitudes and longitudes are the same.

The application will notify the pilot by showing the relevant details in the form of a table such as ICAO, model, coordinates in latitude, longitude, and altitude, their speeds in knots, horizontal and vertical separation in the nautical mile (nm), proximity

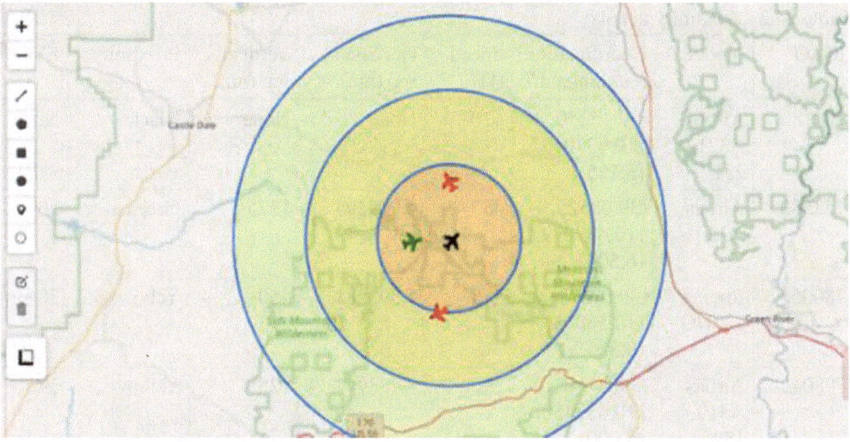

Fig. 24.6 Aircraft in the red region

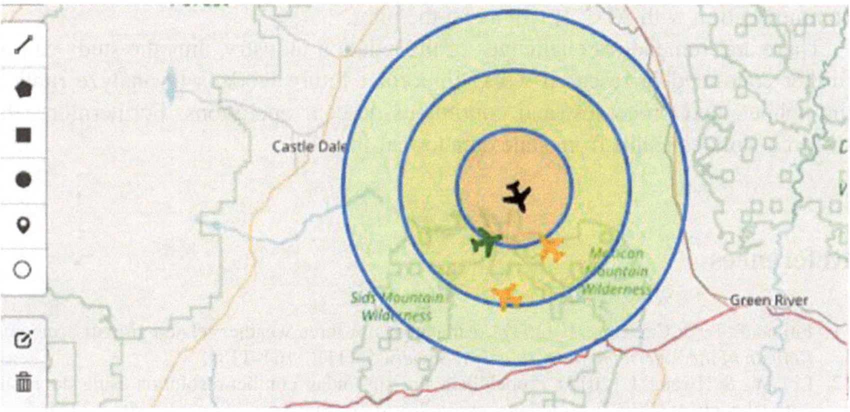

Fig. 24.7 Aircraft in the yellow region

level of the surrounding planes along with the expected time of the collision. Table 24.3 summarizes all the attributes in one place.

24.5 Conclusion and Future Works

The framework can serve as a prototype for the aviation industry. The web-based application is live and deployed in the cloud environment. The proximity is checked to see both the horizontal and vertical separation of the aircraft with the other aircraft.

Table 24.3 Proximity details

ICAO	Model	Coordinates (lat, long, alt)	Speed (kt)	Horizontal sep (nm)	Vertical sep (nm)	Proximity	ETC
780186	Airbus A321 231	(39.15250, –110.637079, 15,125)	410	None	None	Black	None
758550	Airbus A320 216	(39.08525, –110.697079, 16,500)	487	4.909295	1375	Green	20 s
78006C	Boeing 737NG	(39.07571, –110.569153, 15,575)	315	5.590281	450	Yellow	28 s
798042	Airbus A319 132	(39.00746, –110.652450, 15,000)	390	8.723944	125	Yellow	39 s

Then it will notify the alert on the screen, so even in the worst case when there is no communication with ATC. It will assist the pilot.

There are tremendous challenges of the aviation industry, thus the study can be further continued in research work. Important future work is to analyze runway throughput, taxi processes, and continuous descent operations. Furthermore, the research work can shift from stale data to real-time data.

References

1. Fujita, T. T., & Caracena, F. (1977). An analysis of three weather-related aircraft accidents. *Bulletin of the American Meteorological Society, 58*(11), 1164–1181.
2. Liu, W., & Hwang, I. (2013). Probabilistic aircraft midair conflict resolution using stochastic optimal control. *IEEE Transactions on Intelligent Transportation Systems, 15*(1), 37–46.
3. Weick, K. E. (1990). The vulnerable system: An analysis of the Tenerife air disaster. *Journal of management, 16*(3), 571–593.
4. (2020, September 18) *Automatic dependent surveillance – Broadcast*. https://en.wikipedia.org/wiki/Automatic_dependent_surveillance_%E2%80%93_broadcast Accessed September 25, 2020
5. (2020, Januray). *Ins and outs*. https://www.faa.gov/nextgen/equipadsb/capabilities/ins_outs/. Accessed February 14, 2020
6. (2014). How much is the minimum safe distance between two planes in-flight? https://aviation.stackexchange.com/questions/2806/how-much-is-the-minimum-safe-distance-between-two-planes-in-flight. Accessed March 10,2020
7. Whitfield, B. (2017, February 8). *How it works: ADS-B*. https://www.flyingmag.com/how-it-works-ads-b/. Accessed February 14, 2020
8. (2020, February). *Separation aeronautics*. https://en.wikipedia.org/wiki/Separation_(aeronautics). Accessed March 10, 2020
9. (2020, June). *Separation standards*. https://www.airservicesaustralia.com/about-us/our-services/how-air-traffic-control-works/separation-standards/. Accessed August 8, 2020
10. Kuchar, J. K., & Yang, L. C. (2000). A review of conflict detection and resolution modeling methods. *IEEE Transactions on Intelligent Transportation Systems, 1*(4), 179–189.

11. Rekkas, C., & Rees, M. (2008, September). Towards ADS-B implementation in Europe. In *2008 Tyrrhenian International Workshop on Digital Communications-Enhanced Surveillance of Aircraft and Vehicles* (pp. 1–4). IEEE
12. (2020,April). *Basic calculations.* https://www.flightliteracy.com/basic-calculations/. Accessed August 15, 2020

Index of Authors

················

Developments in Biological Standardization

Vol. 51 : Lyon 1981	Immunization of adult birds with inactivated oil adjuvant vaccines
Vol. 52 : Lyon 1981	Herpes virus of man and animal: Standardization of immunological procedures
Vol. 53 : Dublin 1982	Enteric infections in man and animals: Standardization of immunological procedures
Vol. 54 : Athens 1982	Viral hepatitis: Standardization in immunoprophylaxis of infections by hepatitis viruses
Vol. 55 : Copenhagen 1982	Fifth General Meeting of European Society of Animal Cell Technology
Vol. 56 : Algeria 1983	IIIrd International Symposium on Brucellosis
Vol. 57 : Paris 1983	Monoclonal antibodies: Standardization of their characterization and use
Vol. 58 : Budapest 1983	BCG vaccines and tuberculins
Vol. 59 : Geneva 1983	Standardization and control of biologicals produced by recombinant DNA technology
Vol. 60 : Gardone Riviera 1983	Production and exploitation of existing and new animal cell substrates
Vol. 61 : Geneva 1984	Proceedings of the Fourth International Symposium on pertussis
Vol. 62 : Stockholm 1985	Diagnostics and vaccines for parasitic diseases
Vol. 63 : San Francisco 1985	Use and standardization of chemically defined antigens
Vol. 64 : London 1985	Reduction of animal usage in the development and control of biological products
Vol. 65 : Amsterdam 1985	Use and standardization of combined vaccines
Vol. 66 : Baden 1985	Advances in animal cell technology: cell engineering, evaluation and exploitation
Vol. 67 : Melbourne 1986	Standardization in blood fractionation including coagulation factors
Vol. 68 : Geneva, 1986	Cells – Products – Safety
Vol. 69 : London, 1987	Cytokines Laboratory and Clinical Evaluation
Vol. 70 :Arlington, VA, 1988	Continuous Cell Lines as Substrates for Biologicals
Vol. 71 :Utrecht, 1989	Monoclonal Antibodies for Therapy, Prevention and In vivo Diagnosis of Human Disease
Vol. 72 : Annecy, 1989	Progress in Animal Retroviruses
Vol. 73 : Shizuoka, 1990	Pertussis: Evaluation and Research on Acellular Pertussis Vaccines
Vol. 74 : Bethesda, MD, 1990	Biological Product Freeze-Drying and Formulation
Vol. 75 : London, 1990	Virological Aspects of the Safety of Biological Products
Vol. 76 : Bethesda, MD, 1991	Continuous Cell Lines - An International Workshop on Current Issues
Vol. 77 : Annecy, 1991	Standardization of the Immunopharmacology of Natural and Synthetic Immunomodulators
Vol. 78 : Bethesda, MD, 1991	Poliovirus Attenuation: Molecular Mechanisms and Practical Aspects
Vol. 79 : Ploufragan, 1992	The First Steps Towards an International Harmonization of Veterinary Biologicals: 1993 and Free Circulation of Vaccines within the EEC
Vol. 80 : Heidelberg, 1992	Transmissible Spongiform Encephalopathies - Impact on Animal and Human Health
Vol. 81 : Cannes, 1992	Virological Safety Aspects of Plasma Derivatives

Vol. 82: Albany NY, 1993	Recombinant Vectors in Vaccine Development
Vol. 83: Annecy, 1993	Genetic Stability and Recombinant Product Consistency
Vol. 84: Langen, 1993	Non-Target Effects of Live Vaccines
Vol. 85: Santa Fé, NM, 1994	Genetics of Streptococci, Enterococci and Lactococci
Vol. 86: Langen, 1994	Replacement, Reduction and Refinement of Animal Experiments in the Development and Control of Biological Products
Vol. 87: Geneva, 1995	New Approaches to Stabilisation of Vaccines Potency
Vol. 88: Bethesda, MD, 1995	Viral Safety and Evaluation of Viral Clearance from Biopharmaceutical Products
Vol. 89: Rome, 1995	Pertussis Vaccine Trials
Vol. 90: Oslo, 1996	Fish Vaccinology
Vol. 91: San Francisco, CA, 1996	Development of Specifications for Biotechnology Pharmaceuticals Products
Vol. 92: Bergen, 1996	Modulation of the Immune Response to Vaccine Antigens
Vol. 93: Annecy, 1996	Safety of Biological Products Prepared from Mammalian Cell Culture
Vol. 94: Bethesda, MD, 1997	Simian Virus 40 (SV40): A Possible Human Polyomavirus
Vol. 95: Paris, 1997	Preclinical and Clinical Development of New Vaccines
Vol. 96: Washington, DC, 1995	Characterization of Biotechnology Pharmaceutical Products
Vol. 97: London, 1997	Biological Characterization and Assay of Cytokines and Growth Factors
Vol. 98: London, 1997	Inactivated Influenza Vaccines Prepared in Cell Culture
Vol. 99: Strasbourg, 1998	Animal Sera, Animal Sera Derivatives and Substitutes Used in the Manufacture of Pharmaceuticals: Viral Safety and Regulatory Aspects
Vol. 100: Geneva, 1998	A Celebration of 50 Years of Progress in Biological Standardization and Control at WHO
Vol. 101: London, 1998	Alternatives to Animals in the Development and Control of Biological Products for Human and Veterinary Use

Developments in Biologicals

Vol. 102: San Francisco, CA, 1999	Advances in Transfusion Safety
Vol. 103: Annecy, 1999	Physico-Chemical Procedures for the Characterization of Vaccines
Vol. 104: Langen, 1999	Development and Clinical Progress of DNA Vaccines
Vol. 105: Paris, 2000	Progress in Polio Eradication: Vaccine Strategies for the End Game
Vol. 106: Rockville, MD, 2000	Evolving Scientific and Regulatory Perspectives on Cell Substrates for Vaccine Development
Vol. 107: London, 2000	The Design and Analysis of Potency Assays for Biotechnology Products
Vol. 108: Langen, 2001	Advances in Transfusion Safety
Vol. 109: Washington, DC, 2000	Biologics 2000 – Comparability of Biotechnology Products
Vol. 110: Palm Cove, 2001	Orphan Vaccines – Bridging the Gap
Vol. 111: Utrecht, 2001	Advancing Science and Elimination of the Use of Laboratory Animals for Development and Control of Vaccines and Hormones
Vol. 112: Bethesda, MD, 2001	Immunogenicity of Therapeutic Biological Products
Vol. 113: Berlin, 2001	Process Validation for Manufacturing of Biologics and Biotechnology Products
Vol. 114: Ames, IA, 2002	Vaccines for OIE List A and Emerging Animal Diseases
Vol. 115: Bergen, 2002	Laboratory Correlates of Immunity to Influenza – A Reassessment
Vol. 116: Los Angeles, CA, 2003	Development of Therapeutic Cancer Vaccines
Vol. 117: Langen, 2003	Consideration of Alternative Licensing Procedures for Vaccines for Minor Species, Minor Indications and Autogenous/Autologous Products
Vol. 118: Langen, 2003	PDA/EMEA European Virus Safety Forum
Vol. 119: Buenos Aires, 2004	Control of Infectious Animal Diseases by Vaccination

IMPRIMERIE MEDECINE HYGIENE GENEVE SUISSE

juin - 2005